AF616393

20 0238827 9

Current Topics in Microbiology 215 and Immunology

Springer
Berlin
Heidelberg
New York
Barcelona
Budapest
Hong Kong
London
Milan
Paris
Santa Clara
Singapore
Tokyo

Tuberculosis

Edited by T.M. Shinnick

With 46 Figures

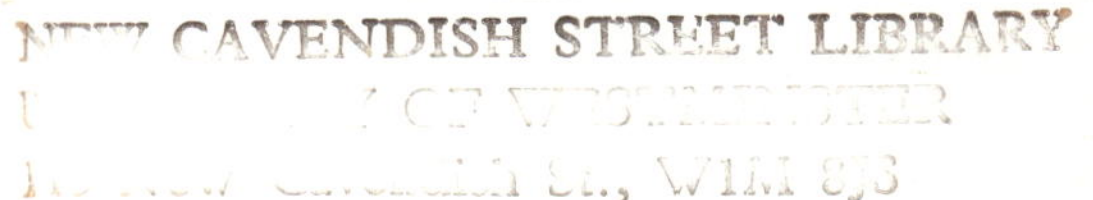

Thomas M. Shinnick, Ph.D.

Chief, Immunology and Molecular Pathogenesis Section
Division of AIDS, STD, and TB Laboratory Research
Centers for Disease Control and Prevention
Department of Health and Human Services
Atlanta, GA 30333
USA

Cover illustration: Tuberculosis results from a long series of interactions between Mycobacterium tuberculosis, macrophages, and the host's immune system, some of which are schematically represented in the cover design. The disease process begins with the inhalation of aerosolized particles containing 1-3 viable tubercle bacilli. Once in the alveolar space in the lower lung, the bacilli are ingested by alveolar macrophages and most are destroyed. However, some survive phagocytosis and replicate intracellularly. The large bacterial burden eventually kills the alveolar macrophages, and the released bacilli are ingested by other alveolar macrophages or by blood borne monocytes. At this early stage of the infection, there is relatively little tissue destruction or cell death. Also, at this stage the bacilli-laden macrophages can leave the initial site of infection to interact with other components of the immune system and to transport the bacilli to the upper regions of the lung. In the upper lobe, the bacilli multiply and granulomas are formed as part of the cellular immune response to the tubercle bacilli. As the infection progresses, the center of the granuloma liquefies to form a caseous necrotic lesion or tubercle – the classic cavitary lesion of tuberculosis. Ultimately, the lesion ruptures into the nearby bronchi, the bacilli discharged through the airway, and the infectious process begins anew.

Cover design: Design & Production, Heidelberg

ISSN 0070–217X
ISBN 3–540–60985–7 Springer-Verlag Berlin Heidelberg New York

Library of Congress Catalog Card Number 15 - 12910
Printed in Germany

Typesetting: Scientific Publishing Services (P) Ltd, Madras
SPIN: 10495419 27/3020/SPS – 5 4 3 2 1 0 – Printed on acid-free paper

Preface

Tuberculosis has plagued mankind since prehistoric times and is still an important source of morbidity and mortality, with particularly devastating effects in developing and tropical countries. Tuberculosis results from an infection with *Mycobacterium tuberculosis,* and the World Health Organization estimates that perhaps as much as one-third of the world's population or approximately 1.9 billion persons are or have been infected with *M. tuberculosis*. Each year, there are 8–10 million new cases of tuberculosis and about 3 million deaths due to it. Indeed, tuberculosis is the leading cause of death in adults due to a single infectious agent and accounts for approximately 26% of all preventable adult deaths in the world. In addition, tuberculosis is an enormous social and economic problem because approximately 95% of new cases occur in developing countries and because about 80% of tuberculosis cases affect persons of child-bearing age and during their most economically productive years (ages 15–59).

Tuberculosis has also re-emerged as an important public health problem in many developed countries. For example, between 1985 and 1992, the number of tuberculosis cases reported to the United States Centers for Disease Control and Prevention increased by more than 20%. Similarly, Austria experienced a 5% increase in tuberculosis cases from 1987 to 1991, Ireland a 9% increase from 1988 to 1991, Denmark a 20% increase from 1987 to 1992, and Italy a 27% increase from 1988 to 1992. In many of these countries, including the United States, the resurgence of tuberculosis has been accompanied by outbreaks of tuberculosis caused by *M. tuberculosis* strains that are resistant to many of the commonly used antituberculosis drugs. Recent advances in our understanding of the molecular basis of the emerging drug resistance in *M. tuberculosis* are discussed by Heym et al.

From the immediate public health perspective, keys to controlling the spread of tuberculosis include the rapid diagnosis of tuberculosis and the prompt initiation of effective chemotherapy and proper infection control procedures. After all, tuberculosis has been a curable and preventable disease for almost four decades, and currently recommended short-course chemotherapy regimens can have cure rates exceeding 90%. In the long run, however, we need to learn more about the biology of *M. tuberculosis* and how it causes disease, which in turn may lead to new vaccines or treatments or intervention strategies. New approaches will be required because the approach of aggressive case finding and chemotherapy with expensive new drugs, which should be effective in controlling tuberculosis in industrialized countries, may not be the answer for developing countries, where as many as half of all adults have been infected with *M. tuberculosis* and where resources are scarce.

However, having half the population infected with *M. tuberculosis* does not mean that half the population will develop tuberculosis. Only about 10% of infected immunocompetent persons will develop active disease during their lifetimes, and some of these cases develop as many as 40–50 years after the initial infection event. This is because many of the clinical manifestations of tuberculosis are actually the result of a long series of interactions between the mycobacterium and the host immune system. Two key aspects of this are that (1) tubercle bacilli survive and replicate within the host's phagocytic cells, primarily within macrophages, and (2) much of the disease process, as well as immunity to disease, is dependent on the host's immune response to the mycobacterium.

The first section of this volume concentrates on the interactions of mycobacteria and host cells. As discussed by Lee et al., the mycobacterial cell wall has a quite complex architecture and plays critical roles in the interaction of *M. tuberculosis* with the environment and host cells. An important aspect of these interactions is the entry of this intracellular pathogen into its host cell, and Schlesinger reviews the various pathways *M. tuberculosis* exploits to enter mononuclear phagocytes. Once inside, *M. tuberculosis* somehow avoids the bactericidal activities of the host cell and grows within the macrophage. The complex interactions of macrophages and mycobacteria are discussed by O'Brien et al. from the viewpoint of the mechanisms macrophages use to kill *M. tuberculosis,* and by Quinn et al. from the point of view of mechanisms the mycobacteria use to survive and replicate within macrophages. The latter chapter is complemented by Hatfull's discussion of the molecular genetic tools available for use in dissecting the biology of *M. tuberculosis.*

The second section of this volume concentrates on the interactions of *M. tuberculosis* with the host's immune system. McMurray et al. describe the various animal models that are available for studying tuberculosis, and Orme reviews our understanding of the immune response to *M. tuberculosis* in the animal models. Next, Barnes and Modlin describe the cellular immune response observed in persons infected with *M. tuberculosis.* Several key aspects of the cellular immune response are then described in detail: (1) Toossi and Ellner describe the curious lack of responsiveness, anergy, which often occurs in tuberculosis patients; (2) Rook and Stanford discuss the deleterious aspects of the immune response of *M. tuberculosis,* which lead to much of the tissue destruction and pathology seen in tuberculosis; and (3) Huebner discusses the current status of the use of BCG as a vaccine to prevent tuberculosis. Finally, Szalay and Kaufmann discuss what we have learned from studies of other intracellular pathogens and how this might further our understanding of tuberculosis, and perhaps, provide some direction for future research on tuberculosis.

The series of events that lead from an infection with *M. tuberculosis* to the development of clinically important tuberculosis is long, complex, and multifaceted. As discussed in this volume, elucidating the details of the disease process will require a multidisciplinary effort combining basic biologic studies on the tubercle bacillus, sophisticated immunologic investigations in animal models, and careful dissection of the human immune response during the various stages of infection and disease. Better understanding of the bacillus and disease process should ultimately lead to the development of new effective treatment or prevention strategies to control this deadly disease.

Atlanta, June 1996 T. Shinnick

List of Contents

List of Contributors

(Their addresses can be found at the beginning of their respective chapters.)

Mycobacterium tuberculosis Cell Envelope

R.E. LEE, P.J. BRENNAN, and G.S. BESRA

1 Introduction

The mycobacterial cell wall is a complex and intriguing mixture of components which sets *Mycobacterium tuberculosis* apart from all other known bacterial species (GOODFELLOW and MINNIKIN 1984). To understand the *M. tuberculosis* cell wall, one must first consider the biology of the tubercle bacillus. Tuberculosis has long been known as a cause of morbidity and mortality worldwide. Indeed it is believed that one third of the word's population is infected with *M. tuberculosis* (SUDRE et al. 1992). Evidence of tuberculosis-like infections date back many thousands of years, and it is very likely that tuberculosis-related infections have plagued humankind since the dawn of civilization. *M. tuberculosis* is primarily an intracellular pathogen which resides within the phagolysosomes of alveolar macrophages. Perhaps as a consequence of this intracellular environment, the highly intricate features of the tubercle bacilli cell wall have undergone extensive evolutionary changes.

Department of Microbiology, Colorado State University, Fort Collins, CO 80523, USA

The reservoir for *M. tuberculosis* is humans. The hereditary theory of transmission has been dismissed. It is now accepted that infection with *M. tuberculosis* develops after contact with infectious persons. *M. tuberculosis* infection is transmitted almost exclusively by the airborne route and the "infectious unit" is a small bacillus-containing particle called a droplet nucleus. When a droplet nucleus containing one or two viable bacilli is inhaled by a non-immune person, it is deposited on the alveolar surface where the bacilli begin to multiply. Initially, the infecting organism meets only limited resistance from the host, as phagocytosis by alveolar macrophages has little effect on the bacilli, which continue to multiply intracellularly in the non-immune host. The earliest evidence of host tissue recognition is a dilation of the capillaries, followed by a migration of polymorphonuclear leukocytes and macrophages into the infected area. After several weeks of infection, the number of leukocytes in the area decreases and the mononuclear cells predominate; these crowd together and contain pale, foamy, cytoplasmic material which is rich in lipids. The resulting unit is called a tubercle, the fundamental lesion of tuberculosis.

The intracellular macrophage phagocyte, home of the tubercle bacilli, is thus of great interest. Intracellular *M. tuberculosis* bacilli are not usually found in direct contact with the components of the vacuoles but are separated by an electron-transparent zone (FRÉHEL et al. 1988). Into this zone are believed to be secreted various proteins and carbohydrates which are part of the putative mechanism allowing intracellular growth. The lack of acidification of these lysosomes, which has also been noted for *M. avium* and implicated for *M. tuberculosis*, severely restricts lysosomal hydrolase activity (RUSSELL et al. 1994). The mycobacteria seem to be able to prevent the fusion of its vacuole with proton-ATPase positive vesicles which are associated with the acidification of vacuoles. *M. tuberculosis* can be grown on a wide range of carbon sources in vitro; thus in vivo it is free to assimilate a wide range of host tissue metabolites (BARCLAY and WHEELER 1989), a process which is aided by the excretion of proteases and lipases through the cell wall and into the vacuole.

Tuberculosis is essentially a disease of the immune system and it is thus pertinent to discuss aspects of the immunology of the disease and how they relate to the cell wall components. There are no mycobacterial exotoxins; much of the tissue damage is believed to originate from the host's immune system trying to combat the disease. The CD4 class of T cells is thought to regulate the immune response to *M. tuberculosis* and is known to be primarily directed towards the secreted or export proteins of the cell wall (ORME et al. 1993). These cells are capable of lysing mononuclear cells which are infected with *M. tuberculosis*. Thus, many latent tuberculosis infections become reactivated when CD4 T cells levels become depleted, as in the case of HIV infection (MILLS and MASUR 1990).

A fundamental challenge in mycobacteriology is to understand the envelope's structure and function under such conditions as phagocytosis and resistance to the intracellular attack mechanisms of the human macrophage. We know that the envelope is responsible for the cell's novel outward physical and chemical characteristics: their small size relative to other bacteria; their hydrophobicity; their acid fast staining; their resistance to chemical injury by acids, alkalis and many disinfectants which kill other bacteria. We believe that the biological ability of *M. tuberculosis* to survive long periods of starvation or aridity without dehydration and even their allergenic and immunogenic properties mainly originate from materials in the cell envelope.

2 The Primary Structure of the Cell Wall

The primary cell wall structure consists of a plasma membrane which is supported by a peptidoglycan backbone against the osmotic pressure of the interior. Attached to the peptidoglycan is an arabinogalactan layer and to which are esterified mycolic acids. Associated with the cell wall are a number of free lipids, glycolipids and proteins (BESRA and CHATTERJEE 1994). This basic cell wall structure of *M. tuberculosis* does not differ greatly from that of other nonpathogenic mycobacteria. Thus a great deal of information has been inferred from the study of the more accessible and fast-growing saprophytic bacteria such as *M. smegmatis, M. phlei, M. scrofulaceum* and *M. chelonae*.

2.1 The Plasma Membrane

The plasma membrane is structurally very important, firstly, as a separation barrier between the inner cytosol of the cell and the "periplasmic" space between the plasma membrane and the rigid peptidoglycan layer, and secondly, as an active site in the cell's anabolic (particularly cell wall biosynthesis) and catabolic functioning by allowing nutrients to flow into the cell. Intensive studies using the mycobacterial plasma membrane of *M. phlei* showed it to posses the characteristic functions of bacterial membranes, such as electron transport and oxidative phosphorylation (RATLEDGE 1982). The plasma membrane of *M. tuberculosis* differs only slightly from that of *M. phlei* and consists of a phospholipid bilayer containing diphosphatidylglycerol (cardiolipin), phosphatidylethanolamine, phosphatidylinositol and the phosphatidylmannoside family (PIMs) (Fig. 1). Minor amounts of phosphatidylglycerol and phosphatidylethanolamine have also been isolated from *M. tuberculosis* (SARMA et al. 1970). Analysis by optical rotations and selective lipase action established that the glyceride unit of these phospholipids possess the D-*lyxo* absolute configuration (OKUYAMA and NOJIMA 1965). The fatty acid components of these phospholipids are well known and consist mainly of octadecanoic (stearic), octadecenoic (oleic) and tuberculostearic (10-methyl stearic) fatty acids esterified to the 1-position of the glyceride (OKUYAMA et al. 1967). Hexadecenoic and hexadecanoic (palmitic) acids are attached to the 2-position. It is worthwhile noting that tuberculostearic

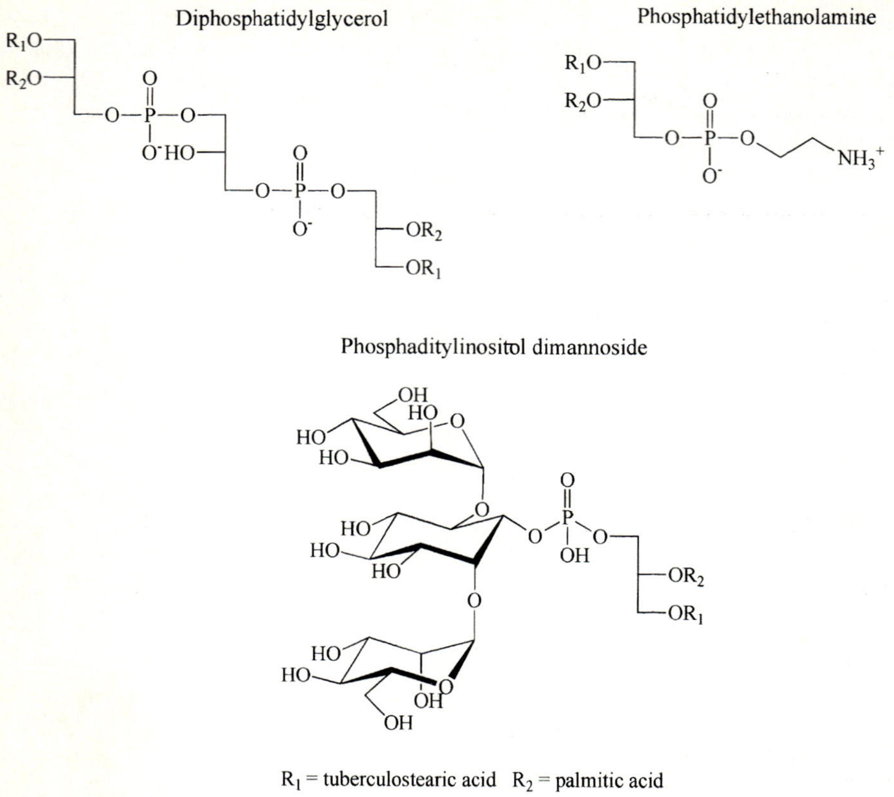

Fig. 1. *Mycobacterium tuberculosis* phospholipids

acids is a good marker for mycobacteria, as it is generally confined to the genus and can be used as a detection marker, especially where it is difficult to use PCR, such as in cerebral tuberculoid infections (BROOKS et al. 1987).

The PIM family is a major plasma membrane component and is closely related to lipomannan (LM) and lipoarabinomannan (LAM) which all share the basic PIM_2 core. Hydrolysis of PIM samples led to the recovery of mannose (in varying quantities), inositol, phosphate, glycerol, tuberculostearic and palmitic acids. The inositol moiety was found to possess no optical rotation and was thus assigned as *myo*-inositol (ANDERSEN 1930). Careful methylation and hydrolysis demonstrated that PIM_2 consisted of inositol (→1) linked to a phosphatidyl diglyceride with the inositol unit being further substituted at the 2- and 6- positions by mannose (LEE and BALLOU 1964). Extension of the mannose chain on the 6-position of the inositol residue by the addition of extra α(1→6) linked mannose units produces the other members of the PIM family up to PIM_5 (LEE and BALLOU 1965), in which the final linkage is α(1→2), and PIM_6 in which the final two linkages are α(1→2) (Fig. 2). Further acylation of the α(1→6) mannose residues attached to the *myo*-inositol ring have been described for the mono and diacyl

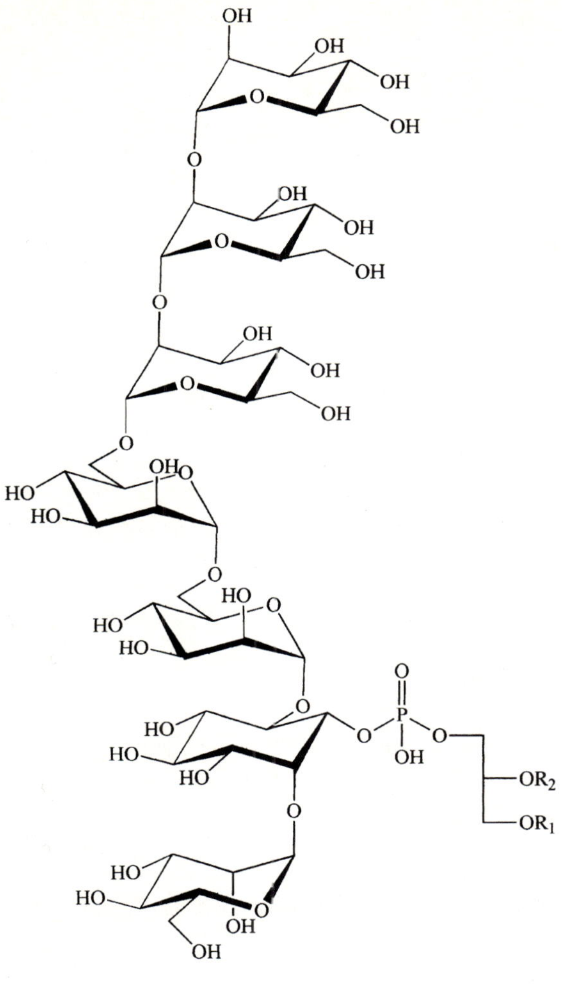

R_1 = tuberculostearic acid R_2 = palmitic acid

Fig. 2. Phosphatidylinositol hexamannoside

phosphatidyl-*myo*-inositol pentamannosides although the exact location has not yet been determined. It has been suggested that the PIM family has an asymmetric distribution within the plasma membrane with the carbohydrate moieties facing into the periplasmic space, as under high resolution electron microscopy it is apparent that the outward facing portion is visibly thicker (BRENNAN and DRAPER 1994). The intracellular location of the PIMs is also indicated by their inability to react with antibodies raised to the whole cells of *M. scrofulaceum* (RIDELL et al. 1986).

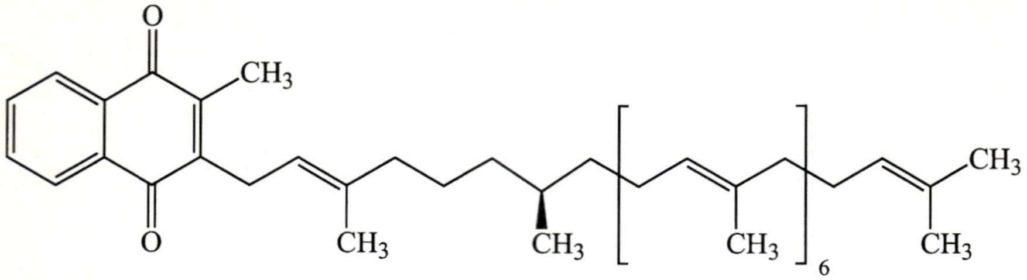

Fig. 3. Menaquinone MK9 (H_2)

Also associated with the plasma membrane are a number of polyterpene based products, thought to be involved in protection against photolytic damage, such as the carotenoids (DAVID 1984) and the menaquinones active in electron transport (MINNIKIN and GOODFELLOW 1980). The carotenoids of mycobacteria are responsible for the characteristic yellow-orange color of photochromagenic mycobacteria such as *M. gordonae* and *M. kanasii*; however, *M. tuberculosis* is not known to be a member of this family as it is generally not exposed to light and the role of carotenoids is probably minor. The isolation of the carotenoids is not trivial due to their high reactivity; however, leprotene has been isolated from *M. tuberculosis* cultures. The menaquinone (MK) family is well known and is found in a wide range of bacteria (GOODFELLOW and MINNIKIN 1984). The *M. tuberculosis* menaquinones systematically known as 2-methyl-polyprenol-1,4-napthoquinone vary in prenoid chain length and saturation at the second prenoid moiety. MK9 (H_2) (Fig. 3) is the most commonly found although MK8 (H_2) and some fully saturated variants are known.

Glycosylphosphopolyprenols involved with cell wall biosynthesis are believed to be plasma membrane associated. The phosphopolyprenols known to be associated with *M. tuberculosis* are phosphodecaprenol and phosphooctahydroheptaprenol. The major glycosylated prenols isolated in mycobacteria thus far include β-D-mannopyranosyl phosphodecaprenol (TAKAYAMA and GOLDMAN 1970), β-D-mannopyranosyl phosphooctahydroheptaprenol (TAKAYAMA et al. 1973), β-D-arabinofuranosyl phosphodecaprenol (WOLUCKA et al. 1994) and 6-*O*-mycolyl-β-D-mannopyranosyl phosphooctahydroheptaprenol (G.S BESRA et al. 1994) (Fig. 4). β-D-Mannopyranosyl phosphodecaprenol and β-D-arabinofuranosyl phosphodecaprenol are believed to be the sugar donors of the mannose and arabinose units, respectively, in arabinogalactan, PIM and lipoarabinomannan (LAM) biosynthesis. The unique 6-*O*-mycolyl-β-D-mannopyranosyl phosphooctahydroheptaprenol is believed to be the product resulting from the key condensation step involved in the biosynthesis of the mycolic acids. Also, it must be assumed that there are a number of glycosylphospho- and pyrophosphoprenol intermediates in peptidoglycan and polysaccharide biosynthesis which have not yet been isolated due to their low concentrations and the high lability of the phosphate, and pyrophosphate prenol linkages.

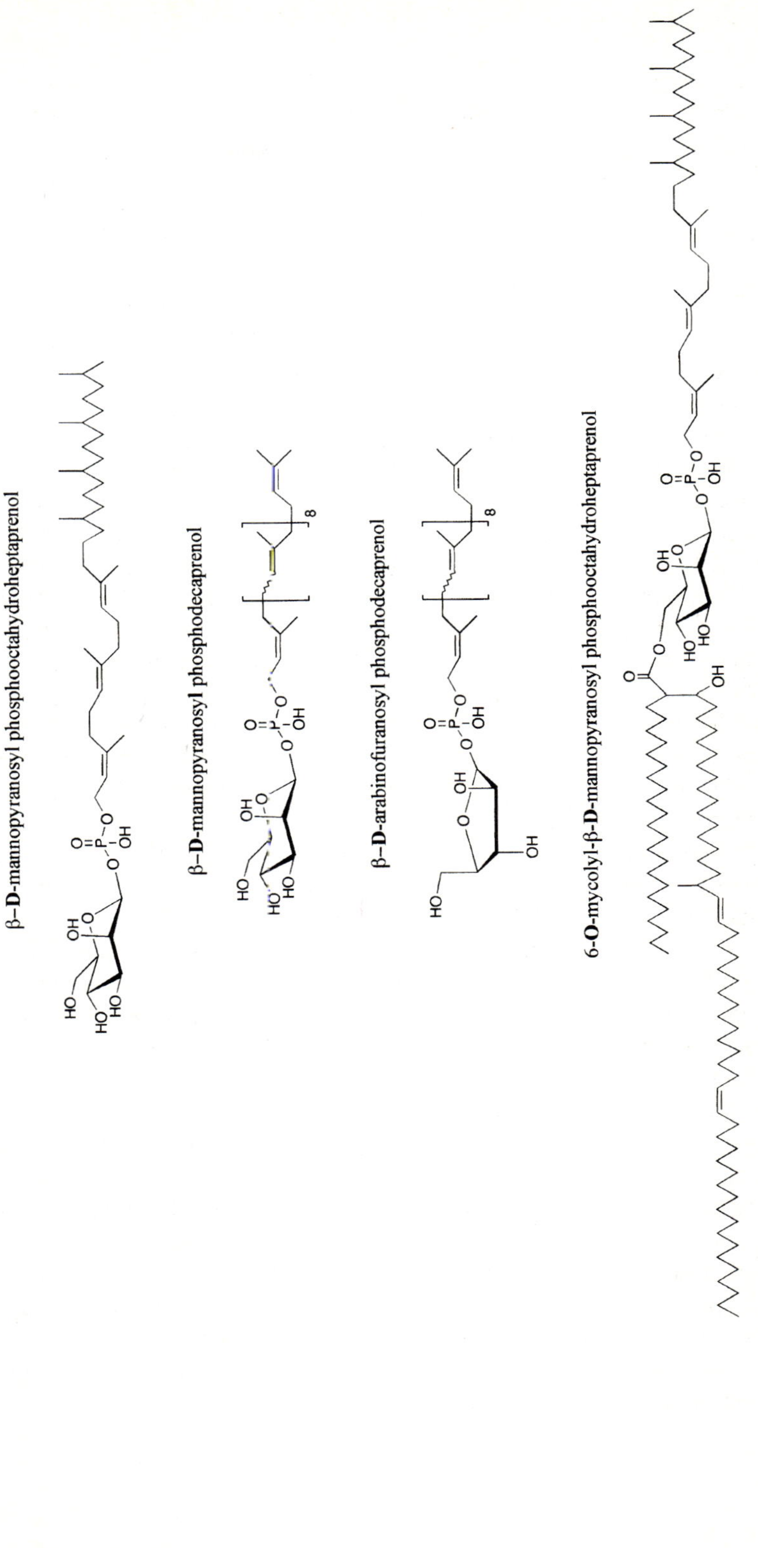

β-**D**-mannopyranosyl phosphooctahydroheptaprenol

β-**D**-mannopyranosyl phosphodecaprenol

β-**D**-arabinofuranosyl phosphodecaprenol

6-**O**-mycolyl-β-**D**-mannopyranosyl phosphooctahydroheptaprenol

Fig. 4. Mycobacterial glycosylphosphopolyprenols

2.2 Lipoarabinomannan: Structure and Location

Lipoarabinomannan and its truncated form lipomannan (LM) (Fig. 5) are thought to be primarily plasma membrane-associated lipopolysaccharides which exhibit a diffuse band between 30 and 35 kDa on SDS-polyacrylamide gel electrophoresis. The actual mass of LAM isolated from *M. bovis* BCG strain by UV laser disorption/ionization mass spectrometry was determined to be 17.4 kDa (VENISSE et al. 1993) and the larger mass on the polyacrylamide gel was attributed to aggregate formation. From hydrolysis experiments LAM is known to contain glycerol, inositol, phosphate, lactate, succinate, palmitate and tuberculostearate. The absolute structure of LAM still remains to be defined; however, the main polysaccharide core structure is known. LAM possess a basic PIM_2 base (HUNTER and BRENNAN 1990) which is one of the reasons why it is believed to be plasma membrane-associated. In LAM the nonreducing terminal of the mannose chain is capped with a polymeric α(1→5) Ara*f* chain (Fig. 6) (CHATTERJEE et al. 1991). The arabinan chain is branched at various sites by the introduction of 3,5-linked α-D-Ara*f* residues. The non-reducing terminals of LAM are very heterogeneous and consist of various linear and branched chain arabinan segments (CHATTERJEE et al. 1993), such as $[t\text{-}\beta\text{-D-Ara}f\text{-}(1\rightarrow2)\text{-}\alpha\text{-D-Ara}f]_2\text{-}3,5\text{-}\alpha\text{-D-Ara}f\text{-}(1\rightarrow5)\text{-}\alpha\text{-D-Ara}f$, and, within the virulent *M. tuberculosis* Erdman strain, these non-reducing segments are capped by mannose residues (CHATTERJEE et al. 1992). The ability of LAM to aggregate and complex to other cell wall components makes its isolation far from trivial. Low availability, large molecular size and heterogeneity of epitopes have hindered the absolute structural determination. The biological and immunological properties of LAM are of great significance and have been studied extensively (BESRA and CHATTERJEE 1994). LAM is known to cause abrogation of T cell activiation, inhibition of-γ-interferon-mediated activation of murine macrophages, inhibition of protein kinase C activity and evocation of a large number of cytokines associated with macrophages such as tumor necrosis factor (TNF). Thus, it has been suggested that LAM mediates many of the clinical manifestations of tuberculosis. The areas of the biomolecule that are responsible for these biological effects are uncertain. LAM does lose many of its serological functions upon deacylation and, interestingly, the closely related LM is biologically inactive.

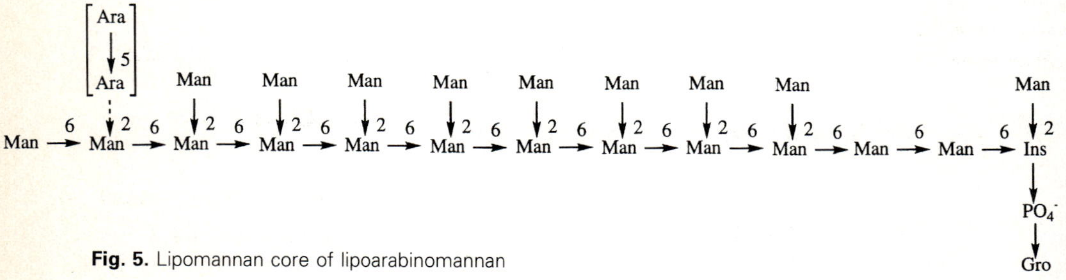

[Ara →5 Ara] ↓2 Man; Man ↓2 Man (×8); Man ↓2 Ins

Man →6 Man →6 Man →6 Man →6 Man →6 Man →6 Man →6 Man →6 Man →6 Man →6 Man →6 Man →6 Ins → PO_4^- → Gro

Fig. 5. Lipomannan core of lipoarabinomannan

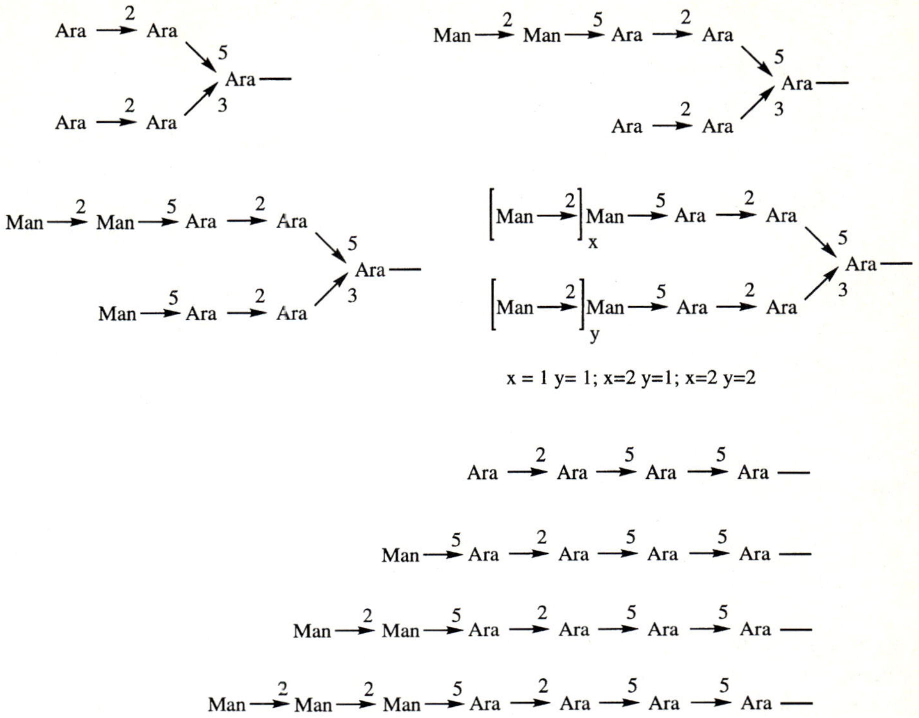

Fig. 6. The nonreducing term nal epitopes of lipoarabinomannan

2.3 Peptidoglycan

The peptidoglycan wall chemotype (IV) was established by the use of selective bacteriolytic enzymes providing a convenient route to the component disaccharide:*N*-acetylglucosamine $\beta(1\rightarrow4)$ *N*-glycolylmuramic acid and the tetrapeptide sequence consisting of L-alanine, D-isoglutamine, *meso*-diaminopimelic acid and D-alanine (Fig. 7) (PETIT and LEDERER 1984). The peptidoglycan units are cross-linked using both the conventional D-alanine to *meso*-diaminopimelic acid linkage and also a *meso*-diaminopimelic acid to *meso*-diaminopimelic acid; indicating that there may be at least two types of transpeptidases present in mycobacteria (Fig. 8). This is clinically significant as transpeptidases are the targets of β-lactam antibiotics.

2.4 The Mycolyl Arabinogalactan Complex

An arabinogalactan polysaccharide is linked through a unique rhamnose glycosyl phosphodiester linkage to the outside of peptidoglycan (McNEIL et al. 1990). The "linker"(Fig. 9) consists of a α-L-Rha*p*-(1$\rightarrow$3)-GlcNAc-1-phosphate which is un-

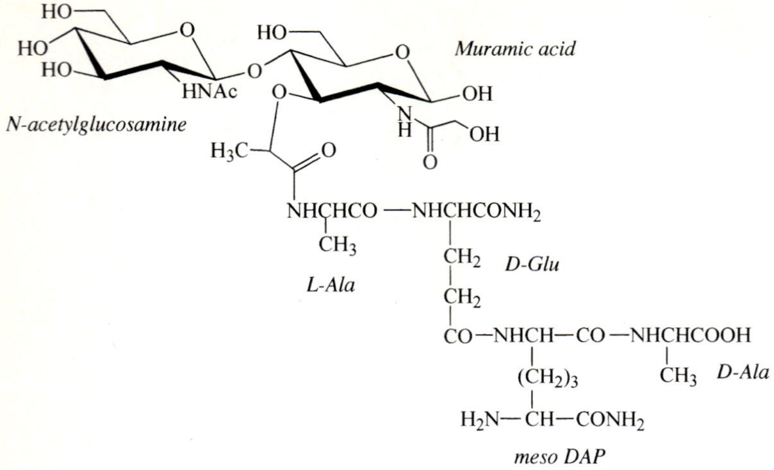

Fig. 7. Peptidoglycan monomer

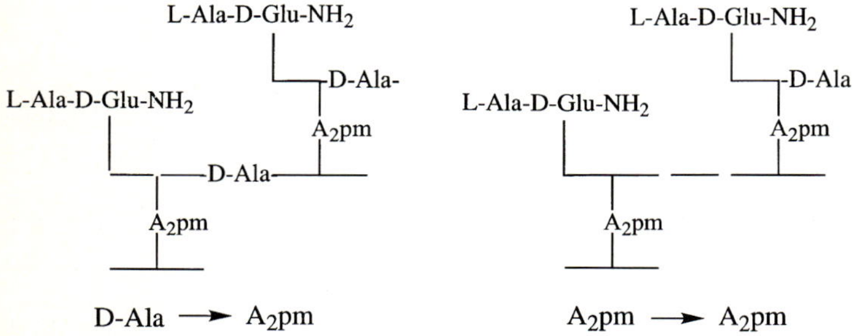

Fig. 8. The two types of interpeptide linkages of the peptidoglycan of mycobacteria

ique to actinomycetes. The phosphate moiety is linked through a phosphodiester linkage to a muramic acid residue on the peptidoglycan backbone securing the reducing galactan terminals of the arabinogalactan polysaccharide.

The arabinogalactan polysaccharide, which is unique to mycobacteria and other actinomycetes, possesses both the furanoid forms of arabinose and galactose. The arabinogalactan polysaccharide consists of a linear strand of $\alpha(1\rightarrow5)$-D-Ara*f* linked to C5 of a $\beta(1\rightarrow6)$D-Gal*f* residue from the alternating $\beta(1\rightarrow5)$, $\beta(1\rightarrow6)$ D-Gal*f* core (Fig. 10) (DAFFÉ et al. 1990). There is thought to be two or three arabinan chains per galactan core. The $\alpha(1\rightarrow5)$-D-Ara*f* arabinan is branched by the introduction of 3,5-α-D-Ara*f* units. The nonreducing terminals of the arabinan are capped with a characteristic hexaarabinose motif [t-β-D-Ara*f*-$(1\rightarrow2)$-α-D-Ara*f*]$_2$-3,5-α-D-Ara*f*-$(1\rightarrow5)$-α-D- Ara*f*]

Esterified to the 5-position of each of the arabinose residues within the hexaarabinosyl motif are the mycolic acids, which are high molecular weight α-

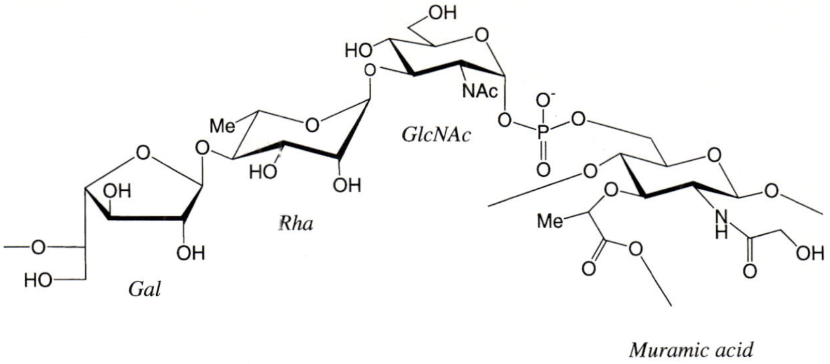

Fig. 9. Rha-GlcNAc linker

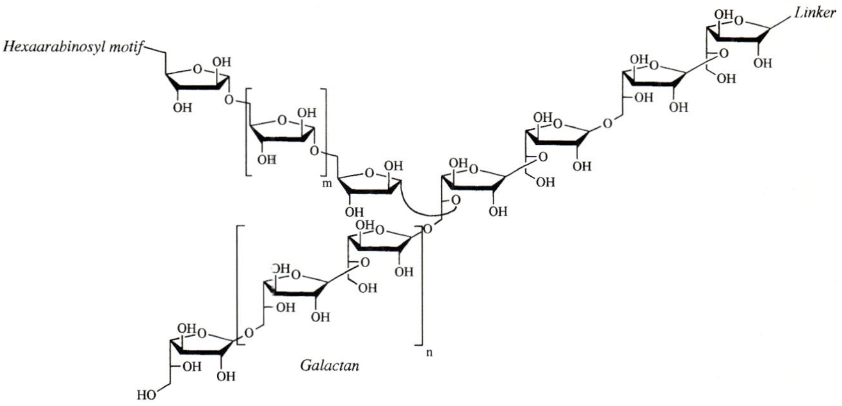

Fig. 10. Arabinogalactan

alkyl-β-hydroxy fatty acics (McNeil et al. 1991) (Fig. 11) and very characteristic of mycobacteria. It is interesting that only two thirds of the pentaarabinosyl motifs are tetramycolated. *M. tuberculosis* contains a variety of homologous mycolic acids differing in substitution of the α-alkyl chains (Fig. 12), which are known to be some of the largest of *Mycobacterium* (Minnikin and Goodfellow 1980). It is believed that some of the outward characteristics of mycobacteria such as the acid fastness and hydrophobicity are due to the high mycolate content of the cell wall. Two families of mycolates are known: (1) α-mycolates possessing no oxygenated functional groups and the β-hydroxy group and (2) the oxygenated mycolates. The major mycolates of *M. tuberculosis* are α-dicyclopropyl, keto and methoxy (Minnikin and Polgar 1967a,b; Qureshi et al. 1978). The absolute stereochemistries of the α-alkyl-β-hydroxy functionalities were assigned as 2R, 3R (Asselineau et al. 1970).

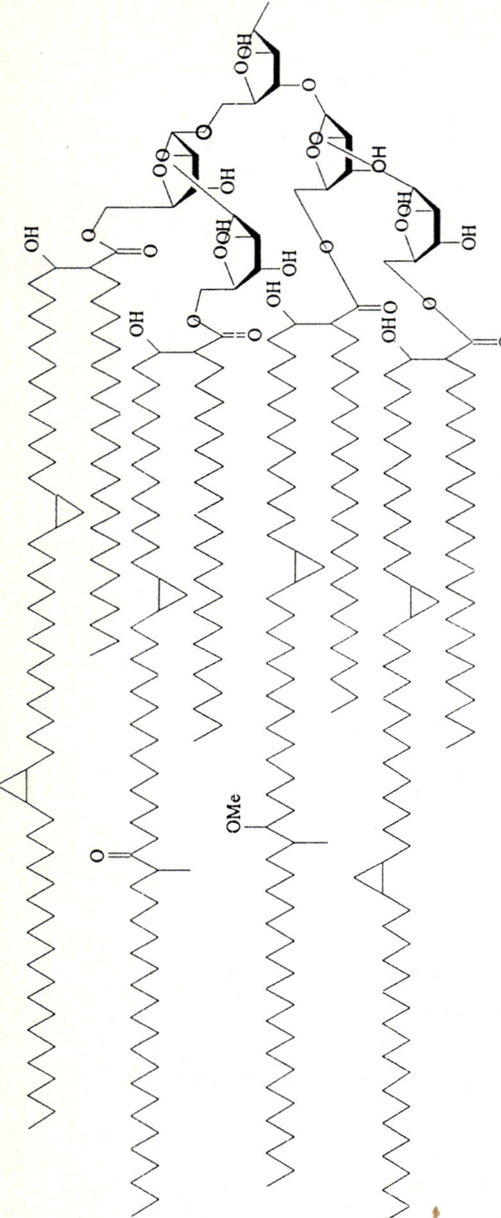

Fig. 11. Tetramycolyl pentaarabinosyl motif

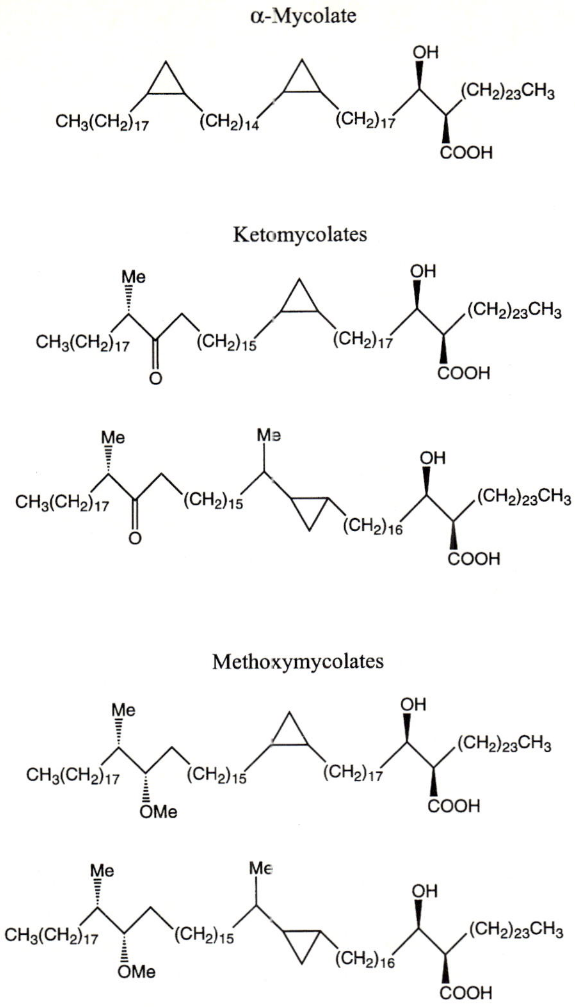

α-Mycolate

Ketomycolates

Methoxymycolates

Fig. 12. *Mycobacterium tuberculosis* mycolates

2.5 The Free Lipids of the Mycobacterial Cell Wall

Associated with the mycolic acid layer are a number of polar and apolar lipids which are thought to extend to the exterior of the cell. There is a great variety in these lipids and their distribution is often strain-dependent. The principal apolar lipids found in all strains of *M. tuberculosis* are the triacylglycerides, which are thought to be an energy reserve as they are generally contained in subcellular vacuoles and are therefore thought not to be associated with the outer cell wall. The next major apolar component of the mycobacterial cell wall capsule is a family of waxes called the phthiocerol dimycocerosates (PDIMs). The PDIMs are composed of multi-methyl branched mycocerosic acids esterified to long chain

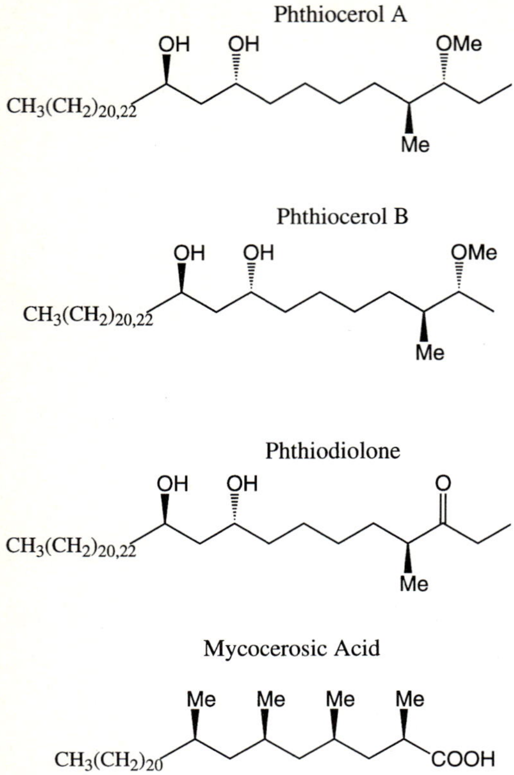

Phthiocerol A

Phthiocerol B

Phthiodiolone

Mycocerosic Acid

Fig. 13. The phthiocerols and mycocerosic acid of *Mycobacterium tuberculosis*

diols called phthiocerols (Fig. 13) (MINNIKIN et al. 1985; DAFFÉ and LANELLE 1988). The family differs in functionality only in the phthiocerol unit. Phthiocerol A is the most abundant, phthiocerol B and phthiodiolone are less common. The PDIMs are thought to be cell wall associated due to their similarity in structure to other more polar lipids and as a result of their highly apolar nature, which predetermines their positioning within the cell.

Major components of the polar lipids are the acylated trehaloses which differ considerably in their acylation patterns to trehalose and in fatty acid content. The major components are the diacyl trehaloses such as the polar acylated trehaloses defined as a complex family of 2,3-di-*O*-acyl trehaloses (BESRA et al. 1992) (Fig. 14). A closely related diacylated trehalose is 6,6′-dimycolyl trehalose (Fig. 15), which has been named cord factor (BLOCK 1950). This is somewhat of a misnomer as the name comes from the appearance of virulent strains of *M. tuberculosis* to grow in serpentine cords, with the original implication that the cord factor causes this cording effect. However 6,6′-dimycolyl is found in many saprophytic non-cording mycobacteria (GOREN 1972). The mycolic acids are esterified to the 6 and 6′ positions of trehalose and consist of the normal distribution of the cell wall bound mycolic acids, i.e., there is no specificity

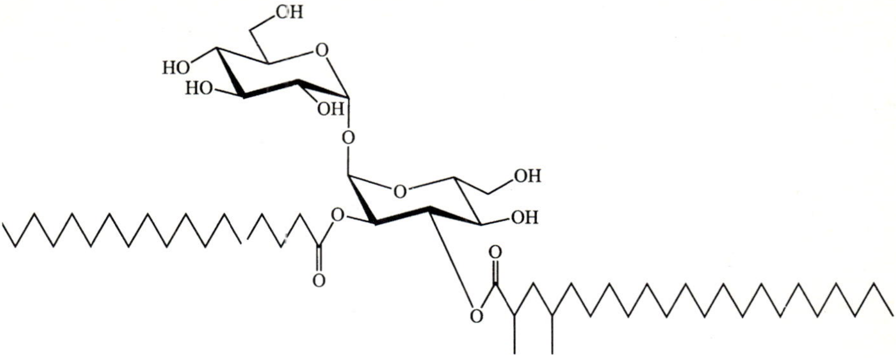

Fig. 14. 2,3-Di-*O*-acyl trehalose

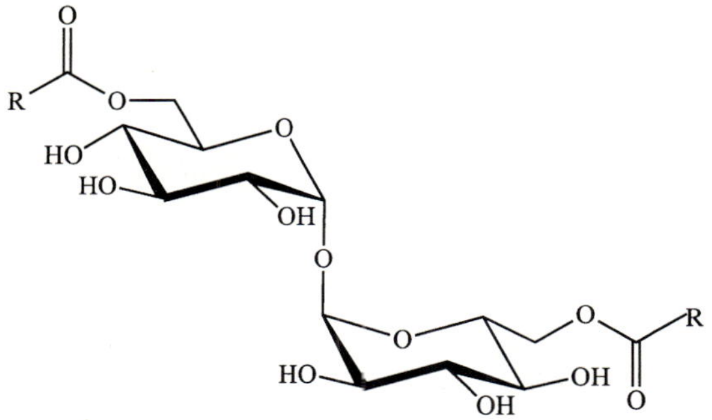

RCOO = Mycolyl

Fig. 15. Trehalose 6,6′-dimycolate

concerning which mycolic acids are bound (STRAIN et al. 1977). Other trehalose based glycolipids include 6-mycolyl trehalose and the larger lipooligosaccharides which are present in *M. tuberculosis* Canetti (DAFFÉ et al. 1991).

The tuberculosis sulfolipids are the most characteristic component of the virulent H37Rv and Erdman strains of *M. tuberculosis* (GOREN 1984) and are not found in the commonly studied attenuated H37Ra and Canetti strains of *M. tuberculosis*. Sulfolipids characteristically bind the cationic dyestuff neutral red and consist of a closely related family of trehalose-2-sulfate esters (Fig. 16) (GOREN et al. 1976). The principal fatty acids which are esterified to a variety of trehalose positions (Table 1) are the phthioceranic, hydroxyphthioceranic and palmitate or stearate fatty acids. There has been some speculation that these charged glycolipids may be involved in intracellular survival by their interaction with phagosomes preventing lysosomal fusion (D'ARCY HART and YOUNG 1988).

The mycosides, i.e., phenolic glycolipids of *M. tuberculosis*, are less abundant than in any other mycobacteria and are limited to *M. tuberculosis* Canetti

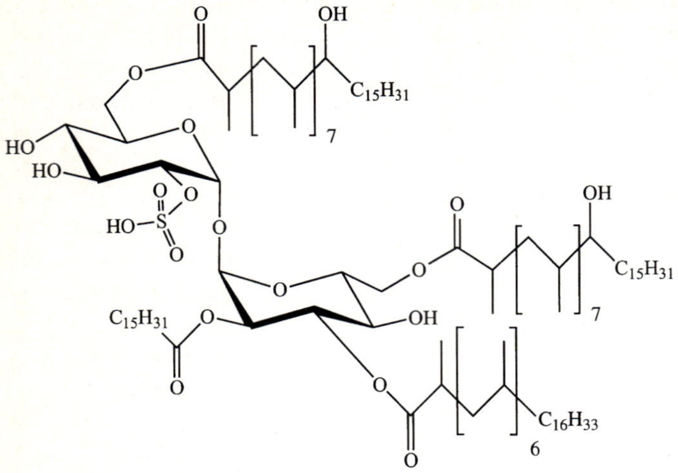

Fig. 16. 2,3,6,6'-Tetra-*O*-acyl trehalose-2-sulfate (SL-I')

strain (Daffé et al 1987). They are related in their lipid core to the phthiocerol dimycocerosates aside from the terminal phenol functionality.

2.6 Cell Wall Proteins and Enzymes

Iron transport and metabolism through the mycobacteria cell wall is well documented (Wheeler and Ratledge 1994). The low abundance, low solubility and high biological need for iron has led to the evolution of a complex iron transport system. Small proteins called exochelins are excreted through the wall and into the medium where they chelate iron. From the exochelins the iron is passed onto hydrophobic lipoprotein mycobactin T (Fig. 17) for passage through the lipid domains of the cell wall.

Table 1. Sulfolipids of *M. tuberculosis*

Sulfolipid	Acylation	Number acyl residues		
		Palmitate/ Stearate	Phthioceranate	Hydroxyphthio ceranate/
SL-I	2,4,6,6'	1	0	3
SL-I'	2,3,6,6'	1	0	3
SL-II	2,3,6,6'	1	1	2
SL-II	2,3,6,6'	1	2	1
SL-III	2,3,6	1	0	2

SL, sulfolipid

Various proteins with a variety of functions are associated with the cell wall and the plasma membrane. Firstly, there are proteins which are needed for cell wall biosynthesis such as glycosyl transferases and transpeptidases. Secondly, proteins are involved in the transport of nutrients into the cell and waste products out of the cell. Finally, there are proteins which are excreted through the cell wall. The slow growth rate and the difficulty in handling cell wall proteins in general and especially those of virulent organisms makes the study of these proteins far from facile. Thus far, little is known about the enzymes involved in cell wall biosynthesis; however, intensive research is now under way, as the isolation and characterization of such enzymes, especially those involved in the biosynthesis of novel mycobacterial components, will hopefully lead to new drug targets. An exception to this was the isolation of a peptidoglycan associated 23 kDa protein showing homology to the outer membrane protein OmpF of *E.coli* (Hirschfield et al. 1990). Mycobacterial transport proteins have been much sought, and it is assumed, due to the normal nature of the mycobacterial cytoplamic membrane, that similar bacterial type porins and transport proteins exist. In fact, a 38 kDa lipoprotein believed to be associated with the plasma membrane has been isolated from *M. tuberculosis* and shows great homology to the well characterized periplasmic phosphate binding protein of *E. coli* (Chang et al. 1994), Importantly a porin type protein has been isolated from the purified cell wall fraction of *M. chelonae* (Trias et al. 1992). It seems fair to assume that very similar porin type molecules exist in other mycobacteria including *M. tuberculosis*. The porin was defined as a 59 kDa protein possessing a water-filled channel of 2.2 mm in diameter with a low specific activity for diffusion compared to other known bacterial porins. Subsequently this porin has been shown to be cation selective due to the positioning of negative charges at its entrance and it is the only type of porin present (Trias and Benz 1993), *M. tuberculosis* is known to excrete many proteins which have been well studied due to their ready

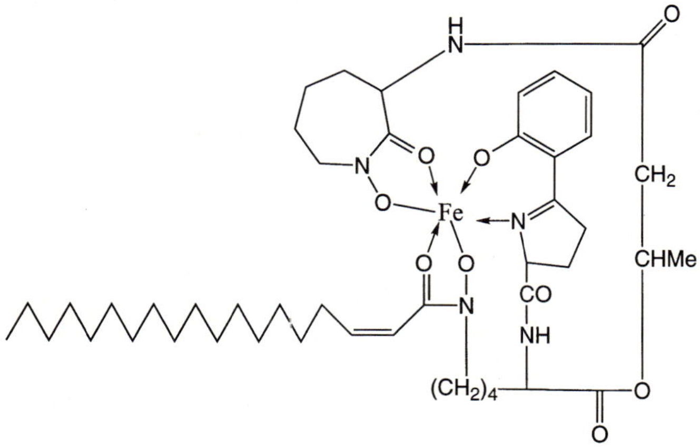

Fig. 17. Mycobactin T

availability in the culture filtrate (Andersen and Brennan 1994). Excreted proteins may be associated with virulence, enabling the tubercle bacilli to survive as an intracellular pathogen. An example of such a protein, the 23 kDa superoxide dismutase enzyme, is believed to be involved in the neutralization of the host's defense mechanisms.

3 Secondary Structure

3.1 Introduction

While the primary structures of the *M. tuberculosis* cell components, as discussed earlier, are well defined, how they combine together to form the cell envelope is less well established. However, due to the important role of the cell wall, it is necessary to understand how all the components fit together in producing their group properties. The prokaryotic cell contains few of the organelles found in a eukaryotic cell, and as a consequence its cell envelope remains the only structure which separates its inner genetic machinery from the outside world, while still permitting a constant exchange of substrates and information between the two. The bacterial cell wall not only provides rigidity against cellular turgor pressure, but it also a key factor in maintaining the shapes which are characteristic of different organisms. This is probably due to the necessity of maintaining a specified cell surface area to the total volume ratio, the latter being dictated by environmental conditions during evolution. We must also remember the law of nature which dictates that, during evolution, whatever may be attained will be attained at the most economical cost, i.e., every structure in a bacterium must serve a purpose in its adaptation to growth. In the above context, mycobacteria are interesting in that they may contain as much as 35% lipid structure in their cell wall (Azuma et al. 1968). Recent studies have unequivocally shown that the various lipids of mycobacteria are not arranged in a haphazard manner; instead the majority of them are arranged in a characteristic fashion contributing to the formation of the outer layer in the trilaminar mycobacterial cell wall (Minnikin 1991).

3.2 Ultrastructure

When the mycobacterial cell envelope is viewed using electron microscopy, a cell wall core consisting of an electron dense peptidoglycan-arabinogalactan layer (3–10 nm) and an electron transparent lipid domain zone (12 nm) can be seen (Fig. 18) (Draper 1982). Also, upon freeze fracture microscopy, mycobacteria clearly possess two distinct cleavage planes (Rastogi 1991). All established knowledge dictates that fragmentation takes place through coherent,

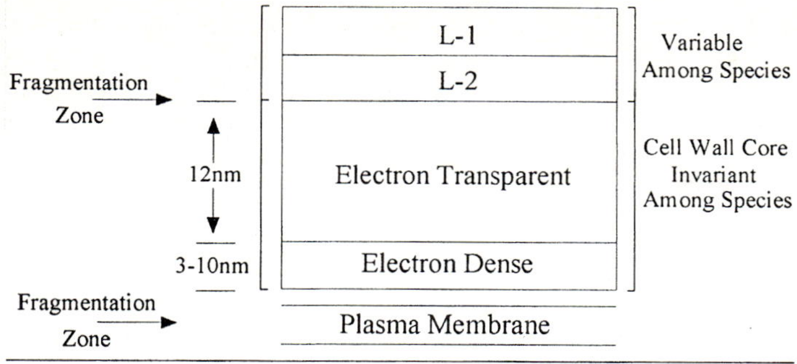

Fig. 18. The mycobacterial cell wall when viewed using electron microscopy

extended nonaqueous domains. The inner cleavage plane corresponds to the plasma membrane, based on conventional polar lipids, and the outer cleavage zone must presumably accommodate the mycolic acids and the free amphiphilic lipids (L-1, L-2). The mycolic acids are covalently bound to the arabinogalactan, so they may form the inner half of an outer membrane and the outer half of the membrane must be occupied by the complex free lipids.

The exact orientation of the mycolic acids within the cell wall is a matter of conjecture. Minnikin initially proposed that the mycolic acids were orientated perpendicular to the cell wall thus forming the inner half of the outer membrane, with the "free" glycolipids postulated to intercalate into the mycolic acids layer leading to the formation of the complete outer layer (L-1, L-2) (MINNIKIN 1982). The model was initially criticized in that there was not enough flexibility in the arabinogalactan matrix to allow the mycolates to form this conformation and not enough free lipid to form the outer layer. Recently, X-ray diffraction studies on the cell wall of *M. chelonae* have demonstrated diffraction patterns which are consistent with a parallel packed arrangement of mycolic acids and other cell wall lipids, as proposed by Minnikin (NIKAIDO et al. 1993).

3.3 Permeability Barrier

The mycobacterial cell wall is notoriously impermeable towards many drugs used in the chemotherapy of bacterial disease (CONNELL and NIKAIDO 1994). The mycolic acids which extend perpendicular to the arabinogalactan/peptidoglycan cell wall and the associated glycolipids which intercalate into the layer are believed to form a nonpermeable bilayer. The tightly packed array is believed to be even more impermeable than the Gram-negative cell wall towards the entrance of lipophilic drugs into the cell. This permeability barrier must also affect the uptake of nutrients into the cell and export of waste products out. Gram-negative bacteria solve this problem by the incorporation of porins into their outer cell wall (NIKAIDO 1992). A similar type porin, a 59 kDa protein, has been isolated from *M.*

chelonae which allows the diffusion of small nonspecific aqueous molecules into reconstructed liposomes (TRAIS et al. 1992). This porin was characterized by low activity and cation selectively due to the presence of negative charges at the entrance and a minimum pore size of 2.2 nm. The low activity (some 20- fold less than the *E. coli* porin) is probably due to the permeability barrier, as, if there was free movement of nutrients through the porin, there would be little biological significance in forming a permeability barrier with the large biosynthetic investment of the mycolic acids. The drug entry pathway for the small hydrophilic antimycobacterial drugs such as isoniazid, pyrazinamide, *p*-aminosalicylic acid and cycloserine is believed to be diffusion through this porin. Other larger more apolar antimycobacterial agents such as rifamycin and clofazimine may diffuse directly through the mycolic acid domain, although it is important to distinguish between the entry into the cell and partition into the lipid domains of the mycobacterial cell wall where the drug is inactive. In fact it has been noted that co-administration of these drugs with ethambutol, which affects cell wall biosynthesis, increases their activity (RASTOGI et al. 1991).

3.4 Lipidology

The unique lipids of the mycobacterial cell wall are the key to understanding the permeability barrier of the cell wall (MINNIKIN 1991), and it must be remembered that these complex lipids represent a considerable biosynthetic investment by the cells. It is highly probable that the mycobacterial lipids behave collectively, with a group behaviour quite different from that of the individual substituents. Chemical systems in physiological conditions are often stabilized by hydrophobic interactions among hydrocarbon chains of component molecules. Chain conformations then become a central issue, raising the following questions: Are they linear or bent. Do kinks occur? Are they transient or permanent. What physical effects do these have on other characteristics, such as ion permeability? Until recently little was known about the physical chemistry of lipids in general and especially lipid aggregates (MENGER 1991).

Knowledge gained from the modelling of simple chemical systems which operate collectively, such as monolayers and micelles, have been widely studied at a chemical level. Monolayer studies using a film balance, which measures the pressure (P_1) exerted by a film of fatty acids or phospholipids as the area it occupies is continually reduced by a moveable barrier, show three phases (Fig. 19) (MENGER 1991). The effect of alkyl substitution can thus be studied, and, as a result, has been shown to increase the size of the molecular area occupied, thus arriving at the "condensed" phase earlier, such as in the case of mycolic acids in which there is an α-substituted alkyl chain (MENGER et al. 1988). The dynamics of lipid orientation within a micellar environment have been studied through a powerful nonintrusive double ^{13}C-NMR labeling technique. The technique was used to conclusively prove that the arrgangement of the alkyl chains

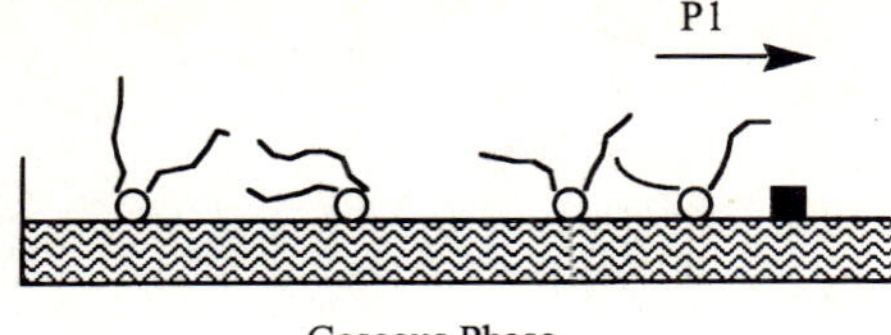

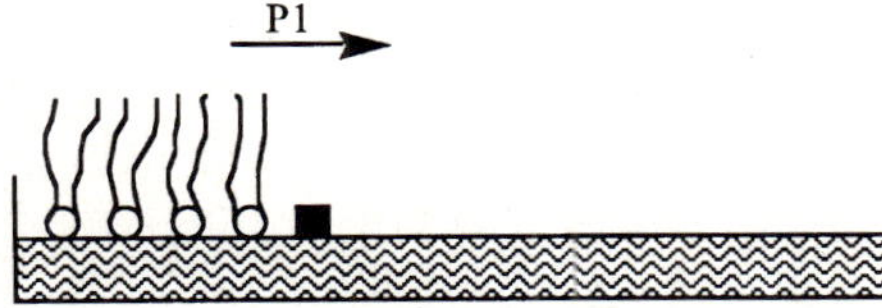

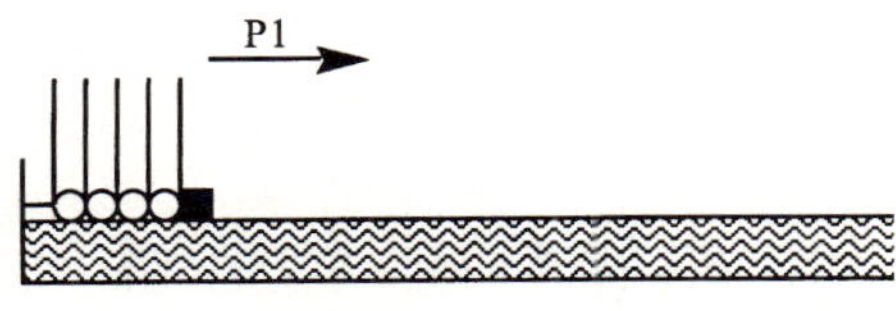

Fig. 19. Lipid monolayer phases

within a micelle is not linear as proposed but has a much more dynamic conformation (MENGER and CARNAHAN 1986).

The selected methylation of the alkyl portion of a phospholipid bilayer has a profound effect on the transition temperature (T_c) between the gel-crystalline phase, in which all the chains are parallel, and the liquid crystalline phase (MENGER et al. 1988). The greatest lowering of T_c is when the methyl branch is in the middle of the alkyl chain, as this exaggerates the natural bend by placing a kink in the chain similar to a *cis*-double bond, thus increasing the entropy of the system. Methyl branching close to the phospholipid head group has less effect since the alkyl chains are held rigidly parallel due to head group interactions (disorder increases down the alkyl chain). Branching at the end of the alkyl chain has little effect due to the high conformational freedom already present. This work can be directly applied to the mycobacterial plasma membrane where tuberculostearic acid (10-methylstearic acid) is commonly found and thus the positioning of the methyl branch is key for the fluidity of the membrane.

The biological activity of many natural compounds is based upon the conformations which they adopt. In only a few instances does nature use rigid active substances such as steroids. Often conformationally flexible substances are used which nonetheless adopt a preferred conformation (HOFFMANN 1992). The

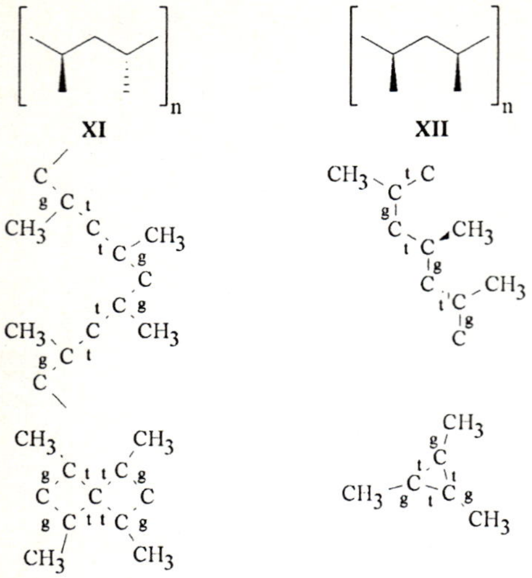

Fig. 20. The conformations adopted by syndiotatic polyproylene

g^+ g^- interaction (also known as the *syn*-pentane interaction) is an important factor in the conformation adopted in polypropiogenic natural substances such as the lipid tails of the mycobacterial outer cell wall glycolipids. The net result of this interaction is to produce characteristic helical chains depending on the relative stereochemistry of the isotactic methyl branches. In both *erythro* and *threo* cases the methyls point outwards (Fig. 20).

4 Discussion

The makeup and structure of the mycobacterial cell envelope has been discussed and has been shown to be a complex multifaceted system. The cell wall structure is certainly involved in the intracellular survival of the *M. tuberculosis* bacilli by providing a thick inert lipid "armour", thus contributing to a permeability barrier protecting it from host-killing mechanisms. Intracellular survival may also be aided by the secretion of active intermediates through the cell wall into the phagosomal space.

How the *M. tuberculosis* cell wall differs from that of its attenuated strains and saprophytic mycobacteria must be addressed. Virulent *M. tuberculosis* contains few components which differ from those of other mycobacteria, the notable exception being the sulfolipids and their possible role in the prevention of the fusion of secondary lysosomes with phagosomes (D'Arcy Hart and Young 1988). Other more subtle differences may exist in the mycolates and the complex free lipids, creating a thicker skin which may be necessary for intracellular survival. Other factors may include regulation of mycolate gene expression and

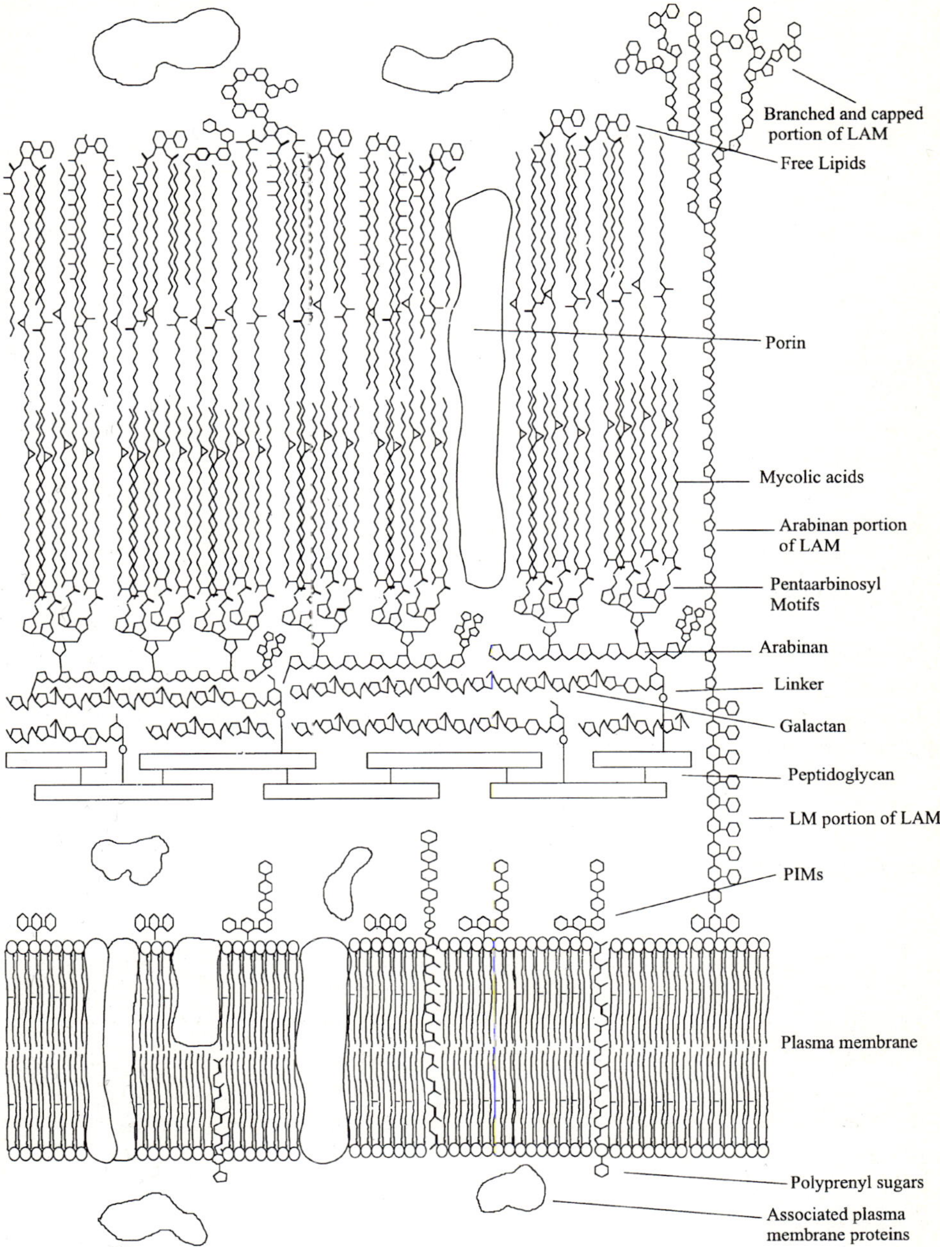

Fig. 21. The *Mycobacterium tuberculosis* cell envelope

lipid biosynthesis upon phagocytosis by *M. tuberculosis*. The slow growth rate may also be key to intracellular survival.

The evidence suggests that the mycolic acids are arranged perpendicular to the arabinogalactan/peptidoglycan cell wall. The X-ray diffraction data prove the parallel packed nature of the outside wall, conclusively dismissing other proposed cell wall structures such as that of WHEELER and RATLEDGE (1988). A large amount of information about the unique mycobacterial cell wall can be inferred from physical chemical data. The large entropy gain in the exclusion of water from the mycolic acid domain must be the driving force behind the packing of the lipids, even with the mycolic acids being tethered to the arabinogalactan. The flexible glycosyl linkages $\alpha(1\rightarrow5)$-D-Ara*f*, $\beta(1\rightarrow6)$ D-Gal*f* in the arabinogalactan are key to the ability of the mycolic acids to migrate and dock into this conformation (CONNELL and NIKAIDO 1994). There is much information to be gleaned in the functionalities of the mycolic acids. We have argued previously that the lipid functionalities serve many important structural purposes and that neither a parallel packed crystalline state or a random mess is desirable. The conserved β-hydroxy functionality, the α-substituted alkyl chain and the regular chain functionalizations are all important in determining structure and fluidity of the mycolyl layer. None of the tuberculosis mycolates contain double bonds; instead, where desaturation initially occurs, these sites are cyclopropanoated. This is a clear adaptation to growth, stopping any possible autoxidation and unwanted and possibly lethal cross-linking; also, the more constrained angles of the cyclopropyl functionality may be needed to maintain fluidity. With the very large mycolates of *M. tuberculosis,* the pertinent question may well be not whether the mycolates form a parallel packed structure with the external lipids intercalating in between, but how the fluidity of the mycolates and the bilayer is maintained. This may explain the presence of free apolar lipids containing polymethyl branching such as the phthiocerol dimycocerosates.

From this we visualize an updated model of the *M. tuberculosis* cell wall (Fig. 21). It must be remembered that the cell envelope is a dynamic system controlling the inflow of nutrients, the excretion of enzymes and proteins involved in intracellular survival and the export of waste products whilst still allowing growth and flexibility.

Acknowledgments. Research conducted by the authors was supported by two grants (AI - 18357 and AI - 38087) from the National Institute of Allergy and Infectious Diseases, National Institutes of Health.

References

Andersen AB, Brennan PJ (1994) Proteins and antigens of *Mycobacterium tuberculosis*. In: Bloom B (ed) Tuberculosis: pathogenesis, protection and control. American Society for Microbiology, Washington DC, pp 307–332

Andersen RJ (1930) The chemistry of the lipids of the tubercle bacilli. XIV. The occurrence of inosite in the phosphatide from the human tubercle bacilli. J Am Chem Soc 52: 1607–1608

Asselineau C, Tocanne G, Tocanne J-F (1970) Stéréochimie des acids mycoliques. Bull Soc Chim Fr 1455–1459

Azuma I, Yamamura Y, Fukushi K (1968) Fractionation of the mycobacterial cell wall. J Bacteriol 96: 1885–1887

Barclay R, Wheeler PR (1989) Metabolism of mycobacteria in tissues. In: Ratledge C, Stanford J, Grange JM (eds) The biology of the mycobacteria, vol 3. Academic, London, pp 37–196

Besra GS, Chatterjee D (1994) Lipids and carbohydrates of *Mycobacterium tuberculosis*. In: Bloom B (ed) Tuberculosis: pathogenesis, protection and control. American Society for Microbiology, Washington DC, pp 285–306

Besra GS, Bolton RC, McNeil MR, Ridell M, Simpson KE, Glushka J, Halbeck HV, Brennan PJ, Minnikin DE (1992) Structural elucidation of a novel family of acyltrehaloses from *Mycobacterium tuberculosis*. Biochemistry 31: 9832–9837

Besra GS, Sievert T, Lee RE, Slayden RA, Brennan PJ, Takayama K (1994) Identification of the apparent carrier in mycolic acid synthesis. Proc Natl Acad Sci USA 91: 12735–12739

Bloch H (1950) Studies on the virulence of tubercle bacilli. Isolation and biological properties of a constituent of virulent organisms. J Exp Med 91: 197–218

Brennan PJ, Draper P (1994) Ultrastructure of *Mycobacterium tuberculosis*. In Bloom B (ed) Tuberculosis: pathogenesis, protection and control. American Society for Microbiology, Washington DC, pp 271–284

Brooks BR, Daneshvar MI, Fast DM, Good RC (1987) Selective procedures for detecting femtomole quantities of tuberculostearic acid in serum and cerebrospinal fluid by frequency pulsed electron capture gas-liquid chromatography. J Clin Microbiol 25: 1201–1206

Chang Z, Choudhary A, Lathigra R, Quiocho FA (1994) The immunodominant 38-kDa lipoprotein antigen of *Mycobacterium tuberculosis* is a phosphate-binding protein. J Biol Chem 269: 1956–1958

Chatterjee D, Bozic CM, McNeil M, Brennan PJ (1991) Structural features of the arabinan component of the lipoarabinomannan of *Mycobacterium tuberculosis*. J Biol Chem 266: 9652–9660

Chatterjee D, Lowell K, Rivoire B, McNeil MR, Brennan PJ (1992) Lipoarabinomannan of *Mycobacterium tuberculosis*: capping with mannosyl residues in some strains. J Biol Chem 267: 6234–6239

Chatterjee D, Khoo K-H, McNeil MR, Dell A, Morris HR, Brennan PJ (1993) Structural definition of the non-reducing termini of mannose-capped LAM from *Mycobacterium tuberculosis* through selective enzymic degradation and fast atom bombardment-mass spectrometry. Glycobiology 3: 497–506

Connell ND, Nikaido H (1994) Membrane permeability and transport in *Mycobacterium tuberculosis*. In: Bloom BR (ed) Tuberculosis: Pathogenesis, protection and control. American Society for Microbiology, Washington DC, pp 333–352

Daffé M, Laneelle MA (1988) Distribution of phthiocerol diester, phenolic mycosides and related compounds in mycobacteria. J Gen Microbiol 134: 2049–2055

Daffé M, Lacave C, Laneelle M-A, Laneelle G (1987) Structure of the major triglycosyl-phenolphthiocerol of *Mycobacterium tuberculosis* strain Canetti. Eur J Biochem 167: 155–160

Daffé M, Brennan PJ, McNeil M (1990) Predominant structural features of the cell wall of *Mycobacterium tuberculosis* as revealed through characterizations of oligoglycosyl alditol fragments by gas chromatography/mass spectrometry and by ^{1}H and ^{13}C NMR analysis. J Biol Chem 265: 6734–6743

Daffé M, McNeil M, Brennan PJ (1991) Novel type-specific lipooligosaccharides from *Mycobacterium tuberculosis*. Biochemistry 30: 378–388

D'Arcy Hart P, Young MR (1988) Polyanionic agents inhibit phagosome-lysosome fusion in cultured macrophages. J Leukoc Biol 43: 179–182

David HL (1984) Carotenoid pigments of the mycobacteria. In: Kubica GP, Wayne LG (eds) The mycobacteria: a sourcebook, part A. Dekker, New York, pp 537–543

Draper P (1982) The anatomy of mycobacteria. In: Ratledge C, Stanford J (eds) The biology of the mycobacteria, vol 1. Academic, London, pp 9–52

Fréhel C, Rastogi N, Benichou J-C, Ryter A (1988) Do test tube grown pathogenic mycobacteria possess a protective capsule? FEMS Microbiol Lett 56: 225–230

Goodfellow M, Minnikin DE (1984) Circumscription of the genus. In: Kubica GP, Wayne LG (eds) The mycobacteria: a sourcebook, part A. Dekker, New York, pp 1–24

Goren MB (1972) Mycobacterial lipids: selected topics. Bacteriol Rev 36: 33–64f

Goren MB (1984) Biosynthesis and structure of phospholipids and sulfatides. In: Kubica GP, Wayne LG (eds) The mycobacteria: a sourcebook, part A. Dekker, New York, pp 379–416

Goren MB, Brokl O, Roller P, Fales HM, Das BC (1976) Sulfatides of *Mycobacterium tuberculosis*: the structure of the principal sulfatide (SL-1). Biochemistry 15: 2728–2735
Hirschfield GR, McNeil M, Brennan PJ (1990) Peptidoglycan-associated polypeptides of *Mycobacterium tuberculosis*. J Bacteriol 172: 1005–1013
Hoffmann RW (1992) Flexible molecules with a defined shape-conformational design. Angew Chem Int Ed Engl 31: 1124–1134
Hunter SW, Brennan PJ (1990) Evidence for the presence of a phosphatidylinositol anchor on the lipoarabinomannan and lipomannan of *Mycobacterium tuberculosis*. J Biol Chem 265: 9272–9279
Lee YC, Ballou C (1964) Structural studies on the myo-inositol mannosides from the glycolipids of *Mycobacterium tuberculosis* and *Mycobacterium phlei*. J Biol Chem 239: 1316–1327
Lee YC, Ballou C (1965) Complete structures of the glycophosholipids of mycobacteria. Biochemistry 4: 1316–1327
McNeil MR, Daffé M, Brennan PJ (1990) Evidence for the nature of the link between arabinogalactan and peptidoglycan of mycobacterial cell walls. J Biol Chem 265: 18200–18206
McNeil MR, Daffé M, Brennan PJ (1991) Location of the mycolyl ester sybstituents in the cell walls of mycobacteria. J Biol Chem 266: 13217–13223
Menger FM (1991) Groups of organic molecules that operate collectively. Angew Chem Int Ed Engl 30: 1086–1099
Menger FM, Carnahan DW (1986) A new method for studying chain conformation - proof of nonradial binding to micelles - chain bending at an enzyme surface. J Am Chem Soc 108: 1297–1298
Menger FM, Wood MG, Zhou QZ, Hopkins HP, Fumero J (1988) Thermotropic properties of synthetic chain-substituted phosphatidylcholines - effect of substituent size, polarity, number, and location on molecular packing in bilayers. J Am Chem Soc 110: 6804–6810
Mills J, Masur H (1990) AIDS related infections. Sci Am 263: 50–57
Minnikin DE (1982) Lipids: complex lipids, their chemistry, biosynthesis and roles. In: Ratledge C, Stanford J (eds) The biology of the mycobacteria, vol 1. Academic, London, pp 95–187
Minnikin DE (1991) Chemical principles in the organization of lipid components in the mycobacterial cell envelope. Res Microbiol 142: 423–427
Minnikin DE, Goodfellow M (1980) Lipid composition in the classification and identification of acid-fast bacteria. In: Goodfellow M, Board RG (eds) Microbiological classification and identification. Academic, London, pp 189–256
Minnikin DE, Polgar N (1967a) The mycolic acids from human and avian tubercle bacilli. J Chem Soc Chem Comm 916–918
Minnikin DE, Polgar N (1967b) The methoxymycolic and ketomycolic acids from human tubercle bacilli. J Chem Soc Chem Comm 1172–1174
Minnikin DE, Dobson G, Goodfellow M, Magnusson M, Ridell M (1985) Distribution of some mycobacterial waxes based on the phthiocerol family. J Gen Microbiol 131: 1375–1381
Nikaido H (1992) Porins and specific channels of bacterial outer membranes. Mol Microbiol 6: 435–442
Nikaido H, Kim S-H, Rosenberg EY (1993) Physical organization of lipids in the cell wall of *Mycobacterium chelonae*. Mol Microbiol 8: 1025–1030
Okuyama H, Nojima S (1965) Studies on hydrolysis of cardiolipin by snake venom phospholipase A. J Biochem 57: 529–537
Okuyama H, Kankura T, Nojima S (1967) Positional distribution of fatty acids in phospholipids of mycobacteria. J Biochem 61: 732–737
Orme IM, Andersen P, Boom WH (1993) T cell response to *Mycobacterium tuberculosis*. J Infect Dis 167: 1481–1497
Petit J-F, Lederer E (1984) The structure of the mycobacterial cell wall (peptidoglycan). In: Kubica GP, Wayne LG (eds) The mycobacteria: a sourcebook, part A. Dekker, New York, pp 301–311
Qureshi N, Takayama K, Jordi HC, Schnoes HK (1978) Characterization of the purified components of a new homologous series of α-mycolic acids from *Mycobacterium tuberculosis* H37Ra. J Biol Chem 253: 5411–5417
Rastogi N (1991) Recent observations concerning structure and function relationships in the mycobacterial cell envelope: elaboration of a model in terms of mycobacterial pathogenicity, virulence and drug resistance. Res Microbiol 142: 464–475
Rastogi N, Goh KS, Labrousse V (1992) Activity of clarithromycin compared with those of other drugs against *Mycobacterium paratuberculosis* and further enhancement of its extra cellular and intracellular activities by ethambutol. Antimicrob Agents Chemother 36: 2843–2846
Ratledge C (1982) Lipids: cell composition, fatty acid biosynthesis. In: Ratledge C, Stanford J (eds) The biology of the mycobacteria, vol 1. Academic, London, pp 53–94

Ridell M, Minnikin DE, Parlett JH, Mattsby-Baltzer I (1986) Detection of mycobacterial lipid antigens by a combination of thin-layer chromatography and immunostanding. Lett Appl Microbiol 2: 89–92
Russell DG et al (1994) Lack of acidification of mycobacterium phagosomes produced by exclusion of the vesicular proton-ATPase. Science 263: 678–681
Sarma GR, Chandramouli V, Venkitasubramanian TA (1970) Occurrence of phosphonolipids in mycobacteria. Biochim Biophys Acta 218: 561–563
Strain SM, Toubiana R, Ribi E, Parker R (1977) Separation of the mixture of trehalose 6,6-dimycolates comprising the mycobacterial glycolipid fraction. Biochem Biophys Res Commun 77: 449–456
Sudre P, Tendam G, A Kochi (1992) Tuberculosis-a global overview of the situation today. Bull WHO 70: 149–159
Takayama K, Goldman DS (1970) Enzymatic synthesis of mannosyl-1-phosphoryl-decaprenol by a cell-free system of *Mycobacterium tuberculosis*. J Biol Chem 245: 6251–6257
Takayama K, Schnoes HK, Semmler EJ (1973) Characterization of the alkaline-stable mannophospholipids of *Mycobacterium smegmatis*. Biochim Biophys Acta 136: 212–221
Trais J, Benz R (1993) Characterization of the channel formed by the mycobacterial porin in lipid bilayer membranes. J Biol Chem 268: 6234–6240
Trais J, Jarlier V, Benz R (1992) Porins in the cell wall of mycobacteria. Science 258: 1479–1481
Venisse A, Berjeaud J-M, Chaurand P, Gilleron M, Puzo G (1993) Structural features of lipoarabinomannan from *Mycobacterium bovis* BCG. J Biol Chem 268: 12401–12411
Wheeler PR, Ratledge C (1988) Metabolism in *Mycobacterium leprae, M. tuberculosis* and other pathogenic mycobacteria. Br Med Bull 44: 547–561
Wheeler PR, Ratledge C (1994) Metabolism of *Mycobacterium tuberculosis*. In: Bloom B (ed) Tuberculosis: pathogenesis, protection and control. American Society for Microbiology, Washington DC, pp 353–385
Wolucka BA, McNeil MR, Hoffmann E, Chojnacki T, Brennan P. J (1994) Recognition of the lipid intermediate for arabinogalactan/arabinomannan biosynthesis and its relation to the mode of action of ethambutol on mycobacteria. J Biol Chem 269: 23328–23335

The Molecular Genetics of *Mycobacterium tuberculosis*

G.F. Hatfull

1 Introduction

The discipline of mycobacterial molecular genetics was effectively born in the late 1980s with the description of methods for the efficient introduction of DNA into mycobacteria and the subsequent development of simple vector systems. Although the slow growth and pathogenicity of *Mycobacterium tuberculosis* present major challenges for the molecular geneticist, many important questions

Department of Biological Sciences, University of Pittsburgh, Pittsburgh, PA 15260, USA can be addressed by using fast-growing mycobacterial relatives (such as *Mycobacterium smegmatis*) and bacteriophages as relevant model systems. The importance of this strategy is illustrated by comparing the time required for obtaining recombinant mycobacteria from transformation experiments, which is typically 3–4 days for *M. smegmatis* and 3–4 weeks for *M. tuberculosis*; bacteriophage plaques on lawns of *M. smegmatis* or *M. tuberculosis* can be visualized in 24 h and 4 days, respectively. While the attenuated anti-tuberculosis vaccine strain bacille Calmette-Guérin (BCG) grows as slowly as *M. tuberculosis*, its use avoids some of the biosafety concerns associated with the cultivation of highly pathogenic strains. The relatively young field of mycobacterial genetics thus includes the use of a number of different mycobacterial systems each of which is, however, relevant to our understanding of *M. tuberculosis*. Our current understanding of these molecular genetic systems and the genetic basis of mycobacterial growth will be briefly reviewed in this chapter.

2 The Genome of *Mycobacterium tuberculosis*

M. tuberculosis contains a single chromosome approximately 4×10^6 bp in length and which in general appears to be similar in structure and organization to the genomes of other eubacterial species; like its other mycobacterial relatives, it has a high G+C content of about 65% (Clark-Curtiss 1990). The genome can be conveniently visualized by pulsed-field gel electrophoresis (PFGE), although few restriction enzymes yield a number of fragments suitable for establishment of a map. Most six-base recognition enzymes yield more than 50 fragments although *Dra* I (cuts at TTTAAA) to produce 35 fragments and has been used to examine polymorphisms among *M. tuberculosis* strains (Cole and Smith 1994). The eight-base A/T recognition enzymes *Pac* I (TTAATTAA) and Swa I (ATTTAAAT) do not cut the *M. tuberculosis* genome at all (Cole and Smith 1994).

Although very few mycobacterial genes have been described it appears as though the genome organization is generally similar to that in other bacteria. For example, a 35 kb region of the *M. tuberculosis* genome has been sequenced which revealed about 30 closely spaced genes transcribed from both strands of the DNA; about half of these genes have significant sequence similarity to previously described genes of known function (Cole and Smith 1994). Taking these observations together with the calculated size of the genome, a reasonable estimate for the total complement of *M. tuberculosis* genes is about 3500.

M. tuberculosis differs from most other bacteria with respect to genes encoding stable RNAs. For example, *M. tuberculosis* contains only a single operon encoding the 16S and 23S ribosomal RNAs, in contrast to the seven operons in *E. coli* (Bercovier et al. 1986); *M. smegmatis* contains two rRNA operons (Bercovier et al. 1986). The number of tRNA genes in any mycobacterial species is not known, but the absence of mycobacterial suppressor strains is consistent with there being a limited repertoire of these genes. It is not clear whether the relatively small number of stable RNA genes is related to the slow growth rate of the mycobacteria.

An effort is underway to determine the entire DNA sequence of the *M. tuberculosis* genome using a multiplex strategy and cosmid derivatives each containing approximately 40 kb of *M. tuberculosis* DNA (COLE and SMITH 1994). These cosmids are first mapped by fingerprinting and hybridization to produce an ordered cosmid map and then sequenced using a high through-put sequencing strategy. Using these methods about one third of the *M. leprae* genome has been sequenced (COLE and SMITH 1994) and it is likely that the *M. tuberculosis* genome will be completely sequenced within the next few years.

2.1 Naturally Occurring Plasmids

Naturally occurring plasmids are not commonly found in clinical isolates of *M. tuberculosis* (FALKINGHAM and CRAWFORD 1994). Thus, while plasmids are known to be carriers of virulence determinants and antibiotic resistance genes in other bacterial systems it is unlikely that they contribute to these phenomena in *M. tuberculosis*. In contrast, plasmids are common inhabitants of *M. avium* and apossible role in virulence or drug resistance cannot be excluded (FRANZBLAU et al. 1986; CRAWFORD and BATES 1986). In the United States it is rare that *M. avium* isolates from AIDS patients do not contain plasmids (JUCKER and FALKINGHAM 1990) although in the United Kingdom only about 50% of isolates carry plasmids (HELLYER et al. 1991), a proportion similar to that observed with nonclinical isolates (MEISSNER and FALKINGHAM 1986). Presumably these plasmids (which vary considerably in size and number) do not encode essential virulence functions but may nevertheless influence the growth or survival of the bacteria in particular environments.

While many mycobacterial plasmids have been described from a variety of non-*M. tuberculosis* sources, only one, pAL5000, has been characterized in detail. This plasmid was isolated from *M. fortuitum*, is relatively small and the complete 4821 bp sequence has been described (LABIDI et al. 1992). The sequences required for autonomous replication are included in a 2.6 kb segment of the plasmid which can function in other mycobacteria including *M. tuberculosis*, BCG and *M. smegmatis* (SNAPPER et al. 1988; RANES et al. 1990) and which has been used to construct mycobacterial cloning vectors (see below).

2.2 Genetic Transformation

Although there were several early reports of mycobacterial transformation the first description of efficient DNA uptake was in 1987 by JACOBS and colleagues. Since little was known about small autonomously replicating plasmids, antibiotic resistance genes, or gene expression signals in the mycobacteria, a systematic approach was required. In the first step, a method was developed for the uptake of mycobacteriophage DNA by *M. smegmatis* spheroplasts using the formation of phage plaques as an assay (Jacobs et al. 1987). By constructing bacteriophage-plasmid hybrids that could replicate in *E. coli* as plasmids and in mycobacteria and phages (shuttle phasmids), any foreign DNA could be introduced. Secondly, shuttle phasmids were constructed with a temperate phage (L1) such that lysogens could be recovered carrying a stable integrated copy of the viral genome in the bacterial chromosome (Snapper et al. 1988). The use of temperate shuttle phasmids thus allowed the introduction of drug resistance genes and the determination of which of these might be useful for direct selection of transformants; one such gene is the *aph* kanamycin resistance gene derived from Tn*903*, although a variety of selectable genetic markers are now available (Hatfull 1993; see below). Finally, with a suitable assay in hand it was possible to show that electroporation was a simple and efficient means of DNA uptake and it is now the preferred method for transformation (Snapper et al. 1988; Ranes et al. 1990).

Although transformation of both fast- and slow-growing mycobacteria is now relatively simple the differences in growth rates greatly affects the timing of the process. Following electroporation of *M. smegmatis* and plating on solid media, transformant colonies can be observed after about 3 days growth at 37 °C. However, transformants of either *M. tuberculosis* or BCG cannot be seen until about 3 weeks of incubation at 37 °C.

3 Mycobacterial Cloning Vectors

3.1 Extrachromosomal Plasmid Vectors

Small plasmid vectors have been constructed using replicons from naturally occurring extrachromosomal plasmids; the most commonly used vectors are derived from the *Mycobacterium fortuitum* plasmid pAL5000, which replicates in both fast and slow-growing mycobacteria and has a copy number of approximately three to five per cell (Stover et al. 1991). These shuttle plasmids are not particularly stable in *M. smegmatis* although their presence can be simply maintained by antibiotic selection. Most of the derivatives are shuttle vectors that can also replicate in *E. coli*, simplifying the purification and manipulation of the DNA; examples of such vectors are pYUB12 (Snapper et al. 1990), pMD31 (Donnelly-Wu et al. 1993) and pRR3 (Ranes et al. 1990). Shuttle vectors have been described that utilize other replicons that function in mycobacteria including plasmids pMSC262 (Goto et al. 1991), RSF1010 (Hermans et al. 1991) and pNG2 (Radford and Hodgson 1991).

pAL5000-derived plasmid vectors efficiently transform the slow-growing mycobacteria such as *M. tuberculosis* and BCG using electrocompetent cells

(>10^5 transformants/μg DNA). However, these plasmids transform *M. smegmatis* mc^{2}6 (a derivative of the type strain ATCC607) very poorly, yielding only one to ten transformants/μg DNA. Jacobs and colleagues reasoned that the transformants that were obtained may arise from mutant cells that were more competent for transformation (SNAPPER et al. 1990). Thus one such transformant was cultured in the absence of antibiotic selection, a plasmid free derivative obtained and tested for an improved transformation frequency. The *M. smegmatis* strain isolated (mc^{2}155) did indeed yield transformants at high frequency (>10^5 transformants/μg DNA) and is the strain of choice for *M. smegmatis* transformation (SNAPPER et al. 1990). The genetic basis and molecular explanation for the efficient plasmid transformation (ept) phenotype have yet to be investigated.

3.2 Integration-Proficient Plasmid Vectors

While extrachromosomal vectors are suitable for many purposes, they are often poorly maintained in the absence of antibiotic selection. In addition, plasmid-borne genes may be expressed differently than when present on the bacterial chromosome especially if their expression is normally regulated. To circumvent these concerns, integration-proficient vectors have been developed in which the entire plasmid DNA is integrated site-specifically into the bacterial chromosome (LEE et al. 1991). These plasmids do not contain a mycobacterial origin of DNA replication but instead carry a segment of DNA containing the attachment site (*attP*) and integrase gene (*int*) from a temperate bacteriophage such as L5 (LEE et al. 1991). When this DNA is introduced into either slow- or fast-growing mycobacteria by electroporation, the *int* gene is expressed and catalyzes a site-specific recombination event between the *attP* site and the bacterial attachment site, *attB* resulting in the insertion of the entire plasmid into the bacterial chromosome; the event can be very efficient and yields a high frequency of transformants (>10^5 transformants/μg DNA; LEE et al. 1991). Transformation is not significantly affected by the *ept* genotype of the *M. smegmatis* strain used and these vectors efficiently transform both mc^{2}6 and mc^{2}155. Moreover, the chromosomal *attB* site overlaps a host tRNA gene that is highly conserved such that these vectors may transform a diverse collection of mycobacterial species. Integration-proficient vectors have been constructed using the *attP-int* components from phages FRAT 1 (HAESELEER et al. 1993) and the Streptomyces plasmid pSAM2 (MARTIN et al. 1991) in addition to phage L5 (LEE et al. 1991).

A characteristic feature of phage integration is that excision is not a simple reversal of the integration reaction and requires a specific phage-encoded protein excisionase. Thus integration-proficient vectors constructed without the phage excisionase gene are stably maintained even in the absence of antibiotic selection and yet still transform with a high efficiency (LEE et al. 1991). This combination of efficiency and stability is important for several genetic approaches including the construction of live recombinant BCG vaccines (STOVER et al. 1991) and in vivo selection of virulent recombinants (PASCOPELLA et al. 1994), in which maintenance of extrachromosomal plasmids by antibiotic selection is not feasible.

3.3 Selectable Markers

A variety of selectable markers for transformation of the mycobacteria has been described. The majority of these are antibiotic resistance genes such as those derived from transposable elements described in other bacteria. The first described and most commonly used of these are the *aph* genes encoding kanamycin resistance derived from Tn*903* and Tn*5*. These function in both the slow- and fast-growing mycobacteria and appear to be expressed from their own signals; spontaneous kanamycin resistance in mycobacteria occurs only at low frequencies.

An important feature of kanamycin is that it is quite stable over the extended periods of incubation needed for growth of BCG and *M. tuberculosis*; antibiotics such as tetracycline can be used for selection in *M. smegmatis* but its poor stability precludes its use for the slow-growing mycobacteria. However, several additional antibiotic-selectable markers for the mycobacteria have been described (HATFULL 1993).

β-Lactam-based selectable schemes are generally not useful for the mycobacteria since most laboratory strains of mycobacteria contain an endogenous β-lactamase and are resistant to penicillins. However, expression of a mutant form of the *M. fortuitum blaF* gene (*blaF**) in *M. smegmatis* confers levels of ampicillin resistance much greater than the endogenous enzyme and could perhaps be used for direct selection (TIMM et al. 1994).

Although antibiotic resistance genes provide good selectable markers for most laboratory experiments, there are some particular situations in which they are not suitable. For example, recombinant BCG intended for use as live human vaccines should ideally be devoid of antibiotic resistance genes to avoid possible dissemination of these to other pathogenic bacteria in vivo (DONNELLY-WU et al. 1993). In addition, the introduction of antibiotic resistance genes into *M. tuberculosis* significantly enhances the biological hazard. One possible alternative selectable marker uses the bacteriophage L5 immunity gene 71 to confer protection from phage infection and has been demonstrated to function well in *M. smegmatis* (DONNELLY-WU et al. 1993); this scheme appears to also work with at least some vectors in BCG (HATFULL 1994). Complementation of auxotrophic mutations may provide additional selectable schemes for the mycobacteria.

4 Recombination

4.1 General Recombination

General or homologous recombination is an important component of any genetic system since it can be used for allele exchange, gene replacement and gene mapping. In *M. smegmatis*, homologous recombination appears to operate similarly to other well-characterized bacterial systems and has been used to integrate nonreplicating plasmids containing *M. smegmatis* DNA into the genome following electroporation (HUSSON et al 1990). In these experiments both of the expected classes of products were observed, gene duplication resulting from a single DNA crossover and gene replacement from a pair of crossover events. Homologous recombination thus appears to be a useful general tool for *M. smegmatis* genetics.

When similar recombination experiments were performed in *M. tuberculosis* and BCG a different result was observed (KALPANA et al. 1991). Following electroporation, transformants were obtained whether or not the plasmid contained sequences homologous to the bacterial genome and transformation was more efficient with linearized DNAs than closed circular DNAs. Further examination indicated that there was a high proportion of illegitimate recombination events (i.e., between nonhomologous DNAs) and less than 1% of the transformants arose from true homologous recombination; this process thus may be more useful for insertional mutagenesis than for gene replacement (KALPANA et al. 1991). More recently a higher frequency of transformants from homologous recombination events (two of ten) was observed at the *uraA* locus of BCG (ALDOVANI et al. 1993) although both were the result of single crossovers. Gene replacement in *M. tuberculosis* and BCG thus remains an elusive goal and until resolved presents a barrier to identifying gene functions by reverse genetics and full usage of information from the genome sequencing project.

While the reasons for the apparently poor frequencies of homologous recombination in *M. tuberculosis* and BCG remain obscure, at least part of the explanation may involve the rather unusual *recA* genes. COLSTON and colleagues have shown that the *recA* genes of *M. tuberculosis* and BCG encode a large protein of 85 kDa which is then processed by a protein splicing event. The final products are a 38 kDa protein with sequence similarity to RecA proteins from other bacteria and a 47 kDa "intein" protein which is removed from the middle of the 85 kDa primary gene product. The processing event appears to be autocatalytic and can occur in *E. coli* as well as mycobacteria (DAVIS et al. 1991, 1992). The significance of this unexpected molecular complication is not obvious, although it has been suggested that it may be related to mycobacterial virulence since only the pathogenic species (*M. tuberculosis, M. microti, M. bovis, and M. leprae*) have intein-containing *recA* genes; the saprophytic species including *M. smegmatis* carry *recA* genes without inteins (COLSTON and DAVIS 1994).

4.2 Transposition

A number of transposable elements have been identified in a variety of mycobacterial species including *M. tuberculosis*; however, all of the transposons described so far are of the class I type, either insertion sequences (ISs) or composite elements (McAdam et al. 1994). These transposable elements are important for two distinct purposes, as a sensitive measure of strain differences that can be used for epidemiological studies and as genetic tools for creating mutations and insertion of foreign DNA. Molecular epidemiology is discussed in the chapter by McCray and Castro and will not be further discussed here.

The use of transposons for insertional mutagenesis requires an element that moves at reasonably high frequencies and an efficient delivery system. Several mycobacterial insertion sequences have been demonstrated to move to new locations including IS*900*, IS*6110* and IS*6120* in *M. smegmatis* and IS*6100* and IS*1096* in *M. smegmatis* and BCG (McAdam et al. 1994); however, like most bacterial transposons the frequency of hopping is typically low. Examination of target specificity shows that IS*6110*, IS*6120* and IS*1096* all select their target sites with little or no sequence specificity suggesting that these could be used as mutagens.

Since the mycobacterial transposons move at low frequency an efficient delivery system is needed. For example, although Gicquel and colleagues (Martin et al. 1990) were able to observe transposition of Tn*610* from nonreplicating plasmids following electroporation, the low frequency of tranposition resulted in only 1–20 events/μg DNA. A more efficient delivery system has been described that uses plasmid vectors that are temperature-sensitive for extrachromosomal replication such that they do not replicate at 39 °C (Guilhot et al. 1994). Transposition of Tn611 (a kanamycin resistant derivative of Tn*610*) from the plasmid onto the chromosome was then detected by selection for kanamycin-resistant colonies at the higher temperature (Guilhot et al. 1994). Characterization of the products showed that the predominant event was insertion of the entire plasmid into the chromosome as a consequence of a replicative transposition process. This delivery system was effective for the isolation of auxotrophic mutants of *M. smegmatis* but has yet to be utilized in *M. tuberculosis* (Guilhot et al. 1994). Since at least in some instances an active pAL5000-derived origin of replication is not tolerated on the mycobacterial chromosome it is also possible that mutants derived from this process may be cold-sensitive for growth.

An alternative strategy for transposon delivery is to use nonreplicating mycobacteriophages, similar to the use of lambda for the efficient delivery of Tn*5* and Tn*10* in *E. coli*. While several mycobacteriophages including the well-characterized L5 (Hatfull and Sarkis 1993) are candidates for such systems, they have yet to be adapted for this purpose. The advantage of such an approach is that phages typically infect their hosts very efficiently (see below) and transposition events can be easily selected even at low transposition frequencies (Jacobs and Bloom 1994).

Genes involved in mycobacterial pathogenicity are not necessarily essential for mycobacterial growth, as demonstrated by the avirulent strains *M. tuberculosis* H37Ra and BCG. However, isolation and characterization of the defective genetic loci in these strains are complicated by their poor definition and probable accumulation of multiple defects (PASCOPELLA et al. 1994). Transposon mutagenesis should provide a powerful system for the isolation of mycobacterial mutants specifically defective in pathogenesis, leading to the characterization of virulence genes and new vaccines strains (JACOBS and BLOOM 1994).

4.3 Site-Specific Recombination

Site-specific recombination involves the reciprocal exchange of DNA via short regions of identity and is catalyzed by the action of specific enzymes. These events are widespread among bacterial species and includes phage integration, transposon cointegrate resolution and phase variation. However, the only examples to date in the mycobacteria are those of phage integration and related systems which have been adapted for vector development as described above. In general the mechanism of phage L5 integration appears to be similar to that described for phage lambda even though there is considerable divergence between the integrase proteins that catalyze the event (LEE et al. 1991). Interestingly, L5 integrase-mediated recombination in vitro exhibits a strong requirement for a protein factor present in extracts of *M. smegmatis* (LEE and HATFULL 1993). This mycobacterial integration host factor (mIHF) shares several properties with the small heat stable proteins of *E. coli*, but neither *E. coli* IHF, HU or crude extracts of *E. coli* can substitute for mIHF in L5 integrase recombination (LEE and HATFULL 1993). The physiological role of mIHF is not known.

5 Gene Expression

5.1 Transcription

Although relatively little is known about gene expression in mycobacteria, the emerging picture shows it to be generally similar to that in other bacteria. For example, the RNA polymerase of *M. smegmatis* has been characterized, shown to be composed of a β, β', α_2 subunit structure and shown to transcribe mycobacterial DNA in vitro (LEVIN and HATFULL 1993). Transcription initiation occurs at specific loci although sequence specificity of mycobacterial promoters and the number of different classes of promoters have yet to be described.

Early anecdotal observations indicated that mycobacterial genes are not expressed well from their own transcriptional signals in *E. coli* (JACOBS et al. 1986; DALE and PATKI 1990). More recently, mycobacterial DNAs with promoter activity were isolated using a reporter gene system and most were indeed found to be much less active in *E. coli* than in *M. smegmatis* (Das Gupta et al. 1993). The transcription initiation sites of only very few mycobacterial promoters have been empirically determined, and thus it is not yet possible to assert that there are differences in promoter specificity between mycobacteria and *E. coli*, although this seems likely. However, it is noteworthy that the BCG hsp60 promoter is recognized by both *E. coli* and *M. smegmatis* RNA polymerase in vitro, although they respond differently to the superhelical state of the DNA template (Levin and Hatfull 1993). Further characterization of the mycobacterial promoters including the sites of transcription initiation and elucidation of consensus sequences is needed before it will be possible to accurately predict transcriptional units from DNA sequence alone.

Analysis of 5′ ends of mycobacterial RNA species can be accurately determined by S1 nuclease protection or primer-extension using in vivo isolated RNA. In many cases, this 5′ end is likely to correspond to the nucleotide used for transcription initiation although it could be the product of a specific processing event. One method of discerning between these possibilities is through use of purified mycobacterial RNA polymerase in an in vitro system in which transcription initiation can be observed in the absence of processing. An in vitro transcription system utilizing *M. smegmatis* RNA polymerase has been described (Levin and Hatfull 1993) but not one derived from *M. tuberculosis*.

Regulated promoters are of interest to the microbial geneticist since they provide the ability to switch genes on and off as desired. Stover et al. (1991) described regulated expression of the *E. coli lacZ* gene when fused to the BCG hsp60 promoter such that expression was induced by heat shock or other stress conditions. Interestingly, induction was only observed when the promoter was integrated into the bacterial chromosome and was constitutively active when on an extrachromsomal plasmid (Stover et al. 1991). Mahenthiralingam et al. (1993) have characterized the acetamidase gene of *M. smegmatis* and demonstrated that its expression is strongly induced by the addition of acetamide although the sequences required for regulated expression are not yet known.

Transcription termination of mycobacterial genes is poorly understood. However, sequences have been observed at the ends of genes (Hatfull and Sarkis 1993) with structural features similar to the Rho-independent class of terminators described in other bacteria (i.e., a stem-loop followed by five to six U residues in the RNA) and are likely to act to terminate transcription. It is not known if mycobacteria also utilize a Rho-dependent mechanism for transcription termination.

5.2 Translation

Little is known about the detailed mechanism of protein synthesis and its regulation in mycobacteria. However, a standard genetic code appears to be used and this is supported by direct comparison of empirically determined amino acids sequences with DNA sequences for more than half of the 64 available codons (HATFULL 1994; TIMM et al. 1994). There is considerable bias in the usage of codons, with a general preference for those with G or C in the third position, an expected behavior given the overall high G+C content of the DNA, although the bias is stronger for some codon groups than others. These codon usage biases are particularly useful since they allow accurate prediction of the location of genes within any segment of mycobacterial DNA sequence (HATFULL 1994).

The signals for initiation of translation appear to be generally similar to those described in other bacteria although in only rather few cases have the initiation codons been confirmed by direct NH_2-terminal amino acid sequencing. Comparison of phage L5 structural protein sequences with their gene sequences indicates that mycobacteria can not only use AUG and GUG for initiation but can also utilize UUG (HATFULL and SARKIS 1993). The codon UUG is used infrequently for initiation in phage L5 (five genes from a total of 85; HATFULL and SARKIS 1993) and has yet to be described for any host genes; AUG and GUG appear to be used with similar frequency for both phage and host genes (HATFULL and SARKIS 1993).

The mechanism for selection of the initiation codon appears to be similar to that in other bacteria in which a specific region within the 3′ end of the 16S ribosomal RNA pairs with a ribosome binding site (Shine-Dalgarno sequence) within the mRNA. This ribosome binding site is usually related to the sequence 5′-AGGAGGA located 3–12 bases upstream of the initiation codon and appears to be associated with many mycobacterial genes. Moreover the sequence of the 16S rRNA of several mycobacterial species has been determined and the pairing sequence is highly conserved. Interestingly, at least one departure of this organization has been described for the *blaF* gene of *M. fortuitum*, in which the proposed initiation codon is located at the extreme 5′ end of the transcript (TIMM et al. 1994), although the actual initiation codon used has not been demonstrated by direct amino acid sequencing. This suggests an alternative mechanism for translation initiation although the organization is not unprecedented and has been observed previously in the *cI* gene of phage lambda and some *Streptomyces* genes (STROHL 1992).

5.3 Reporter Gene Systems

The measurement of gene expression and its regulation is facilitated through the use of reporter genes whose activity can be readily detected and assayed. In general these reporter genes can be used in two separate ways, as a selection or color-based assay on solid media or as a quantitative assay of liquid cultures. Typically, the reporter gene can be coupled to a regulatory region of interest by either transcriptional or translational fusion to a known coding region; the results of such experimental strategies do, however, have to be interpreted with caution since the precise nature of the fusion can greatly influence the activity.

Several reporter genes have been described for use in the mycobacteria. For example, the *lacZ* gene of E. *coli* has been expressed from mycobacterial expression signals and activity detected by conversion of the β-galactosidase substrate X-Gal (added to solid medium) to a blue color (BARLETTA et al. 1992; TIMM et al. 1994). The amount of β-galactosidase produced by the fusion strain can be assayed quantitatively using a standard Miller assay; however, the cells must be efficiently disrupted for these assays and failure to do so can result in considerable variation among samples. Expression of the *E. coli phoA* gene (encoding an alkaline phosphatase) in *M. smegmatis* can also be detected by a color change in solid media producing a blue color in the presence of the indicator 5-bromo-4-chloro-3-indoxyl phosphate (XP) and assayed quantitatively (TIMM et al. 1994). Since PhoA activity is dependent on export from the cell it has the potential for use as a reporter gene for studying mycobacterial protein export (TIMM et al. 1994). A third reporter is the CAT gene (encoding chloramphenicol acetyltransferase) which can confer resistance to chloramphenicol when fused to mycobacterial promoters (DAS GUPTA et al. 1993); CAT activity can also be quantitatively measured following lysis of the cells.

Genes encoding luciferase enzymes have also been used as reporters in mycobacteria. For example, the enzyme derived from fireflies (*Pnotinus pyralis*) has been expressed in both *M. smegmatis, M. tuberculosis* and BCG (JACOBS et al. 1993; COOKSEY et al. 1993) and the *Vibrio harveyi* luciferase enzyme in *M. smegmatis* (ANDREW and ROBERTS 1993). The firefly luciferase is encoded by a single gene (FF*lux*) whereas the *Vibrio* enzyme requires two genes *luxA* and *luxB*. However, the expression of both enzymes can be readily detected in a simple quantitative assay in which the substrate (luciferin for FFlux and decanal for Vibrio LuxAB) is added directly to cultures of cells. The substrates are taken up by the mycobacteria and enzyme activity monitored by detection of light, eliminating the need for cell lysis.

The simplicity of the assay to detect mycobacterial luminescence and its responsiveness to antibiotics suggest that it may be a powerful assay for finding new antimycobacterial drugs. The effect of large numbers of potential drugs on the growth of mycobacteria can be screened in a microtiter plate format and light output measured over a period of a few hours or days, in contrast to the weeks and months required by more traditional methods (JACOBS et al. 1993; COOKSEY et al. 1993).

6 Mycobacteriophages

Bacteriophages are important components of any bacterial genetic system. Their relatively small genome size simplifies their characterization, and their genetic constitution and regulation usually reflect the hosts that they exploit. More than 250 mycobacteriophages have been described infecting a wide variety of mycobacterial species (for review see HATFULL and JACOBS 1994). Some of these phages have broad host ranges and can infect both fast- and slow-growing mycobacteria (e.g., D29), whereas other have restricted host ranges (e.g., DS6A is highly specific for species of the *M. tuberculosis* complex). This host specificity has been exploited for the phage typing of mycobacterial isolates (SNIDER et al. 1984).

Mycobacteriophages play an indispensable role in the development of mycobacterial genetic systems since their relatively rapid reproduction can experimentally compensate somewhat for the otherwise slow growth of their hosts. Two mycobacteriophages, I3 and L5, have become the focus of more detailed studies because of their specific attributes; I3 is a generalized transducing phage and L5 is the only mycobacteriophage demonstrated to be a true temperate phage.

6.1 Mycobacteriophage I3

Mycobacteriophage I3 was first described by SUNDARAJ and RAMAKRISHNAN in 1971 and is of particular interest since it is a generalized transducing phage of the mycobacteria. It has a large hexagonal head containing a genome of approximately 145 kb and long contractile tail. The genome is circularly permuted and packaged by a head-fill mechanism but is unusual in that it contains random single-stranded interruptions of about six to ten bases in each of the strands (REDDY and GOPINATHAN 1986). I3 has been shown to transduce both auxotrophic and drug resistance markers in *M. smegmatis*.

6.2 Mycobacteriophage L5

Mycobacteriophage L5 is the best characterized of the phages that infect the mycobacteria (HATFULL 1994). Morphologically, it is composed of an icosahedral head containing a linear double-stranded DNA genome and a long flexible tail, features that are common to a large group of bacteriophages including lambda of *E. coli*. The host range of L5 is somewhat unclear although it efficiently forms plaques on lawns of *M. smegmatis*; L1 (which is virtually identical to L5; LEE et al. 1991) appears to at least inject its DNA into BCG and may productively infect some substrains of BCG (SNAPPER et al. 1988). D29, a phage which shares very good sequence similarity with L5, efficiently infects *M. smegmatis*, BCG and *M. tuberculosis*.

L5 is the only mycobacteriophage that has been shown to be a true temperate virus with the capability to form stable lysogens containing an integrated prophage as well as undergoing lytic growth (SNAPPER et al. 1988). Temperate phages are particularly attractive for study since the process of site-specific integration, regulation of gene expression and the lytic-lysogenic switch involve an especially intimate association between the virus and its host.

6.2.1 The L5 Genome

Phage L5 virions contain a linear double-stranded DNA genome 52 297 bp in length whose DNA sequence has been determined (HATFULL and SARKIS 1993). The ends of the genome contain short single-stranded 3′ extensions that are complementary such that they can ligate together following injection into the cell. There is no extensive DNA sequence comparison between L5 and other bacteriophages with the exception of its close relatives L1 and D29 (HATFULL 1994).

The attachment site (*attP*) that is used for integration of the genome during lysogeny is situated close to the center of the L5 genome dividing the genome into a left and right arm. There appear to be few noncoding regions (with the exception of the regions closest to *cos*) and 88 closely spaced putative genes have been identified, including three tRNA genes (HATFULL and SARKIS 1993). The genes in the right arm are transcribed in the leftwards direction and those in the left arm are transcribed rightwards (with the exception of gene 33). Protein sequence comparison indicates that gp44 is a DNA polymerase and gp33 is the phage integrase but no other obvious gene functions could be inferred.

6.2.2 L5 Gene Expression

Since L5 is a true temperate phage, infection has two possible outcomes, lytic growth resulting in lysis of the cell and release of progeny phage, or lysogeny, a parasitic state in which the viral genome is integrated into the bacterial genome and the lytic functions are switched off. The relative simplicity of the L5 genome and ease of manipulation make it a good model system for understanding gene expression and how its regulation can influence the biological outcome. In addition, powerful regulated promoters are useful tools which can be exploited for the expression of foreign genes in recombinant vaccine construction.

Two sets of L5 genes are expressed during lytic growth; at early times the right arm genes are synthesized and presumably are required for DNA replication; at later times the structural and assembly products encoded by left arm genes are made (HATFULL and SARKIS 1993). As a temperate phage, L5 has the unusual feature of strongly inhibiting host gene expression during lytic growth (HATFULL and SARKIS 1993). The detailed mechanisms of these regulatory events have yet to be described but they are presumably reversible to enable the formation of viable lysogens.

Lysogeny of L5 is maintained by the action of gp71 (DONNELLY-WU et al. 1993). The properties of gp71 are similar to those of other phage repressors and gp71 and mutants defective in gp71 are unable to form stable lysogens (DONNELLY-WU et al. 1993). Moreover, temperature-sensitive clear plaque mutants have been isolated which contain mutations within gene 71; lysogens of these mutants formed at low temperature are thermoinducible, such that when the temperature is elevated there is a rapid induction of lytic growth (DONNELLY-WU et al. 1993). Other clear plaque mutants of L5 are able to form stable lysogens but at a lower frequency than wild-type L5 implicating a second gene required for establishment of lysogeny; this locus maps to gene region 72–72 of the chromosome and is functionally analogous to *cIII* of phage lambda (DONNELLY-WU et al. 1993). This activity may play a role in regulating the activity or expression of gp71 in the lytic-lysogenic decision.

6.3 Other Mycobacteriophages

Many other mycobacteriophages have features of interest but have not yet been studied in great detail. For example, phage TM4, which reportedly was derived from lysogenic *M. avium*, has been used for the construction of shuttle phasmids and luciferase reporter phages (see below). It has a genome of approximately 50 kb with cohesive ends but does not appear to form stable lysogens in *M. smegmatis*. The simplicity with which it was possible to generate TM4 shuttle phasmids, in which an *E. coli* cosmid vector is inserted randomly into the TM4 genome, suggests that a reasonable proportion of the genome may be nonessential for viral growth (HATFULL and JACOBS 1994). Other phages with restricted host ranges such as DS6A, which only forms plaques on species of the *M. tuberculosis* complex, may be of interest for investigating the specific nature of phage-host interactions. Phage D29, a close relative of L5, has been shown to adsorb to *M. leprae* cells but productive infection has not yet been demonstrated (DAVID et al. 1984).

6.4 Luciferase Reporter Mycobacteriophages

The study of viruses as a model system for understanding and manipulating their hosts is a traditional and perhaps rather obvious strategy. However, the special role of mycobacteriophages in the analysis of slow-growing hosts is illustrated by the development of luciferase reporter mycobacteriophages that have potential as important clinical tools for the diagnosis and drug susceptibility testing of *M. tuberculosis* clincial isolates.

The slow growth of *M. tuberculosis* severely complicates its cultivation in the clinical microbiology laboratory. While the presence of *M. tuberculosis* can usually be inferred from the presence of acid-fast bacilli in sputum samples, this assay is not particularly sensitive and provides no information regarding drug susceptibility. However, with the dramatic increase in the appearance of drug resistance and multiple drug resistant (MDR) *M. tuberculosis* in recent years, there are many geographical areas where it cannot be assumed that standard antibiotic regimens will be effective against tuberculosis. It is therefore necessary to empirically determine the drug susceptibility profiles of clinical isolates by determining the effect of antibiotics on the growth of the bacilli. Traditional microbiological methods can require more than 12 weeks to obtain this information, while the BACTEC system can reduce this time to about 2 weeks under optimal conditions (HEIFETS 1991).

Luciferase reporter phages have the potential to further reduce the time required for drug susceptibility testing. The idea is simple and relies on the use of bacteriophages for efficient and specific delivery of a reporter gene such as luciferase (ULITZER and KUHN 1987; JACOBS et al. 1993); if the activity of the reporter gene can be detected with good sensitivity then it should be detectable even if only a small number of bacilli are present, eliminating the need for time-consuming growth of *M. tuberculosis*. Moreover, since luciferase activity requires both expression of the gene and available ATP, light production is expected to be sensitive to the presence of antibiotics. However, if the bacteria are resistant to an antibiotic then its presence should have little effect on light production. Thus an ideal embodiment of the assay would be one in which the reporter phage is added directly to a sputum sample which is then divided into aliquots and antibiotics added; detection of light in a luminometer a few hours later should reveal whether the organism is sensitive (lights-out) or resistant (lights-on) for any given drug. While this scheme has yet to be perfected, preliminary observations suggest this is a feasible approach.

The first luciferase reporter phage developed was based on phage TM4. Construction of the recombinant phage was accomplished using the shuttle phasmid approach in which a cosmid containing the FF*lux* gene fused to the BCG hsp60 promoter was inserted randomly into the TM4 genome (JACOBS et al. 1993). Unfortunately, since little is known about the genome structure of TM4 the location and orientation of FF*lux* with respect to phage genes is not known. Infection of either *M. smegmatis*, BCG or *M. tuberculosis* in liquid culture resulted in a burst of light output that peaked about 3 h after infection and was dependent on the presence of both phage and bacteria. Moreover, incubation of the cells with antibiotics prior to the phage infection strongly inhibited the production of light unless the bacterial strain was resistant to that antibiotic. This general behavior was observed with all three mycobacterial species and strongly supports the feasibility of the luciferase reporter phage approach (JACOBS et al. 1993).

A potential limitation of the TM4::FF*lux* phage is its poor sensitivity of detection. If fewer than about 10^5 bacilli were present then the light output would be barely detectable above background and a large proportion of clinical samples would go undetected. It is thus desirable to elevate the sensitivity perhaps by increasing the level of gene expression or by using other mycobacteriophages. A better understanding of the biology of TM4 might provide clues as to why the sensitivity is poor relative to that observed when the same hsp60 promoter-FF*lux* fusion is present in the cell on an extrachromosomal plasmid (JACOBS et al. 1993). Other important questions must also be addressed such as whether such phage infections can be performed directly in clinical samples (such as sputum) before luciferase reporter mycobacteriophages find their way into the clinical microbiology laboratory.

7 Concluding Remarks

In this brief review of the molecular genetics of mycobacteria and *M. tuberculosis* I have described several areas that have attracted interest in recent years. Not surprisingly, given the relative youth of the discipline, there are many important aspects of mycobacterial genetics that have yet to be investigated. While the genetic basis of pathogenicity and drug resistance will continue to be a top priority, the clinical importance of the mycobacteria warrants thorough investigation of all aspects of their biology. For example: How is the mycobacterial genome replicated and how is replication regulated? What role does the timing of replication and cell division play in their characteristic slow growth? What are the genetic determinants for biosynthesis of the mycobacterial cell wall? What metabolic systems are used and how are they regulated?

The study of the molecular genetics of the mycobacteria over the past few years indicates that they will share many common features with their better-characterized bacterial relatives. However, the informative example of RecA protein splicing in *M. tuberculosis* (Colston and Davis 1994) dictates that assumptions drawn from existing bacterial dogmas should be applied to the mycobacteria with extreme caution. With the availability of a sophisticated genetic system there should be few barriers to obtaining a full understanding of these pathogens, and a rich reward of novel strategies for the prevention, diagnosis and cure of mycobacterial diseases can be expected.

References

Aldovani A, Husson RH, Young RA (1993) The *uraA* locus and homologous recombination in *Mycobacterium bovis* BCG J Bacteriol 175: 7282–7289

Andrew PW, Roberts IS (1993) Construction of a bioluminescent mycobacterium and its use for assay of antimycobacterial agents. J Clin Microbiol 31: 2251–2254

Barletta RG, Kim DD, Snapper SB, Bloom BR, Jacobs WR Jr (1992) Identification of expression signals of the mycobacteriophages Bxb1, L1 and TM4 using the *Escherichia-Mycobacterium* shuttle plasmids pYUB75 and pYUB76 designed to create translational fusions to the *lacZ* gene. J Gen Microbiol 138: 23–30

Bercovier H, Kafri O, Sela S (1986) Mycobacteria possess a surprisingly small number of ribosomal RNA genes in relation to the size of their genome. Biochem Biophys Res Commun 136: 1136–1141

Clark-Curtiss JE (1990) Genome structure of mycobacteria. In: McFadden J (ed) Molecular biology of the mycobacteria. Academic, London, pp 77–96

Cole ST, Smith DR (1994) Toward mapping and sequencing the genome of *Mycobacterium tuberculosis*. In: Bloom BR (ed) Tuberculosis: pathogenesis, protection and control. American Society for Microbiology, Washington, pp 227–238

Colston MJ, Davis EO (1994) Homologous recombination, DNA repair, and mycobacterial *recA* genes. In: Bloom BR (ed) Tuberculosis: pathogenesis, protection and control. American Society for Microbiology, Washington, pp 217–226

Cooksey RC, Crawford JT, Jacobs WR Jr, Shinnick TM (1993) A rapid method for screening antimicrobial agents for activity against a strain of *Mycobacterium tuberculosis* expressing firefly luciferase. Antimicrob Agents Chemother 37:1348–1352

Crawford JT and Bates JH (1986) Analysis of plasmids in *Mycobacterium avium-intracellulare* isolates from persons with acquired immunodeficiency syndrome. Am Rev Respir Dis 3: 949–951
Dale JW, Patki A (1990) Mycobacterial gene expression and regulation. In: McFadden J (ed) Molecular biology of the mycobacteria. Academic, London, pp 173–198
Das Gupta SK, Bashyam MD, Tyagi AK (1993) Cloning and assessment of mycobacterial promoters by using a plasmid shuttle vector. J Bacteriol 175: 5186–5192
David H, Clement F, Clavel-Sere S, Rastogi N (1984) Abortive infection of *Mycobacterium leprae* by the mycobacteriophage D29. Int J Lepr 52: 515–523
Davis EO, Sedgwick SG, Colston MJ (1991) Novel structure of the *recA* locus of *Mycobacterium tuberculosis* implies processing of the gene product. J Bacteriol 173: 5653–5662
Davis EO, Jenner PJ, Brooks PC, Colston MJ, Sedgwick SG (1992) Protein splicing in the maturation of the *M. tuberculosis* RecA protein: a mechanism for tolerating a novel class of intervening sequences. Cell 71: 201–210
Donnelly-Wu M, Jacobs WR, Hatfull GF (1993) Superinfection immunity of mycobacteriophage L5: applications for genetic transformation of mycobacteria. Mol Microbiol 7: 407–417
Falkingham JO, Crawford JT (1994) Plasmids. In: Bloom BR (ed) Tuberculosis: pathogenesis, protection and control. American Society for Microbiology, Washington DC, pp 185–198
Franzblau SG, Takeda T, Nakamura M (1986) Mycobacterial plasmids: screening and possible relationship to antibiotic resistance in *Mycobacterium avium*/*Mycobacterium intracellulare*. Microbiol Immun 30: 903–907
Goto Y, Taniguchi H, Udou T, Mizuguchi Y, Tokunaga T (1991) Development of a new host vector system in mycobacteria. FEMS Microbiol Lett 83: 277–282
Guilhot CB, Otal I, van Rompaey I, Martin C, Gicquel B (1994) Efficient transposition in mycobacteria: construction of *M. smegmatis* insertional mutant libraries. J Bacteriol 176: 535–539
Haeseleer F, Pollet J-F, Haumont M, Bollen A, Jacobs P (1993) Stable integration and expression of the *Plasmodium falciparum* circumsporozoite protein coding sequence in mycobacteria. Mol Biochem Parasitol 57: 117–126
Hatfull GF (1993) Genetic transformation of Mycobacteria. Trends Microbiol 1: 310–314
Hatfull GF (1994) Mycobacteriophage L5: a toolbox for tuberculosis. ASM News 60: 255–260
Hatfull GF, Jacobs JR Jr (1994) Mycobacteriophages: cornerstones of mycobacterial research. In: Bloom BR (ed) Tuberculosis: pathogenesis, protection and control. American Society for Microbiology, Washington, pp 165–183
Hatfull GF, Sarkis G (1993) DNA sequence, structure and gene expression of mycobacteriophage L5: a phage system for mycobacterial genetics. Mol Microbiol 7: 395–405
Heifets L (1991) Drug susceptibility tests in the managment of chemotherapy in tuberculosis. In: Heiferts L (ed) Drug susceptibility in the chemotherapy of mycobacterial infections. CRC Press, Boca Raton, pp 89–122
Hellyer TJ, Brown IN, Dale JW, Easmon CSF (1991) Plasmid analysis of *Mycobacterium avium-intracellulare* (MAI) isolated in the United Kingdom from patients with and without AIDS. J Med Microbiol 34: 225–231
Hermans J, Martin C, Huijberts GNM, Goosen T, de Bont JAM (1991) Transformation of Mycobacterium aurum and Mycobacterium smegmatis with the broad host range Gram-negative cosmid vector pJRD215. Mol Microl 5: 1561–1566
Husson RA, James BE, Young RA (1990) Gene replacement and expression of foreign DNA in mycobacteria. J Bacteriol 172: 519–524
Jacobs WR Jr, Bloom BR (1994) Molecular genetic strategies for identifying virulence determinants of *Mycobacterium tuberculosis*. In: Bloom BR (ed) Tuberculosis: pathogenesis, protection and control. American Society for Microbiology, Washington DC, pp 253–268
Jacobs WR Jr, Docherty MA, Curtiss R III, Clark-Curtiss JE (1986) Expression of *Mycobacterium leprae* genes from a *Streptococcus mutans* promoter in *E. coli* K-12. Proc Natl Acad Sci USA 83: 1926–1930
Jacobs WR, Tuckman M, Bloom BR (1987) Introduction of foreign DNA into mycobacteria using a shuttle phasmid. Nature 327: 532–535
Jacobs WR, Udani R, Barletta R, Chan J, Kalkut G, Sonse G, Kieser T, Sarkis G, Hatfull GF, Bloom BR (1993) Rapid assessment of drug susceptibilities of *Mycobacterium tuberculosis* by means of luciferase reporter phages. Science 7: 819–822
Jucker MT, Falkingham JO (1990) Epidemiology of infection of nontuberculous mycobacteria. IX. Evidence for two DNA homology groups among plasmids in *M. avium, M. intracellulare* and *M. scrofulaceum*. Am Rev Respir Dis 142: 858–862

Kalpana GV, Bloom BR, Jacobs WR Jr (1991) Insertional mutagenesis and illegitimate recombination in mycobacteria. Proc Natl Acad Sci USA 88: 5433–5437

Labidi A, Mardis E, Roe BA, Wallace RJ (1992) Cloning and DNA sequence of the *Mycobacterium fortuitum* var. *fortuitum* plasmid pAL5000. Plasmid 27: 130–140

Lee MH, Hatfull GF (1993) Mycobacteriophage L5 integrase-mediated site-specific recombination in vitro. J Bacteriol 175: 6836–6841

Lee MH, Pascopella L, Jacobs WR, Hatfull GF (1991) Site-specific integration of mycobacteriophage L5: integration-proficient vectors for *Mycobacterium smegmatis*, *Mycobacterium tuberculosis*, and bacille Calmette-Guérin. Proc Natl Acad Sci USA 88: 3111–3115

Levin M, Hatfull GF (1993) *Mycobacterium smegmatis* RNA polymerase: DNA supercoiling, action of rifampicin and mechanism of rifampicin resistance. Mol Microbiol 8: 277–285

Mahenthiralingam E, Draper P, Davis EO, Colston MJ (1993) Cloning and sequencing of the gene which encodes the highly inducible acetamidase of *Mycobacterium smegmatis*. J Gen Microbiol 139: 575–583

Martin CJ, Timm CJ, Rauzier J, Gomez-Lus R, Davis J, Gicquel B (1990) Transposition of an antibiotic resistance element in mycobacteria. Nature 345: 739–743

Martin CP, Mazodier E, Mediola MV, Gicquel B, Smokvina T, Thompson CJ, Davis J (1991) Site-specific integration of the *Streptomyces* plasmid pSAM2 in *Mycobacterium smegmatis*. Mol Microbiol 5: 2499–2502

McAdam RA, Guilhot C, Gicquel B (1994) Transposition in mycobacteria. In: Bloom BR (ed) Tuberculosis: pathogenesis, protection and control. American Society for Microbiology, Washington, pp 199–216

Meissner PS, Falkingham JO (1986) Plasmid DNA profiles as epidemiological markers for clinical and environmental isolates of *Mycobacterium avium*, *Mycobacterium* intracellulare, and *Mycobacterium scrofulaceum*. J Infect Dis 153: 325–331

Pascopella L, Collins FM, Martin JM, Lee MH, Hatfull GF, Bloom BR, Jacobs WR Jr (1994) Use of in vivo complementation in *Mycobacterium tuberculosis* to identify a genomic fragment associated with virulence. Infect Immun 62: 1313–1319

Radford AJ, Hodgson ALM (1991) Construction and characterization of a *Mycobacterium-Escherichia coli* shuttle vector. Plasmid 25: 149–153

Ranes MG, Rauzier J, LaGranderie M, Gheorghiu M, Gicquel B (1990) Functional analysis of pAL5000, a plasmid from *Mycobacterium fortuitum*: construction of a "mini" mycobacterium-*Escherichia coli* shuttle vector. J Bacteriol 172: 2793–2797

Reddy AB, Gopinathan KP (1986) Presence of random single-stranded gaps in mycobacteriophage I3 DNA. Gene 44: 227–234

Snapper SB, Lugosi L, Jekkel A, Melton RE, Kieser T, Bloom BR, Jacobs WR (1988) Lysogeny and transformation in mycobacteria: stable expression of foreign genes. Proc Natl Acad Sci USA 85: 6987–6991

Snapper SB, Melton RE, Mustafa S, Kieser T, Jacobs WR Jr (1990) Isolation and characterization of efficient plasmid transformation mutants of *Mycobacterium smegmatis*. Mol Microbiol 4: 1911–1919

Snider DE, Jones WD Jr, Good RC (1984) The usefulness of phage typing *Mycobacterium tuberculosis* isolates. Am Rev Respir Dis 130: 1095–1099

Stover CK, de la Cruz VF, Fuerst TR, Burlein JE, Benson LA, Bennett LT, Bansal GP, Young JF, Lee MH, Hatfull GF, Snapper S, Barletta RG, Jacobs WR Jr, Bloom BR (1991) New use of BCG for recombinant vaccines. Nature 351: 456–460

Strohl WR (1992) Compilation and analysis of DNA sequences associated with apparent streptomycete promoters. Nucleic Acids Res 20: 961–974

Sundaraj CV, Ramakrishnan T (1971) Transduction in Mycobacterium smegmatis. Nature 228: 280–281

Timm J, Perilli MG, Duez C, Trias J, Orefici G, Fattorini L, Amicosante G, Oratore A, Joris B, Frere JM, Pugsley AP, Gicquel B (1994) Transcription and expression analysis, using *lacZ* and *phoA* gene fusions, of *Mycobacterium fortuitum* β-lactamase gene cloned from a natural isolate and high-level β-lactamase producer. Mol Microbiol 12: 491–504

Ulitzer S and Kuhn J (1987) Introduction of lux genes into bacteria: a new approach for specific determinants of bacteria and their antibiotic susceptibilities, In: Sclomerick J, Andersen RR, Kapp A, Ernst M, Woods WG (eds) Bioluminescence and chemiluminescence: new perspectives. Wiley, New York, pp 463–472

Mechanisms of Drug Resistance in *Mycobacterium tuberculosis*

B. Heym[1], W. Philipp[2], and S.T. Cole[2]

1 Introduction

Fifty years ago Selman Waksman and his colleagues discovered streptomycin and, for the first time, provided a chemotherapeutical approach to tuberculosis treatment as an alternative to fresh air, diet and physical exercise. Soon after the initiation of the chemotherapy era with streptomycin it was recognized that not all cases of tuberculosis could be cured due to the emergence of resistant mutants (Mitchison 1950). In the following years many other drugs were discovered, some of them too toxic for clinical use, but others like isoniazid, pyrazinamide, ethambutol or rifampin were very active (Forbes et al. 1965; Fox 1951; Tsukamura et al. 1958; Yeager et al. 1952), and physicians realised that the association of several drugs given simultaneously prevented the emergence of resistant mutants (American Thoracic Society 1986; O'Brien 1993). When tu-

[1]Hôpital Ambroise Paré, 9, avenue Charles de Gaulle, 92104 Boulogne Cedex, France
[2]Unité de Génétique Moléculaire Bactérienne, Institut Pasteur, 28 rue du Docteur Roux, 75724 Paris Cedex 15, France

berculosis results from infection with drug susceptible strains of *M. tuberculosis,* the success rate of short course chemotherapy is close to 100% provided that the regimen is strictly adhered to by both the physician and the patient (American Thoracic Society 1992; Grosset 1989; Hopewell 1994; Iseman 1994; Iseman et al. 1993).

Thus, until fairly recently, it was a common belief that tuberculosis no longer represented a major public health problem in the industrialized countries despite being the leading cause of human morbidity and mortality due to an infectious disease worldwide (Bloom and Murray 1992; Murray et al. 1990). The recrudescence of tuberculosis can be attributed to many changes in society such as the decline of adequate control programs, increasing homelessness and extensive poverty, but the principal cause is undoubtedly the AIDS pandemic (Murray et al. 1990). Indeed, in the last few years many HIV-infected individuals have been contaminated with *M. tuberculosis*, rapidly developed disease and died (Barnes et al. 1991; Daley et al. 1992; Frieden et al. 1993; Iseman 1994; Snider and Roper 1992). One of the most disturbing features of AIDS-related tuberculosis has been the emergence of strains of *M. tuberculosis* that show resistance to some, or all, of the front-line anti-tuberculous drugs (Edlin et al. 1992; Frieden et al. 1993; Iseman 1994; Snider and Roper 1992).

Many studies have been carried out to explain the modes of action of, or the mechanisms of resistance to, the different anti-tuberculous drugs but knowledge about their targets and the resistance mechanisms of *M. tuberculosis* has remained scarce. Our understanding of the mechanisms of resistance to several anti-tuberculous agents has improved only recently with the development of genetic tools to study the mycobacteria and the availability of powerful molecular tools such as the polymerase chain reaction (Jacobs et al. 1987; Saiki et al. 1988).

Resistance to isoniazid (iso-nicotinic acid hydrazide) and rifampin, the two drugs comprising the backbone of the current tuberculosis chemotherapy program (Centers for Disease Control 1993), has been studied extensively. It has been shown that rifampin resistance stems from missense mutations in the *rpoB* gene encoding the DNA-dependent RNA polymerase, the primary target of rifampin (Honoré and Cole 1993; Imboden et al. 1993; Telenti et al. 1993a; Cole 1994). Resistance to isoniazid is associated with at least two independent mechanisms, the more common of which involves mutation or inactivation of the *katG* gene encoding the HPI-type catalase-peroxidase (Zhang et al. 1992, 1993; Heym et al. 1993, 1994b). Mutation of a second gene, *inhA*, can also result in resistance to isoniazid and this is generally accompanied by cross-resistance to the second-line antituberculous drug ethionamide (Banerjee et al. 1994). Resistance to streptomycin and fluoroquinolones results from missense mutations in the genes encoding certain ribosomal subunits and DNA gyrase, respectively (Finken et al. 1993; Takiff et al. 1994).

2 Streptomycin

Streptomycin, a broad spectrum antibiotic of the aminoglycoside family, was the first drug used in the treatment of tuberculosis. Found in 1943 by Selman Waksman, who was studying the bactericidal activities of soil, it was first employed, with excellent success, in 1944 to treat tuberculosis patients (AYVAZIAN 1993). As it has been used since the early beginnings of tuberculosis chemotherapy, and then mostly as monotherapy, the resistance rate to streptomycin is the highest among anti-tuberculous drugs in clinical *M. tuberculosis* strains nowadays (CANETTI and GROSSET 1961; MITCHISON 1950). Because of the absence of an oral form of streptomycin and its toxicity for the kidney and the ear, and, more importantly, because more active and less toxic drugs like isoniazid and, later, ethambutol and rifampin were discovered, streptomycin was more or less excluded as a front-line drug for the treatment of tuberculosis in the late 1960s (GROSSET 1989). Since the mid-1980s, when the recrudescence of tuberculosis and the emergence of multidrug resistant strains were observed, streptomycin has regained importance in the treatment of tuberculosis (CENTERS FOR DISEASE CONTROL 1993; O'BRIEN 1993).

The site of action of streptomycin is the small, or 30S, subunit of the ribosome, especially the ribosomal protein S12 and the 16S rRNA (GARVIN et al. 1974; WEISBLUM and DAVIES 1968). With regard to the 16S rRNA (Fig. 1) it has been established in *Escherichia coli* that the bases between positions 903 and 910 are responsible for the binding of streptomycin to the ribosome (MOAZED and NOLLER 1987), whereas the reaction of streptomycin with the protein S12 seems to interfere in the translation process (GALE et al. 1981). Streptomycin acts at several stages in protein synthesis and its main effects appear to be the inhibition of initiation of mRNA translation, misreading of the genetic code, and aberrant proof-reading by the bacterial ribosome (FUNATSU and WITTMANN 1975; MOAZED and NOLLER 1987; WALLACE and DAVIS 1973).

Resistance to streptomycin has been studied in many bacteria and plants and has been shown to result from three different mechanisms. Firstly, missense mutations in the *rpsL* gene, encoding the S12 protein, have been shown to confer streptomycin resistance by affecting the binding of streptomycin to the ribosome or by changing the ribosomal conformation so that the effects of the drug when bound are diminished. About 70% of resistant clinical isolates of *M. tuberculosis* have an A-G transition in codon 43 of the *rpsL* gene which leads to a Lys-Arg substitution (BÖTTGER 1994; HEYM et al. 1994a). The same mutation has been identified in other bacteria and plants as conferring streptomycin resistance (FUNATSU and WITTMANN 1975; GALILI et al. 1989). Other missense mutations have been found in the *rpsL* gene of resistant strains of *M. tuberculosis* and result in the substitution of Arg-43 by Thr and Lys-88 by Arg. These substitutions are encountered much less frequently (MEIER et al. 1994).

The second mechanism involves changes to the *rrs* gene (KEMPSELL et al. 1992; NOLLER 1984), encoding 16S rRNA. Two regions of this molecule that are

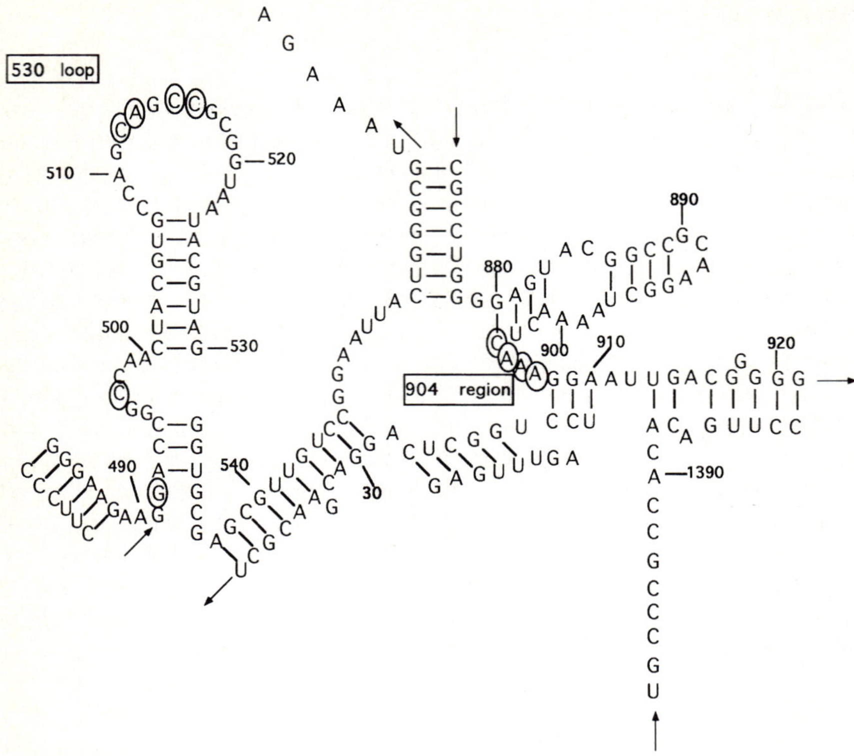

Fig. 1. Secondary structure of the 16S rRNA of *M. tuberculosis:* section of the 530 loop and the 904 region where most of the mutations conferring streptomycin resistance were localized. Bases which give rise to resistance are *circled*

well conserved in all bacteria, the region around nucleotide 904 and the 530 loop (Fig. 1), are implicated in streptomycin resistance. Mutation of the cytosine at position 904 to adenine or guanosine, or adenine 905 to guanosine has been found in streptomycin resistant isolates of *M. tuberculosis*, as described previously in *E. coli* or *Euglena gracilis* (HONORÉ and COLE 1994; MOAZED and NOLLER 1987; MONTANDON et al. 1986). Moreover, it is interesting to note that in *E. coli* streptomycin protects cytosine 912 from the action of alkylating agents or nucleases. The second domain affected by mutations conferring streptomycin resistance is the 530 loop (FINKEN et al. 1993). The mutations of this domain in *E. coli* were found at positions 523 and 525, and, recently, in *M. tuberculosis* at position 491 and 512 corresponding to positions 501 and 522 in *E. coli*, respectively (FINKEN et al. 1993). Mutations in *rrs,* conferring clinically relevant streptomycin resistance, have only been described in *M. tuberculosis* (BÖTTGER 1994) probably because this organism, unlike most other eubacteria, contains a single *rrn* operon.

Little is known about the third means of streptomycin resistance, the existence of which was revealed by the finding that in some resistant isolates neither the *rpsL* nor the *rrs* genes were mutated (Böttger 1994; Honoré and Cole 1994) and that the minimal inhibitor concentration (MIC) is much lower (<80 μg/ml) than for strains known to have mutations in these genes (>640 μg/ml) (Honoré and Cole 1994). Streptomycin resistance in other eubacteria also results from production of aminoglycoside-modifying enzymes or reduced antibiotic uptake. This was first discussed many years ago by Mitchison, who identified three different classes of streptomycin-resistant tubercle bacilli showing low level (MICs of 4–32 μg/ml), intermediate level (MICs of 64–512 μg/ml), and high level (MICs>1000 μg/ml) resistance (Mitchison 1950, 1951). The latter two classes probably harbored mutations in the ribosomal genes. As streptomycin-inactivating enzymes are generally encoded by plasmids or transposons (Benveniste and Davies 1973) which have not yet been found in *M. tuberculosis* (Crawford and Bates 1979; Falkinham and Crawford 1994; Martin et al. 1990), and no cross-resistance with other aminoglycosides is known, this resistance mechanism seems unlikely (Crawford and Bates 1979; Martin et al. 1990). By contrast, perturbation or blocking of streptomycin uptake (Beggs and Andrews 1976; Bryan et al. 1977; Kanner and Gutnick 1972) as a result of altered permeability of the mycobacterial cell wall, which constitutes an effective barrier to all kinds of antibacterial agents, could be responsible (Jarlier and Nikaido 1990; Trias and Benz 1993) and might account for the low and intermediate levels of resistance (Honoré and Cole 1994; Mitchison 1950).

3 Isoniazid

Isonicotinic acid hydrazide or isoniazid (Fig. 2) was synthesized for the first time in 1912 by two Czech biochemists (Meyer and Mally 1912). Its powerful antituberculous activity was detected in 1951 and, of the pathogenic bacteria, only the members of the *M. tuberculosis* complex, *M. tuberculosis*, *M. bovis*, *M. africanum* and *M. microti* are susceptible to isoniazid (Fox 1951). Its activity on

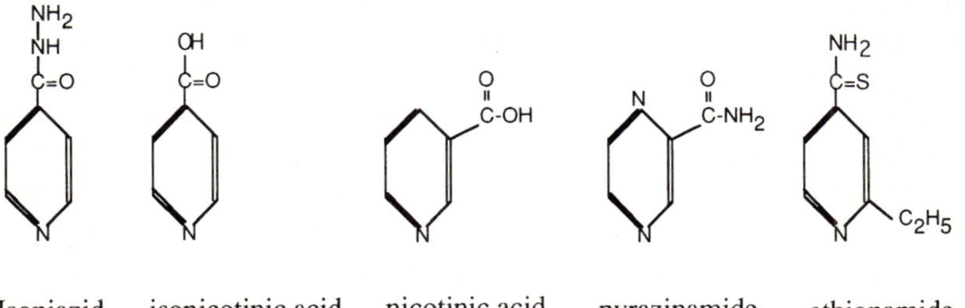

Fig. 2. Chemical structure of isoniazid and a putative active derivative, isonicotinic acid, which could be used instead of nicotinic acid in the biosynthesis of nicotinamide adenine dinucleotide (NAD). Molecules with related structures like pyrazinamide and ethionamide are also presented

these species is excellent, with MICs of about 0.02–0.05 μg/ml. The introduction of isoniazid was thus a milestone in the chemotherapy of tuberculosis (Bernstein et al. 1952; Steenken et al. 1952).

Since that time, the mode of action of isoniazid has been the subject of intensive studies and the drug seems to have pleiotropic effects that are temporally distinct. Although the synthesis of proteins and nucleic acids is affected in the later stages (Herman and Weber 1980; Tsukamura and Tsukamura 1963; Wimpenny 1967; Youatt 1969), one of the first consequences is the loss of acid-fastness of the tubercle bacilli, probably as a result of the inhibition of the synthesis of mycolic acids, the long-chained, unsaturated fatty acids found in the cell wall (Barclay et al. 1953; Cohn et al. 1954). As these are confined essentially to the mycobacteria they represent a selective target for drugs (Davidson and Takayama 1979; Winder 1982).

Shortly after the introduction of isoniazid, resistant strains were isolated from patients treated by isoniazid monotherapy. It was observed that some of the highly resistant strains (MIC>50 μg/ml) had lost their catalase-peroxidase activity and showed attenuated virulence in the guinea-pig model (Buck et al. 1952; Middlebrook 1952; Peizer and Widelock 1955). In 1992, the relationship between isoniazid resistance and loss of catalase-peroxidase activity was explained by the identification of one of the targets, catalase-peroxidase, a heme-containing enzyme encoded by the *katG* gene of *M. tuberculosis* (Zhang et al. 1992), which apparently converts isoniazid to a toxic derivative.

In some highly isoniazid-resistant strains that had completely lost their catalase-peroxidase activity the *katG* gene had been deleted from the chromosome (Heym et al. 1993; Zhang et al. 1993). This observation predicted that mutations affecting expression of the *katG* gene or lowering the catalytic activity of the protein should also result in isoniazid resistance. Mutations belonging to the latter group have recently been identified and characterized in clinically isoniazid-resistant isolates of *M. tuberculosis* (Heym et al. 1994b). Interestingly, most of the mutations identified were of the missense type. The positions of the amino acid substitutions in some of the isoniazid-resistant clinical isolates were consistent with there being a modification of the active site of the enzyme and these strains had lowered or absent catalase and peroxidase activity. In another set of strains the amino acid substitutions were localized at the putative heme-binding site of the protein and in these cases the enzymatic activity was also much lowered (Heym et al. 1994b).

The most frequently found amino acid substitution in the KatG protein was the replacement of Arg-463 by Leu. As it is located in the COOH-terminal of the enzyme, it is not expected to be in the vicinity of the peroxidase active site. It is thus conceivable that it represents a substrate-binding site and may define a domain where isoniazid interacts with the catalase-peroxidase enzyme. Most strikingly, an Arg residue is found at position 463 only in the *M. tuberculosis* enzyme, in contrast to the catalase-peroxidase (HPI) enzymes of *M. intracellulare*, *M. bovis* BCG and *Bacillus stearothermophilus*, where this position is occupied by a Leu residue (Heym et al. 1993; Loprasert et al. 1988; Morris et al. 1992). In the case of *M. bovis* BCG there is a strong correlation between this amino acid substitution and reduced isoniazid susceptibility (HEYM et al. 1994b). HPI consists of two identical subunits of 80 kDa and, in *E. coli,* has been shown to play a protective role against noxious oxidative agents, like H_2O_2, which accumulate during oxidative respiration (SHOEB et al. 1985; LOEWEN et al. 1985; TRIGGS-RAINE et al. 1988). HPI is bi-functional with regard to its catalytic activities and in its catalase mode converts $2H_2O_2$ to $2H_2O$ and O_2, whereas in its peroxidase mode it accepts electrons from a variety of organic electron donors and uses them to reduce H_2O_2 to H_2O (CLAIBORNE and FRIDOVITCH 1979). The peroxidase activity of this enzyme seems to be necessary to activate isoniazid to a toxic substance in the bacterial cell but the nature of the active derivative is still obscure (ZHANG et al. 1993). Iso-nicotinic acid is one of the products of the reaction and was proposed as the active compound many years ago. Iso-nicotinic acid is an analogue of nicotinic acid (Fig. 2) and could replace it in nicotinamide biosynthesis, thus leading to the production of iso-nicotinamide adenine-dinucleotide or iso-NAD and to the perturbation of many NAD-dependent metabolic pathways (SEYDEL et al. 1976; KRÜGER-THIEMER et al. 1975). Recently, JOHNSSON and SCHULTZ (1994) studied the reaction of isoniazid with purified HPI from *M. tuberculosis* and found that this results in the production of iso-nicotinic acid, iso-nicotinamide and pyridine-4-carboxaldehyde as well as a number of highly reactive species including an acyl radical, peracid and aldehyde capable of attacking nucleophilic groups in proteins.

A link between isoniazid resistance, the synthesis of mycolic acids and the NAD pool was found in 1994 with the identification of a second target of isoniazid, the putative fatty acid synthase InhA, encoded by the *inhA* gene (BANERJEE et al. 1994). The overexpression of *inhA* leads to low level isoniazid resistance and is accompanied by cross-resistance to the second-line antituberculous drug ethionamide which has a structure quite similar to isoniazid (Fig. 2). The nucleotide sequence of *inhA* predicts a product that shares 40% identity with the enterobacterial *envM* enzyme associated with fatty acid, phospholipid and lipopolysaccharide biosynthesis (BERGLER et al. 1992; TURNOWSKY et al. 1989; BANERJEE et al. 1994). A possible role for InhA in mycolic acid production was suggested by the results of experiments using cell-free extracts of strains of *M. smegmatis* harboring the cloned *inhA* gene, or an *inhA* missense mutation conferring isoniazid and ethionamide resistance, (described below) to program mycolic acid synthesis in vitro, as this was unaffected by addition of isoniazid. It is currently believed that this mechanism of resistance results from titration of isoniazid, or its active derivative, as a consequence of overproduction of the InhA protein or of alterations in the InhA protein.

Like the putative EnvM fatty acid synthase, the InhA protein probably uses NAD or NADH as a cofactor, which could explain isoniazid susceptibility. It is quite conceivable that the enzymatic activity of InhA would be inhibited by the incorporation of iso-NAD, produced as a result of the action of HPI on isoniazid, or by one of the radicals described by JOHNSSON and SCHULTZ (1994), and this in turn would lead to a block in mycolic acid synthesis and loss of acid-fastness.

Ethionamide- and isoniazid-resistant mutants of *M. smegmatis and M. bovis* have been selected in vitro and shown to harbor a missense mutation in the *inhA* gene, changing the Ser codon at position 94 to Ala. The same mutation has been described in the *E. coli envM* gene and confers a protective effect against diazaborine, an antibacterial agent affecting the biosynthesis of lipids (BERGLER et al. 1992). In none of the clinical isolates studied which showed resistance to ethionamide and, at a low level, to isoniazid, was this mutation found (HEYM et al. 1994a,b). In contrast, a putative regulatory mutation located in the control region of the *inhA* gene, was found in nearly all these strains, and could be responsible for the overexpression of the *inhA* gene (KAPUR et al. 1994). Further work is required to confirm this however.

A series of experiments performed with various mutants in the oxidative responses of *E. coli* or *Salmonella typhimurium*, microorganisms that do not produce mycolic acids, suggested that isoniazid may also act on other targets (ROSNER 1993; ROSNER and STORZ 1994). Strains lacking the redox-sensitive regulatory protein OxyR, or alkylhydroperoxide reductase encoded by the OxyR-controlled *ahpCF* genes, showed increased susceptibility to isoniazid, albeit to very high levels (ROSNER and STORZ 1994), as do certain *E. coli* strains containing the *M. tuberculosis katG* gene (ZHANG et al. 1992). H_2O_2 potentiates this effect and also renders an isoniazid-resistant mutant of *M. smegmatis,* lacking katG, more susceptible to the drug suggesting that nonenzymatic activation of isoniazid may also occur (ROSNER and STORZ 1994).

Although much progress has been made toward understanding the mechanism of action of isoniazid a number of questions remain to be answered. Perhaps the most intriguing of these is why *M. tuberculosis*, of all the mycobacteria, is so exquisitely susceptible to the drug? This is particularly puzzling, as the *katG* and *inhA* genes, and their proteins, are well conserved amongst mycobacteria (BANERJEE et al. 1994; HEYM et al. 1993) even those which are naturally resistant to isoniazid. It also seems unlikely that this is due to defects in the OxyR regulon as *oxyR* and *ahpC* genes have been found in *M. tuberculosis* (our unpublished results). Perhaps it could be explained by differences in the uptake of isoniazid or the fact that *M. tuberculosis* is unusual among mycobacteria as it lacks the HPII catalase encoded by the *katE* gene. One possible consequence of this could be higher intracellular concentrations of H_2O_2 leading to stronger activation of the drug. Another plausible explanation could be provided by the HPI of *M. tuberculosis* itself, as it may contain some features which favor interaction with the drug. Hopefully, a solution to this riddle will be found in the next few years.

4 Rifampin

Rifampin, a lipophilic ansamycin, was introduced in tuberculosis therapy in 1967 and is highly active against mycobacteria as it diffuses rapidly across the hydrophobic cell envelope. Rifampin is the key component of antituberculous therapeutic regimens and its use has greatly shortened the duration of chemotherapy necessary for the successful treatment of drug-susceptible tuberculosis (GROSSET and LEVANTIS 1983; GROSSET 1989). Patients infected with a rifampin-resistant strain of *M. tuberculosis* generally have a rather poor prognosis, particularly because rifampin resistance is often associated with resistance to other front-line drugs (SMALL et al. 1993; FRIEDEN et al. 1993).

In vitro rifampin is one of the most potent inhibitors of bacterial DNA-dependent RNA polymerase known (GALE et al. 1981). RNA polymerase is a complex oligomer and consists of a core enzyme containing four major polypeptide chains ($\alpha 2\beta\beta'$). These can associate with another subunit, σ, which confers specificity for recognition of the correct promoter sites for the initiation of transcription in the DNA template, thus forming the holoenzyme that is essential for bacterial life. The four subunits α, β, β', σ are encoded by the genes *rpoA, rpoB, rpoC* and *rpoD*, respectively (BURGESS et al. 1987). Rifampin inhibits RNA polymerase by covalently binding to the β-subunit, encoded by the *rpoB* gene, which is involved in chain initiation and elongation (WEHRLI 1983; OVCHINNIKOV et al. 1983). The *rpoB* gene of *M. tuberculosis* has been cloned and sequenced and its product, like that of *M. leprae,* is highly similar to the *E. coli* RpoB protein, except for two large insertions in the latter case (HONORÉ and COLE 1993; MILLER et al. 1994).

Resistance of *M. tuberculosis* to rifampin has been well studied (HONORÉ and COLE 1993; KAPUR et al. 1994; IMBODEN et al. 1993; TELENTI et al. 1993a; MILLER et al. 1994). In virtually all strains of *M. tuberculosis* the resistant phenotype was attributable to mutations in a short region, consisting of 27 codons situated near the center of *rpoB* known to bear mutations that confer rifampin resistance in *E. coli* (JIN and GROSS 1989; OVCHINNIKOV et al. 1983). Most of the mutations were missense mutations, but insertions and small deletions also occurred (TELENTI et al. 1993a). All these genetic modifications resulted in a modified enzyme that is still functional but has a largely impaired affinity for rifampin (Fig. 3). Two sites are more frequently mutated – His526 to Tyr and Ser531 to Leu – and represent more than 70% of the mutations found in clinical isolates (IMBODEN et al. 1993; TELENTI et al. 1993a). Genetic evidence that RNA polymerase is the target of rifampin is available in the form of heterologous complementation experiments using rifampin-susceptible and resistant strains of *M. smegmatis* containing various *M. tuberculosis rpoB* alleles (MILLER et al. 1994).

Interestingly, the majority of the mutations identified in *M. tuberculosis* do not appear to confer a significant growth disadvantage, in contrast to the situation in *E. coli* where such mutations are often pleiotropic and result in slower growth probably due to inefficient transcription initiation or to termination defects (JIN and GROSS 1989). This discrepancy may be related to the fact that *M. tuberculosis*, being naturally slow-growing, can tolerate a less active transcriptional apparatus as its growth rate may be limited at another level.

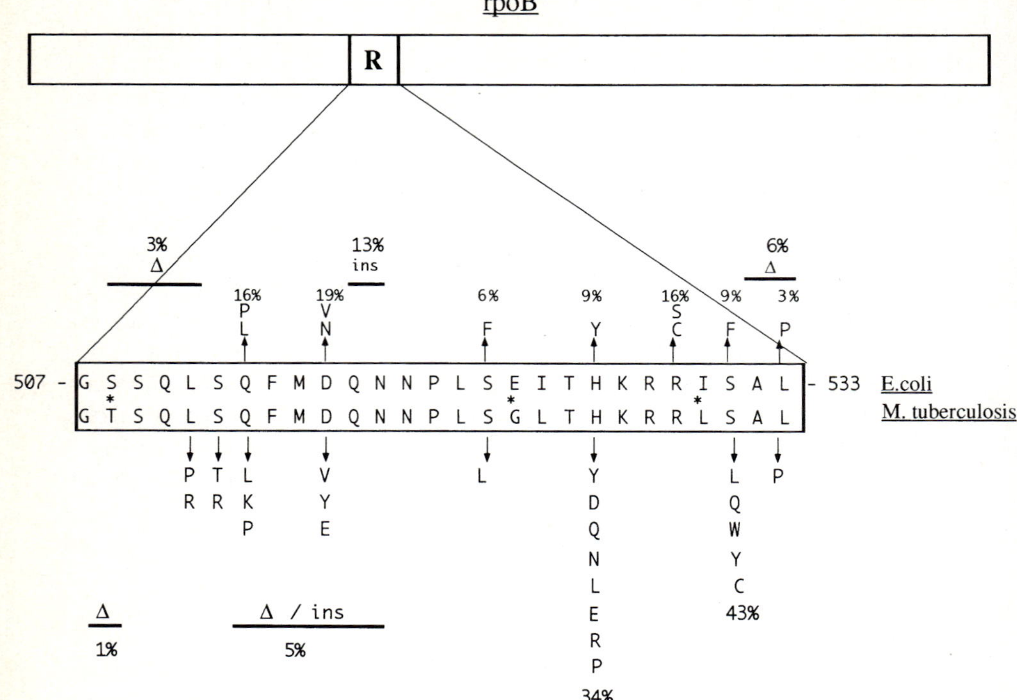

Fig. 3. Region of the *rpoB* gene encoding the β -subunit of the DNA-dependent RNA polymerase, the primary target of rifampicin, and localization (*R*) of mutations conferring rifampicin resistance

5 Pyrazinamide

The antituberculous activity of pyrazinamide was detected in 1952 (Yeager et al. 1952), but its description was overshadowed by the dramatic entry into chemotherapy of isoniazid. Moreover, the results of early studies with pyrazinamide were rather enigmatic, because the drug, particularly in combination with isoniazid, showed good activity in clinical trials but not in vitro (Tarshis and Weed 1953). Furthermore, it was potent in the mouse model but much less active in the guinea pig model (Steenken et al. 1957). The discrepancy between the in vivo and in vitro activity has been explained by the fact that the activity of pyrazinamide depends upon an acid pH in the environment. It was shown that the same strain was inhibited by 250 μg/ml of the drug at pH 7, by 53 μg/ml at pH 6 and by 15 μg/ml at pH 5 (McDermott and Tompsett 1954) and that pyrazinamide showed excellent activity against *M. tuberculosis* grown in rabbit macrophages (Mackaness et al. 1956). Because of its powerful activity in an acidic environment it was proposed that pyrazinamide should be particularly active on intracellular bacilli or in recent caseous lesions (Steenken et al. 1957; Schwartz and Moyer 1954; McDermott and Tompsett 1954). However, it has been shown recently that phagosomes containing *M. avium* are probably not acidic (Sturgill-Koszycki et al. 1994).

Currently, pyrazinamide is one of the essential drugs in short course chemotherapy of tuberculosis (Grosset 1989; Centers for Disease Control 1993) yet essentially nothing is known about its site of action or the mechanisms of resistance. Konno et al. (1967) found that pyrazinamide-susceptible strains of *M. tuberculosis* possess a pyrazinamidase activity capable of deaminating pyrazinamide to pyrazinoic acid, whereas pyrazinamide-resistant isolates, and the naturally resistant *M. bovis* strains, could not convert the drug. The structure of pyrazinamide is quite similar to that of isoniazid (Fig. 2) and the proposed activation intracellularly by pyrazinamidase to give pyrazinoic acid is analogous to the presumed activation of isoniazid to isonicotinic acid by HPI; however, these seem to be two completely independent mechanisms, as cross-resistance has never been observed (Krüger-Thiemer et al. 1975; Seydel et al. 1976). The nature of the pyrazinamidase enzyme remains obscure.

6 Ethambutol

Ethambutol is an effective and specific drug that is part of the standard regimens for the treatment of tuberculosis. It is bacteriostatic and has no effect on the viability and metabolism of nongrowing cells. As for mechanisms of action, effects on nucleic acid metabolism (Forbes et al. 1965), mycolic acid synthesis, phospholipid metabolism, and arabinogalactan synthesis have been described (Kilburn and Greenberg 1977; Takayama and Kilburn 1989). From the work of Takayama and Sareen it seems that the cell wall, and more specifically mycolic acid synthesis, is the primary target of ethambutol (Saaren and Khuller 1990; Takayama and Kilburn 1989). Among the effects observed was the inhibition of the transfer of precursor molecules in mycolic acid synthesis (mycolic-acetyl-trehalose) from the cytoplasm to the cell wall, the accumulation of trehalose-, mono- and dimycolates in the cell and the inhibition of the synthesis of arabinogalactan from D-arabinose (Takayama and Kilburn 1989). Other investigators have suggested that inhibition of glucose metabolism may be involved (Silve et al. 1993), whereas a resistant mutant of *M. tuberculosis*, H37Ra, had less phospholipids and unsaturated fatty acids and more arabinose, galactose, hexosamine and mycolic acids than the ethambutol susceptible strain (Saaren and Khuller 1990). In conclusion, both the action and the target of ethambutol remain unclear.

7 Fluoroquinolones

With the rising incidence of isoniazid- and rifampin-resistant strains there has been much interest in the use of quinolones as antituberculous agents, and various fluoroquinolones have demonstrated good activity against *M. tuberculosis* in vitro (FENLON and CYNAMON 1986; YEW et al. 1990). Among the newer molecules, ofloxacin and ciprofloxacin seem to be the most active (COLLINS and UTTLEY 1985; TRIMBLE et al. 1987). Ofloxacin has been used in some clinical studies because of its better absorption and longer half-life. One of the most recently developed quinolones, sparfloxacin, is approximately two to three times more potent than ciprofloxacin or ofloxacin (RASTOGI and GOH 1991; WITZIG and FRANZBLAU 1993) and showed excellent activity in vivo (LALANDE et al. 1993).

The principal target of the quinolones is DNA gyrase, a type II DNA topoisomerase that is composed of two A and two B subunits encoded by the genes *gyrA* and *gyrB*, respectively (HOOPER and WOLFSON 1993; WANG 1985). Mutations in the putative fluoroquinolone-binding region of the A subunit have been found to confer high-level resistance in several bacterial species (ORAM and FISHER 1991; GOSWITZ et al. 1992; YOSHIDA et al. 1990), whereas mutations in the B subunit tend to confer lower-level resistance (WOLFSON and HOOPER 1989; YOSHIDA et al. 1990). In a recent study, it was shown that high-level ciprofloxacin resistance in *M. tuberculosis* is associated with a limited number of missense mutations around codon 90 of the *gyrA* gene corresponding to the mutations found in other bacterial species (GOSWITZ et al. 1992; ORAM and FISHER 1991; TAKIFF et al. 1994). As described for streptomycin, a second, low-level resistance mechanism is also operational in a minority of strains, with normal DNA gyrase genes, but has not yet been characterized (TAKIFF et al. 1994).

8 Other Drugs

Second-line drugs used for the treatment of tuberculosis are capreomycin, kanamycin, viomycin, para-aminosalicylic acid, thiacetazone and ethionamide. These drugs, in general, are much more toxic than front-line drugs and are only used to treat cases resulting from infection with *M. tuberculosis* strains that are resistant to one or more of the front-line drugs, especially those responsible for multidrug-resistant tuberculosis. With the exception of ethionamide, which shares a target (InhA) with isoniazid, little or nothing is known about the targets or resistance mechanisms involved (WINDER 1982).

9 Multidrug Resistance of *Mycobacterium tuberculosis*

Several recent studies indicate that resistance to various anti-tuberculous agents results from alterations to chromosomal genes encoding the drug targets. Thus, multidrug resistance does not stem from the acquisition by *M. tuberculosis* of a transposable element, or a plasmid, carrying drug resistance determinants (FINKEN et al. 1993; HONORÉ and COLE 1994; MEIER et al. 1994; NAIR et al. 1993; TAKIFF et al. 1994; TELENTI et al. 1993a; ZHANG et al. 1992), but rather appears to result from the stepwise acquisition of new mutations in the genes for different drug targets. A number of operational difficulties like inadequate prescription of chemotherapy, poor compliance or an insufficient number of active drugs in the regimen may be responsible for their selection (GOBLE et al. 1993; ISEMAN 1994).

This interpretation is supported by a comprehensive study of the molecular basis of multidrug resistance (HEYM et al. 1994a) and is nicely illustrated by the findings with a multidrug-resistant strain (resistant to isoniazid, rifampin, streptomycin, ethambutol, ethionamide) isolated from an AIDS patient in Paris. When the genes known to be associated with drug resistance were examined in this strain, a missense mutation was found in the HPI gene (Arg-463 to Leu) and a mutation in the *inhA* locus was detected. These would confer isoniazid and ethionamide resistance. Rifampin and streptomycin resistance could be attributed to the most commonly occurring mutations in the *rpoB* (Ser-531 to Leu) and *rpsL* (Lys-43 to Arg) genes, respectively. The clinical records showed that the patient, who represented a case of primary resistance, did not respond to treatment and that second-line drugs, including ofloxacin, were later prescribed. After the patient's death from other AIDS-related causes it was found that treatment with the fluoroquinolone, the only active drug in the regimen, had led to the emergence of a strain with a mutation in *gyrA* (Asp-94 to Asn).

Moreover, in virtually all cases of drug resistance, exactly the same mutation, or combination of mutations, was found in the various strains of *M. tuberculosis* examined, irrespective of previous treatment, the patients'HIV status or the clinical manifestations (HEYM et al. 1994a). Therefore, as postulated many years ago, it appears that random mutations conferring resistance occur naturally during microbial replication (CANETTI and GROSSET 1961; MITCHISON 1950, 1951). These mutations arise independently and are not induced by drugs but strains harboring them may be selected by incorrect drug use (CENTERS FOR DISEASE CONTROL 1993; HEYM et al. 1994a; ISEMAN et al. 1993; SMALL et al. 1993).

10 Rapid Detection of Drug Resistance

With the recrudescence of tuberculosis and the emergence of multidrug-resistant strains, rapid detection of resistance is more necessary than ever. The first significant progress was made several years ago with the implementation of radiometric susceptibility testing by the BACTEC system. This enables susceptibility testing to be accomplished in 7–10 days, compared to 4–6 weeks with the proportional method (HEIFETS 1988; SEWELL et al.1993; SNIDER et al. 1981; TENOVER et al. 1993).

Progress in mycobacterial molecular genetics has resulted in three new techniques that are currently undergoing clinical evaluation. Two of these involve screening mutational hotspots in the genes encoding drug targets by using the polymerase chain reaction followed by analysis of the single strand conformational polymorphism (PCR-SSCP) or by automated DNA sequence analysis. The third is a biological assay using luciferase reporter phages (JACOBS et al. 1993; KAPUR et al. 1994; TELENTI et al. 1993b).

The principle of PCR-SSCP is based on the fact that the two denatured strands of a PCR-amplified DNA molecule adopt stable intramolecular conformations and that genetic modifications lead to conformational changes that can be easily recognized by their altered electrophoretic mobility compared to the wild-type pattern (HEYM et al. 1994a; HONORÉ and COLE 1993, 1994; TELENTI et al. 1993a). Automated DNA sequence analysis requires a greater level of technological sophistication but yields definitive and unambiguous results (KAPUR et al. 1994). Both these methods are rapid, giving results in under 48 h, and can be performed on minimally grown cultures or rich sputum samples (TELENTI et al. 1993b; IMBODEN et al. 1993; KAPUR et al. 1994), but suffer from complexity as simultaneous screening of several drug targets is required.

The luciferase reporter phage is an ingenious tool for evaluating viability and involves infection of mycobacterial cells with a phage carrying the firefly luciferase gene. In the presence of ATP, found only in living organisms, luciferase produces light from its substrate, luciferin (JACOBS et al. 1993). If mycobacteria are treated with drugs prior to infection with the reporter phage, light is only produced by viable or drug-resistant cells. Although exquisitely simple this method is not without drawbacks and, in its present state, appears to be more suited to screening for new antituberculous compounds (COOKSEY et al. 1993) than for evaluating drug susceptibility of clinical isolates.

11 Chromosomal Mapping of the Genes Involved in Drug Resistance of *Mycobacterium tuberculosis*

As shown above, tremendous progress has been made in the last 3 years in identifying the genes involved in drug resistance. As part of the *M. tuberculosis* genome project these genes were localized on the integrated map of the chromosome (COLE and SMITH 1994) consisting of positioned cosmid clones and *DraI* macro-restriction fragments.

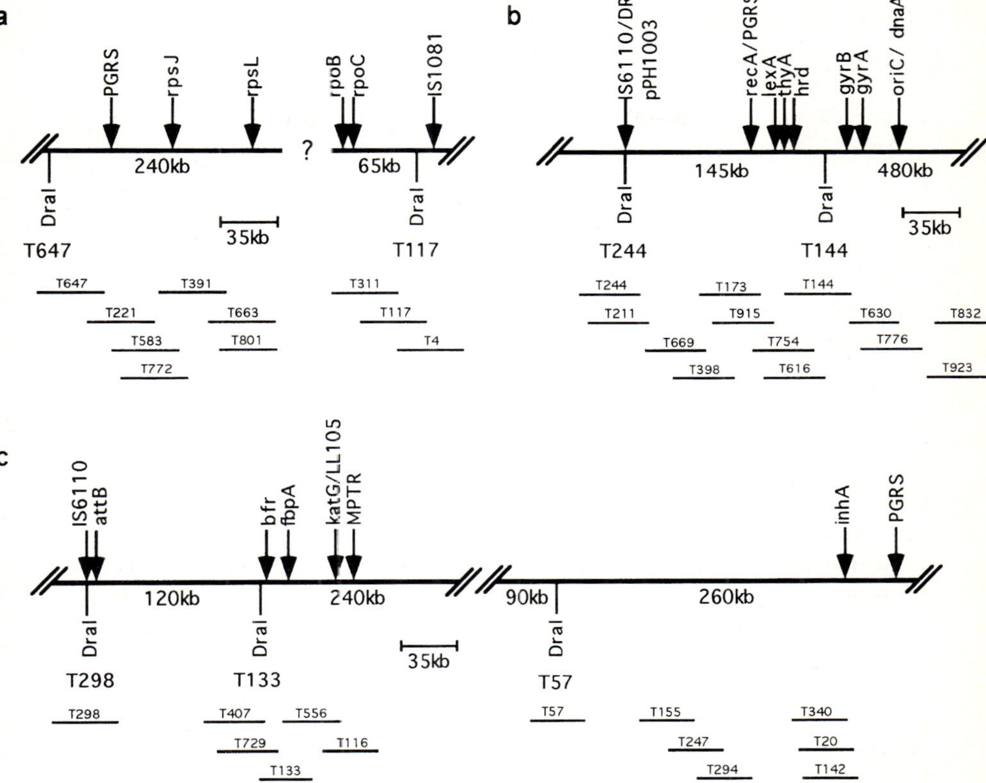

Fig. 4a–c. Organization is strain H37 Rv of the regions with genes involved in resistance or susceptibility to **a** streptomycin and rifampicin, **b** fluoroquinolones and **c** isoniazid. **a** PGRS (polymorphic GC rich repetitive sequence), *rpsJ* and *rpsL* are localized on the same DraI fragment of 240 kb, and, if the situation is the same as in *M. laprae* (indicated by the?), the 65 kb DraI fragment carrying *rpoB* and *rpoC* follows. One copy of the insertion sequence IS*1081* is found about 70 kb upstream of the *rpoBrpoC* locus. **b** One of 16 copies of IS*6110*, flanked by the direct repeat sequence DRr, is about 160 kb upstream of the *gyrA* and *gyrB* genes. Between the IS element and the DNA gyrase genes are two genes implicated in DNA recombination and repair, *recA* and *lexA*, close to genes encoding a putative thymidylate synthetase, *thyA*, and two RNA polymerase sigma factors, *hrd*. The exact orientation of *oriC* (origin of replication) and *dnaA* (initiation of DNA biosynthesis) is not known, because both probes hybridized on the same 5 kb BamHI fragment of cosmid T776. **c** MPTR (major polymorphic tandem repeat) is located downstream of *katG*, whereas *fbpA* (fibronectin binding protein) and *bfr* (bacterioferritin) are found about 40–50 kb upstream of *katG*. LL105 is an anoymous clone of a λgt11 library (kindly provided by . Andersen). *attB* is the putative attachment site for pSAM2, directly followed by one copy of IS*6110*. The *inhA* gene and another PGRS are situated at the end of a 260kb DraI fragment

The *rpsL* gene is located on a 240 kb *DraI* fragment (Fig. 4a) near the *rpsJ* gene, coding for the S10 ribosomal protein, similar to the situation found in other bacteria and *M. leprae* (Cole and Smith 1994; Honoré et al. 1993). The Rif operon of *M. tuberculosis*, containing the *rpoB* and *rpoC* genes, the former of which encodes the target of rifampin, is located on another *DraI* fragment of 65 kb (Fig. 4a) which means that it is further from the *rpsL* locus than in *M. leprae*, where these two loci are only separated by about 20 kb (Honoré et al. 1993).

As in *M. leprae* and other bacterial species, the genes *gyrA* and *gyrB*, whose products are targets of fluoroquinolones, are localized closed to the putative origin of replication, *oriC*, and the *dnaA* gene, one of the genes involved in replication (Smith et al. 1991). This whole region was identified on a 480 kb *DraI* fragment (Fig. 4b).

The genes which are involved in susceptibility or resistance to isoniazid, *katG* and *inhA*, respectively (Banerjee et al. 1994; Heym et al. 1993; Zhang et al. 1992) are separated by more than 100 kb (Fig. 4c). The *katG* locus is relatively close to *fbpA*, encoding fibronectin binding protein A, and *bfr*, coding for bacterioferritin (Pessolani et al. 1994). Upstream of the *katG* gene is a region rich in MPTR (major polymorphic tandem repeats, Hermans et al. 1992) sequences which might play a role in the rearrangement of this chromosomal site, since it was found in some highly isoniazid-resistant isolates that a large fragment harboring the *katG* gene was deleted (Zhang et al. 1993). The *inhA* gene was found at one end of a large continuous stretch of overlapping clones (contig) on a 260 kb *DraI* fragment where no other known genes have been localized to date (Fig. 4c).

Acknowledgements. Work from this laboratory received financial support from the Association Francaise Raoul Follereau, the STD 3 Program of the European Community (Grant TS3*CT93*0243) and the National Institute of Allergy and Inflectious Diseases (A137015).

References

American Thoracic Society (1986) Treatment of tuberculosis and tuberculous infection in adults and children. Am Rev Respir Dis 134: 355–363

American Thoracic Society (1992) Control of tuberculosis in the United States. Am Rev Respir Dis 146: 1623–1633

Association Française Raoul Follereau, the STD3 Program of the European Community (Grant TS3*CT93*0243) and the National Institute of Allergy and Infectious Diseases (A137015).

Ayvazian LF (1993) History of tuberculosis. In: Reichman LB, Hershfield ES (eds) tuberculosis-a comprehensive international approach. Dekker, New York, pp 1–20

Banerjee A, Dubnau E, Quemard A, Balasubramanian V, Um KS, Wilson T, Collins D, de Lisle G, Jacobs WR Jr. (1994) *inhA*, a gene encoding a target for isoniazid and ethionamide in *Mycobacterium tuberculosis*. Science 263: 227–230

Barclay WR, Ebert RH, Koch-Weser D (1953) Mode of action of isoniazid, part I. Am Rev Tuberc 67: 490–496

Barnes PF, Bloch AB, Davidson PT, Snider DE (1991) Tuberculosis in patients with human immunodeficiency virus infection. N Engl J Med 324: 1644–1650

Beggs WH, Andrews FA (1976) Inhibition of dihydrostrepomycin binding to *Mycobacterium smegmatis* by monovalent and divalent cation salts. Antimicrob Agents Chemother 9: 393–396

Benveniste R, Davies J (1973) Mechanisms of antibiotic resistance in bacteria. Annu Rev Biochem 42: 471–506

Bergler H, Högenauer G, Turnowsky F (1992) Sequences of the *envM* gene and two mutated allels in *Escherichia coli*. J Gen Microbiol 138: 2093–2100

Bernstein J, Lott WA, Steinberg BA, Yale HL (1952) Chemotherapy of experimental tuberculosis. Am Rev Tuberc 65: 357–364

Bloom BR, Murray JL (1992) Tuberculosis: commentary on a reemergent killer. Science 257: 1055–1064

Böttger EC (1994) Resistance to drugs targeting protein synthesis in mycobacteria. Trends Microbiol 2: 416–421

Bryan LE, van den Elzen HM (1977) Effects of membrane-energy mutations and cations on streptomycin and gentamicin accumulation by bacteria: a model for entry of streptomycin and gentamicin in susceptible and resistant bacteria. Antimicrob Agents Chemother 12: 163–177

Buck M, Schnitzer RJ (1952) The development of drug resistance of *M. tuberculosis* to isonicotinic acid hydrazide. Am Rev Tub 65: 759–760

Burgess RR, Erickson B, Gentry D, Gribstov M, Hager D, Lesley S, Strickland M, Thompson N (1987) Bacterial RNA polymerase subunits and genes. RNA polymerase and the regulation of transcrption. Elsevier, New York, pp 3–15

Canetti G, Grosset J (1961) Teneur des souches sauvages de Mycobacterium tuberculosis en variants résistants à l'isoniazide et en variants résistants à la streptomycine sur milieu de Loewenstein-Jensen. Ann Inst Pasteur 101: 28

Centers for Disease Control (1993) Initial therapy for tuberculosis in the era of multidrug resistance: recommendations of the advisory council for the elimination of tuberculosis. MMWR 42 (RR-7): 1–8

Claiborne A, Fridovitch I (1979) Purification of the *o*-dianisidine peroxidase from *Escherichia coli*. J Biol Chem 254: 4245–4252

Cohn ML, Kovitz C, Oda U, Middlebrook G (1954) Studies on isoniazid and tubercle bacilli. II. The growth requirements, catalase activity, and pathogenic properties of isoniazid-resistant mutants. Am Rev Tuberc 70: 641–669

Cole ST (1994) Drug resistance mechanisms employed by *Mycobacterium tuberculosis*. Trends Microbiol 2: 412–415

Cole ST, Smith DR (1994) Toward mapping and sequencing the genome of *Mycobacterium tuberculosis*. In: Bloom BR (ed) Tuberculosis: pathogenesis, protection, and control. American Society of Microbiology, Washington DC, pp 227–238

Collins CH, Uttley AHC (1985) *In vitro* susceptibility of mycobacteria to ciprofloxacin. J Antmicrob Chemother 16: 575–580

Cooksey RC, Crawford JT, Jacobs WR, Shinnick TM (1993) A rapid method for screening antimicrobial agents for activities against a strain of *Mycobacterium tuberculosis* expressing firefly luciferase. Antimicrob Agent Chemother 37: 1348–1352

Crawford JT, Bates JH (1979) Isolation of plasmids from mycobacteria. Infect Immun 24(3): 979–981

Daley CL, Small PM, Schecter GF, Schoolnik GK, McAdam RA, Jacobs WR, Hopewell PC (1992) An outbreak of tuberculosis with accelerated progression among persons infected with the human immunodeficiency virus. An analysis using restriction-fragment-length polymorphisms. N Engl J Med 326: 231–235

Davidson LA, Takayama K (1979) Isoniazid inhibition of the synthesis of monounsaturated long-chain fatty acids in *Mycobacterium tuberculosis* H37Ra. Antimicrob Agent Chemother 16: 104–105

Edlin BR, Tokars JI, Grieco MH, Crawford JT, Williams J, Sordillo EM, Ong KR, Kilburn JO, Dooley SW, Castro KG, Jarvis WR, Holmberg SD (1992) An outbreak of multidrug-resistant tuberculosis among hospitalized patients with the acquired immunodeficiency syndrome. N Engl J Med 326: 1514–1521

Falkinham JO III, Crawford JT (1994) Plasmids. In: Bloom BR (ed) Tuberculosis: pathogenesis, protection, and control. American Society for Microbiology, Washington DC, pp 185–198

Fenlon CH, Cynamon MH (1986) Comparative *in vitro* activities of ciprofloxacin and other 4-quinolones against *Mycobacterium tuberculosis* and *Mycobacterium intracellulare*. Antimicrob Agent Chemother 29: 386–388

Finken M, Kirschner P, Meier A, Wrede A, Böttger EC (1993) Molecular basis of streptomycin resistance in *Mycobacterium tuberculosis*: alterations of the ribosomal protein S12 gene and point mutations within a functional 16S ribosomal RNA pseudoknot. Mol Microbiol 9: 1239–1246

Forbes M, Kuck NA, Peets EA (1965) Effect of ethambutol on nucleic acid metabolism in *Mycobacterium smegmatis* and its reversal by polyamines and divalent cations. J Bacteriol 89: 1299–1305

Fox HH (1951) Synthetic tuberculostatics show promise. Chem Eng News 29: 3963–3964
Frieden TR, Sterling T, Pablos-Mendez A, Kilburn JO, Cauthen GM, Dooley SW (1993) The emergence of drug-resistant tuberculosis in New York City. N Engl J Med 328: 521–526
Funatsu G, Wittmann HG (1975) Location of amino acid replacements in protein S12 isolated from *Escherichia coli* mutants resistant to streptomycin. J Mol Biol 68: 547–550
Gale EF, Cundliffe E, Reynolds PE, Richmond MH, Waring MJ (1981) The moelcular basis of antibiotic action. Wiley, London
Galili H, Fromm H, Aviv D, Edelman M, Galun E (1989) Ribosomal protein S12 as a site of streptomycin resistance in *Nicotinia* chloroplasts. Mol Gen Genet 218: 289–292
Garvin RT, Biswas DK, Gorini L (1974) The effects of streptomycin or dihydrostreptomycin binding to 16S RNA or to 30S ribosomal subunits. Proc Natl Acad Sci USA 71: 3814–3818
Goble M, Madsen LA, Waite D, Ackerson L, Horsburgh R (1993) Treatment of 171 patients with pulmonary tuberculosis resistant to isoniazid and rifampin. N Engl J Med 328: 527–532
Goswitz JJ, Willard KE, Fasching CE, Perterson LR (1992) Detection of *gyrA* gene mutations associated with ciprofloxacin resistance in methicillin resistant *Staphylococcus aureus*: analysis by polymerase chain reaction and direct DNA sequencing. Antimicrob Agents Chemother, 36: 1166–1169
Grosset J, Levantis S (1983) Adverse effects of rifampin. Rev Infect Dis 5 [Suppl 3]: S440–S446
Grosset JH (1989) Present status of chemotherapy for tuberculosis. Rev Infect Dis 11 [Suppl 2]: S347–S352
Heifets L (1988) Qualitative and quantitative drug-susceptibility tests in mycobacteriology. Am Rev Respir Dis 137: 1217–1222
Herman RP, Weber MM (1980) Isoniazid interaction with tyrosine as a possible mode of action of the drug in mycobacteria. Antimicrob Agents Chemother 17: 170–178
Hermans PWM, van Soolingen D, van Embden JDA (1992) Characterization of a major polymorphic tandem repeat in *Mycobacterium tuberculosis* and its potential use in the epidemiology of *Mycobacterium kansasii* and *Mycobacterium* gordonae. J Bacteriol 174: 4157–4156
Heym B, Zhang Y, Poulet S, Young D, Cole ST (1993) Characterization of the *katG* gene encoding a catalase-peroxidase required for the isoniazid susceptibility of *Mycobacterium tuberculosis*. J Bacteriol 175: 4255–4259
Heym B, Honoré N, Truffot-Pernot C, Banerjee A, Schurra C, Jacobs WR Jr, van Embden JDA, Grosset JH, Cole ST (1994a) The implications of multidrug-resistance for the future of short course chemotherapy of tuberculosis: a molecular study. Lancet 344: 293–298
Heym B, Alzari PM, Honoré N, Cole ST (1994b) Missense mutations in the catalase-peroxidase gene, *katG*, are associated with isoniazid resistance in *Mycobacterium tuberculosis*. Mol Microbiol 15: 235–245
Honoré N, Cole ST (1993) The moelcular basis of rifampin resistance in *Mycobacterium leprae*. Antimicrob Agents Chemother 37: 414–418
Honoré N, Cole ST (1994) Streptomycin resistance in mycobacteria. Antimicrob Agents Chemother 38: 238–242
Honoré N, Bergh S, Chanteau S, Doucet-Populaire F, Eiglmaier K, Garnier T, Georges C, Launois P, Limpaiboon T, Newton S, Niang K, del Portillo P, Ramesh GR, Reddi P, Ridel PR, Sittisombut N, WU-Hunter S, Cole ST (1993) Nucleotide Sequence of the first cosmid from the *Mycobacterium leprae* genome project: structure and function of the Rif-Str regions. Mol Microbiol 7: 207–214
Hooper DC, Wolfson JS (1993) Mechanisms of quinolone action and bacterial killing. Quinolone antimicrobial Agents. American Society for Microbiology, Washington DC, pp 53–76
Hopewell PC (1994) The cure: organization and administration of therapy for tuberculosis. In: Porter JDH, McAdam KPW (eds) Tuberculosis - back to the future. Wiley, Chichester, pp 99–120
Imboden P, Cole S, Bodmer T, Telenti A (1993) Detection of Rifampin Resistance Mutations in *Mycobacterium tuberculosis* and *M. leprae*. In: Persing DH, Smith TF, Tenover FC, White TJ (eds) Diagnostic moelcular microbiology. Principles and applications. American Society for Microbiology, Washington DC, pp 519–526
Iseman MD (1994) Evolution of drug-resistant tuberculosis: a tale of two species. Proc Natl Acad Sci USA 91: 2428–2429
Iseman MD, Cohn DL, Sbarbaro JA (1993) Directly observed treatment of tuberculosis. N Engl J Med 328: 576–578
Jacobs WR Jr, Tuckman M, Bloom BR (1987) Introduction of foreign DNA into mycobacteria using a shuttle plasmid. Nature 327: 532–535
Jacobs WR Jr, Barletta RG, Udani R, Chan J, Kalkut G, Sosne G, Kieser T, Sarkis GJ, Hatfull GF, Bloom BR (1993) Rapid assessment of drug susceptibilities by means of luciferase reporter phages. Science 260: 819–822

Jarlier V, Nikaido H (1990) Permeability barrier to hydrophilic solutes in *Mycobacterium chelonei*. J Bacteriol 172: 1418–1423

Jin DJ, Gross CA (1989) Characterization of the pleiotropic phenotypes of rifampin-resistant *rpoB* mutants of *Escherichia coli*. J Bacteriol 171: 5229–5231

Johnsson K, Schultz PG (1994) Mechanistic studies of the oxidation of isoniazid by the catalase-peroxidase from *Mycobacterium tuberculosis*. J Am Chem Soc 116: 7425–7426

Kanner BI, Gutnick DL (1972) Use of neomycin in the isolation of mutants blocked in energy conservation in *Escherichia coli*. J Bacteriol 111: 287–289

Kapur V, Hamrick MR, LI L-L, Plikaytis BB, Shinnick TM, Telenti A, Jacobs WR, Banerjee A, Cole ST, Yam WC, Clarridge JE, Kreiswirth BN, Musser JM (1994) Application of automated DNA sequencing strategies to problems of *Mycobacterium* diagnostics: speciation by characterization of polymorphisms in the gene (*hsp65*) encoding a 65-kilodalton heat shock protein, and a rapid an unambiguous identification of mutations associated with antibiotic resistance in *Mycobacterium tuberculosis*. JAMA (submitted)

Kempsell KE, Ji Y-E, Estrada-G ICE, Colston MJ, Cox RA (1992) The Nucleotide Sequence of the Promoter, 16S rRNA and Spacer Region of the Ribosomal RNA Operon of *Mycobacterium tuberculosis* and Comparison with *Mycobacterium leprae* Precursor rRNA. J Gen Microbiol 138: 1717–1727

Kilburn JO, Greenberg J (1977) Effect of ethambutol on the viable cell count in *Mycobacterium smegmatis*. Antimicrob Agent Chemother 11: 534–540

Konno K, Feldmann FM, McDermott W (1967) Pyrazinamide susceptibility and amidase activity of tubercle bacilli. Am Rev Respir Dis 95: 461–469

Krüger-Thiemer E, Kröger H, Nestler HJ, Seydel JK (1975) Bakteriologische Grundlagen der Chemotherapie der Tuberkulose. Mykobakterien und mykobakterielle Krankheiten. Fischer, Jena, pp 251–288

Lalande V, Truffot-Pernot C, Paccaly-Moulin A, Grosset J, Ji B (1993) Powerful bactericidal activity of sparfloxacin (AT-4140) against *Mycobacterium tuberculosis* in mice. Antimicrob Agent Chemother 37: 407–413

Loewen PC, Switala J, Triggs-Raine BL (1985) Catalases HPI and HPII in *Escherichia coli* are induced independently. Arch Biochem Biophys 243: 144–149

Loprasert S, Negoro S, Okada H (1988) Thermostable peroxidase from *Bacillus stearothermophilus*. J Gen Microbiol 134: 1971–1976

Mackaness GB (1956) The intracellular activation of pyrazinamide and nicotinamide. Am Rev Tuberc Pulm Dis 74: 718–728

Martin C, Ranes M, Gicquel B (1990) Plasmids, antibiotic resistance, and mobile genetic elements in mycobacteria. In: McFadden J (ed) The molecular biology of the mycobacteria. Surrey University Press, London, pp 120–138

McDermott W, Tompsett R (1954) Activation of pyrazinamide and nicotinamide in acidic environments *in vitro*. Am Rev Tuberc 71: 748–754

Meier A, Kirschner P, Bange F-C, Vogel U, Böttger EC (1994) Genetic alterations in streptomycin-resistant *Mycobacterium tuberculosis*: mapping of mutations conferring resistance. Antimicrob Agents Chemother 38: 228–233

Meyer H, Mally J (1912) Über Hydrazinderivate der Pyridincarbonsäuren. Monatshefte Chem 33: 393–414

Middlebrook G (1952) Sterilization of tubercle bacilli by isonicotinic acid hydrazide and the incidence of variants resistant to the drug in vitro. Am Rev Tuberc 65: 765–767

Miller LP, Crawford JT, Shinnick TM (1994) The *rpoB* gene of *Mycobacterium tuberculosis*. Antimicrob Agent Chemother 38: 805–811

Mitchison DA (1950) Development of streptomycin resistant strains of tubercle bacilli in pulmonary tuberculosis. Thorax 5: 144–161

Mitchison DA (1951) The segregation of streptomycin-resistant variants of *Mycobacterium tuberculosis* into groups with characteristic levels of resistance. J Gen Microbiol 5: 596–604

Moazed D, Noller HF (1987) Interaction of antibiotics with functional sites in 16S ribosomal RNA. Nature 327: 389–394

Montandon P-E, Wagner R, Stutz E (1986) *E. coli* ribosomes with a C912 to U base change in the 16S rRNA are streptomycin resistant. EMBO J 5: 3705–3708

Morris SL, Nair J, Rouse DA (1992) The catalase-peroxidase of *Mycobacterium intracellulare*: nucleotide sequence analysis and expression in *Escherichia coli*. J Gen Microbiol 138: 2363–2370

Murray CJL, Styblo K, Rouillon A (1990) Tuberculosis in developing countries: burden, intervention and cost. Bull Int Un Tuberc 65: 6–24

Nair J, Rouse DA, Bai G-H, Morris SL (1993) The *rpsL* gene and streptomycin resistance in single and multiple drug-resistant strains of *Mycobacterium tuberculosis*. Mol Microbiol 10: 521–527

Noller H (1984) Structure of ribosomal RNA. Annu Rev Biochem 53: 119–162

O'Brien RJ (1993) The treatment of tuberculosis. In: Reichman LB, Hershfield ES (eds) Tuberculosis - a comprehensive international approach. Dekker, New York, pp 207–240

Oram M, Fisher LM (1991) 4-Quinolone resistance mutation in the DNA gyrase of *Escherichia coli* clinical isolates identified using the polymerase chain reaction. Antimicrob Agent Chemother 35: 387–389

Ovchinnikov YA, Monastyrskaya GS, Guriev SO, Kalinina NF, Sverdlof ED, Gragerov AI, Bass IA, Kiver IF, Moiseyeva EP, Igumnov VN, Mindlin SZ, Nikiforov VG, Khesin RP (1983) RNA polymerase rifampicin resistance mutations in *Escherichia coli:* sequence changes and dominance. Mol Gen Genet 190: 344–348

Peizer LR, Widelock D (1955) The correlation of rate of catalase activity,Guinea pig virulence, and isoniazid resistance of tubercle bacilli from specimens of patients under isoniazid therapy. Am Rev Respir Dis 72: 246–251

Pessolani MCV, Smith DR, Rivoire B, McCormick J, Hefta SA, Cole ST, Brennan P (1994) Purification, characterization, gene sequence and significance of a bacterioferritin from *Mycobacterium leprae*. J Exp Med 180: 319–327

Rastogi N, Goh KS (1991) *In vitro* activity of the new difluorinated quinolone sparfloxacin (AT -4140) against *Mycobacterium tuberculosis* compared with activities of ofloxacin and ciprofloxacin. Antimicrob Agent Chemother 35: 759–764

Rosner JL (1993) An alternative to *Mycobacterium tuberculosis* for studying isoniazid resistance: *oxyR* regulon mutants of *Escherichia coli*. Antimicrob Agent Chemother 37: 2251–2253

Rosner JL, Storz G (1994) Peroxides and isoniazid-susceptibility of *Escherichia coli* and *Mycobacterium smegmatis*. Antimicrob Agent Chemother 38: 1829–1833

Saaren M, Khuller GK (1990) Cell wall and membrane changes associated with ethambutol resistance in *Mycobacterium tuberculosis* H37Ra. Antimicrob Agent Chemother 34: 1773–1776

Saiki RK, Gelfand DH, Stoffel S, Scharf SJ, Higuchi R, Horn GT, Mullis KB, Erlich HA (1988) Primer-directed enzymatic amplification of DNA with thermostable DNA polymerase. Science 239: 487–491

Schwartz WS, Moyer RE (1954) The chemotherapy of pulmonary tuberculosis with pyrazinamide used alone and in combination with streptomycin, para-aminosalicylic acid, or isoniazid. Am Rev Tuberc 70: 413–429

Sewell DL, Rashad AL, Rourke WJ, Poor SL, McCarthy JAC, Pfaller MA (1993) Comparison of the septi-chek AFB and BACTEC systems and conventional culture for recovery of mycobacteria. J Clin Microbiol 31: 2689–2691

Seydel JK, Schaper K-J, Wempe E, Cordes HP (1976) Mode of action and quantitative structure-activity correlations of tuberculostatic drugs of the isonicotinic acid hydrazide type. J Med Chem 19: 483–491

Shoeb HA, Bowman BU Jr, Ottolenghi AC, Merola AJ (1985) Evidence for the generation of active oxygen by isoniazid treatment of extracts of *Mycobacterium tuberculosis* H37Ra. Antimicrob Agent Chemother 27: 404–407

Silve G, Valero-Guillen P, Quemard A, Dupont M, Daffe M, Laneelle G (1993) Ethambutol inhibition of glucose metabolism in mycobacteria: a possible target of the drug. Antimicrob Agent Chemother 37: 1536–1538

Small PM, Shafer RW, Hopewell PC, Singh SP, Murphy MJ, Desmond E, Sierra MF, Schoolnik GK (1993) Exogenous reinfection with multidrug-resistant *Mycobacterium tuberculosis* in patients with advanced HIV infection. N Engl J Med 328: 1137–1144

Smith DW, Tee TW, Baird C, Krishnapillai V (1991) Pseudomonad replication origins: a paradigm for bacterial origins? Mol Microbiol 5: 2581–2587

Snider DE Jr, Roper WL (1992) The new tuberculosis. N Engl J Med 326: 703–705

Snider DE Jr, Good RC, Kilburn JO, Laskowski LF Jr, Lusk RH, Marr JH, Reggiardo Z, Middlebrook G (1981) Rapid drug-susceptibility testing of *Mycobacterium tuberculosis*. Am Rev Respir Dis 123: 402–406

Steenken W Jr, Meade GM, Wolinsky E, Coates EO Jr (1952) Demonstration of increased drug resistance of tubercle bacilli from patients treated with hydrazines of isonicotinic acid. Am Rev Tuberc 65: 754–758

Steenken W Jr, Wolinsky E, Smith MM, Montalbine V (1957) Further observations on pyrazinamide alone and in combination with other drugs in experimental tuberculosis. Am Rev Tub Pulm Dis 76: 643–659

Sturgill-Koszycki S, Schlesinger PH, Chakraborty P, Haddix PL, Collins HL, Fok AK, Allen RD, Gluck SL, Heuser J, Russell DG (1994) Lack of acidification in *Mycobacterium* phagosomes produced by exclusion of the vesicular proton ATPase. Science 263: 678–681

Takayama K, Kilburn JO (1989) Inhibition of synthesis of arabinogalactan by ethambutol in *Mycobacterium smegmatis*. Antimicrob Agent Chemother 33: 1493–1499

Takiff HE, Salazar L, Guerrero C, Philipp W, Huang WM, Kreisworth B, Cole S, Telenti A, Jacobs WR (1994) Cloning and nucleotide sequence of the *Mycobacterium tuberculosis gyrA* and *gyrB* genes, and characterization of quinolone resistance mutations. Antimicrob Agents Chemother 38: 773–780

Tarshis MS, Weed WA Jr (1953) Lack of significant *in vitro* sensitivity of *Mycobacterium tuberculosis* to pyrazinamide on three different solid media. Am Rev Tuberc 67: 391–395

Telenti A, Imboden P, Marchesi F, Lowrie D, Cole ST, Colston MJ, Matter L, Schopfer K, Bodmer T (1993a) Detection of rifampicin-resistance mutations in *Mycobacterium tuberculosis*. Lancet 341: 647–650

Telenti A, Imboden P, Marchesi F, Schmidheini T, Bodmer T (1993b) Direct, automated detection of rifampicin-resistant *Mycobacterium tuberculosis* by polymerase chain reaction and single-strand conformation polymorphism analysis. Antimicrob Agents Chemother 37: 2054–2058

Tenover FC, Crawford JT, Huebner RE, Geiter LJ, Horsburgh CR, Good RC (1993) The resurgence of tuberculosis: is your laboratory ready? J Clin Mircobiol 31: 767–770

Trias J, Benz R (1993) Characterization of the channel formed by the mycobacterial porin in lipid bilayer membranes. J Biol Chem 268: 6234–6240.

Triggs-Raine BL, Doble BW, Mulvey MR, Sorby PA, Loewen PC (1988) Nucleotide sequence of *katG*, encoding catalase HPI of *Escherichia coli*. J Bact 1970: 4415–4419

Trimble KA, Clark RB, Sanders WE Jr, Frankel JW, Cacciatore R, Valdez H (1987) Activity of ciprofloxacin against mycobacteria *in vitro*: comparison of BACTEC and macrobroth dilution methods. J Antimicrob Chemother 19: 617–622

Tsukamura M, Tsukamura S (1963) Isotopic studies on the effect of isoniazid on protein synthesis of mycobacteria. Jpn J Tuberc 11: 14–17

Tsukamura M, Hasimoto M, Noda Y (1958) Transformation of isoniazid and streptomycin resistance in *Mycobacterium avium* by the desoxyribonucleate derived from isoniazid- and streptomycin-double-resistant cultures. Am Rev Respir Dis 81: 403–406

Turnowsky F, Fuchs K, Jeschek C, Högenauer G (1989) *envM* genes of *Salmonella typhimurium* and *Escherichia coli*. J Bacteriol 171: 6555–6565

Wallace BJ, Davis BD (1973) Cyclic blockade of initiation sites by streptomycin-damaged ribosomes in *Escherichia coli*: an explanation for dominance of sensitivity. J Mol Biol 75: 377–390

Wang JC (1985) DNA topoisomerases. Annu Rev Biochem 54: 665–697

Wehrli W (1983) Rifampin: mechanisms of action and resistance. Rev Infect Dis 5: S407–S411

Weisblum B, Davies J (1968) Antibiotic inhibitors of the bacterial ribosome. Bacteriol Rev 32: 493–528

Wimpenny JWT (1967) Effect of isoniazid on biosynthesis in *Mycobacterium tuberculosis var. bovis* BCG. J Gen Microbiol 47: 379–388

Winder FG (1982) Effect of isoniazid on biosynthesis in *Mycobacterium tuberculosis var. bovis* BCG. J Gen Microbiol 47: 379–388

Witzig RS, Franzblau SG (1993) Susceptibility of *Mycobacterium kansasii* to ofloxacin, sparfloxacin, clarithromycin, azithromycin, and fusidic acid. Antimicrob Agent Chemother 37: 1997–1999

Wolfson JS, Hooper DC (1989) Bacterial resistance to quinolones. Rev Infect Dis 11: S960–S968

Yeager RL, Munroe WGC, Dessau FI (1952) Pyrazinamide (aldinamide) in the treatment of pulmonary tuberculosis. Am Rev Tuberc 65: 523–534

Yew WW, Kwan Sy-L, Ma WK, Khin MA, Chau PY (1990) In vitro activity of ofloxacin against *Mycobacterium tuberculosis* and its clinical efficacy in multiply resistant pulmonary tuberculosis. J Antimicrob Chemother 26: 227–236

Yoshida H, Bogaki M, Nakamura M, Nakamura S (1990) Quinolone resistance-determining region in the DNA gyrase *gyrA* gene of *Escherichia coli*. Antimicrob Agent Chemother 34: 1271–1272

Youatt J (1969) A review of the action of isoniazid. Am Rev Respir Dis 99: 729–749

Zhang Y, Heym B, Allen B, Young D, Cole S (1992) The catalase-peroxidase gene and isoniazid resistance of *Mycobacterium tuberculosis*. Nature 358: 591–593

Zhang Y, Garbe T and Young D (1993) Transformation with *katG* restores isoniazid-sensitivity in *Mycobacterium tuberculosis* isloates resistant to a range of drug concentrations. Mol Microbiol 8: 521–524

Entry of *Mycobacterium tuberculosis* into Mononuclear Phagocytes

L.S. Schlesinger

1 Introduction

Mycobacterium tuberculosis is one of a diverse group of intracellular pathogens that survives the process of entry into mammalian cells. Multiplication within the cell is critical for the pathogenesis of disease by these host-adapted microbes. For *M. tuberculosis*, the major host cell reservoir is the mononuclear phagocyte, including both monocytes and macrophages. These cells, along with polymorphonuclear leukocytes, are considered professional phagocytes in that they have adapted specifically to engulf and destroy bacteria or other foreign particulate matter. Given the potent microbiocidal mechanisms that these cells possess, *M. tuberculosis* has adapted strategies to circumvent adverse host

Department of Internal Medicine, The University of Iowa, Department of Veterans Affairs Medical Center, 200 Hawkins Dr., SW54-GH, Iowa City, IA 52242, USA cellular responses during and after entry. The earliest interaction between *M. tuberculosis* and the mononuclear phagocyte is binding of the bacterium to the cell surface and subsequent internalization. Specific receptor-ligand interactions mediate this internalization and the outcome of one interaction may be different than the outcome of another (Fig. 1). For example, it can be only attachment of the bacterium, which for *M. tuberculosis* may be less important for disease pathogenesis, or it can be ingestion (phagocytosis), which in turn can lead to either intracellular survival or death of the bacterium. Recent studies provide evidence for the importance of specific receptor-ligand pathways (single or multiple) and cooperativity between receptors in dictating the fate of phagocytosed intracellular pathogens (Joiner et al. 1990; Mosser and Edelson 1987; Drevets et al. 1992; Hoepelman and Tuomanen 1992; Hieny et al. 1992). Apart from receptor-ligand interactions, the fate of the pathogen is dependent upon the microbe's unique surface molecules, which either by themselves or by binding to specific host molecules can influence the host cell response during entry. This chapter will discuss first differences between monocytes and macrophages, including their major receptors for phagocytosis. It will then focus on the known receptor-ligand interactions that mediate phagocytosis of M. tuberculosis and other pathogenic mycobacteria. Finally, it will discuss molecules that have been found to regulate the phagocytic pathway of *M. tuberculosis* and the potential consequences of phagocytosis. Emphasis will

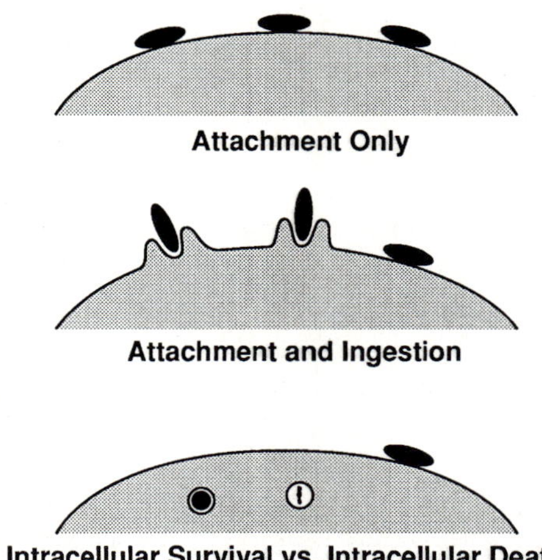

Outcome Dependent Upon:

1. Host Cell Studied
2. Receptor-Ligand Pathway (Single or Multiple, i.e. Cooperativity)
3. Unique Characteristics of Surface Molecules on the Microbial Pathogen

Fig. 1. Factors influencing the outcome of the initial receptor-ligand interaction between the intracellular pathogen and mononuclear phagocytes

be placed on human cells in primary culture, since human beings serve as the natural reservoir for this pathogen.

2 Mononuclear Phagocytes

Mononuclear phagocytes originate from a common precursor cell in the bone marrow but differentiate in tissues into long-lived resident macrophages with a variety of phenotypes that are dictated by local tissue factors. In the absence of an inflammatory stimulus the numbers of resident macrophages greatly exceed the numbers of newly arriving and differentiating monocytes. Thus, predominantly resident macrophages ingest newly arriving mycobacteria as would occur during initial infection of healthy individuals (reviewed in YOUMANS 1979). However, as mycobacteria resist killing and form a persistent focus of infection, the numbers of mononuclear phagocytes greatly increase due to an influx of monocytes from the bloodstream. These monocytes may ingest *M. tuberculosis* shortly after their arrival or may differentiate in unique fashion in the presence of inflammatory mediators from neighboring cells prior to ingesting *M. tuberculosis*. Thus, the receptor-ligand interactions between *M. tuberculosis* and the phagocyte may differ depending on the tissue site, the degree of cell differentiation, and the presence of inflammatory mediators. One early study using guinea pig macrophages illustrates this point (MAXWELL and MARCUS 1968). Adherence of the virulent H37Rv strain and the attenuated H37Ra strain of *M. tuberculosis* to resident macrophages was comparable using peritoneal macrophages but was greater for H37Rv using alveolar macrophages. In the same study, after immunization with *M. bovis* BCG, adherence of bacteria to alveolar macrophages was markedly increased for H37Rv but was reduced for H37Ra when compared to resident macrophages.

During in vitro culture the functional capabilities of freshly isolated human monocytes change. This change is a model for studying mature macrophage function (JOHNSON et al. 1977; NAKAGAWARA et al. 1981) and has been utilized in in vitro studies of phagocytosis of *M. tuberculosis* and other pathogenic mycobacteria. Such monocyte-derived macrophages (MDMS) exhibit altered phagocytic capacity (NEWMAN et al. 1980; WRIGHT and SILVERSTEIN 1982), increased expression of existing surface receptors, and expression of new receptors (SPEERT and SILVERSTEIN 1985; MYONES et al. 1988; HOGG et al. 1986). Human alveolar macrophages are specialized phagocytes and are the first cells encountered by *M. tuberculosis* as it enters the lung. Therefore, their interaction with *M. tuberculosis* has particular pathophysiologic relevance. Several factors unique to alveolar macrophages that may impact on the phagocytosis and killing of mycobacteria include (reviewed in FELS and COHN 1986): an increased basal respiratory rate in the oxygenated tissues of the lung, poor bactericidal activity, altered opsonin requirements for uptake of other bacteria, surface-bound immunoglobulin, and the exposure to surfactant proteins which can enhance receptor-mediated phagocytosis (TENNER et al. 1989).

As one illustration of the importance of studying both monocytes and macrophages in assays of phagocytosis, there is a greater than sixfold increase in phagocytosis of *M. leprae* by MDMS compared to monocytes under conditions in which the bacteria to phagocyte ratio is kept the same (SCHLESINGER and HORWITZ 1991a).

3 Major Surface Receptors for Phagocytosis on Monocytes and Macrophages: Role in the Phagocytosis of Intracellular Pathogens Other Than Mycobacteria

Major mononuclear phagocyte receptors mediating phagocytosis of intracellular pathogens include Fc receptors, complement receptors (CRS: CR1, CR3, and CR4), and the mannose receptor (MR) (Table 1) (HORWITZ 1988). In the absence of specific immune antibody, CR and MR play major roles. CR1 (CD35), a single membrane glycoprotein, binds C3- and C4-coated particles and also regulates complement activation ROSS 1986). CR3 (CD11b/CD18) and CR4 (CD11c/CD18; p150, 95) are heterodimers (α and β chain) and belong to the family of β-2 integrins (leukocyte integrins) (ARNAOUT 1990). In addition to containing binding sites for C3–opsonized particles, CR3 and CR4 are involved in cell-cell and cell-

Table 1. Mononuclear phagocyte receptors mediating phagocytosis[a]

Receptor	*Major ligand*
Fc receptors	Fc domain of IgG
FcRI	
FcRII	
FcRIII	
Complement receptors	
CR1	C3b (also C4B, C3bi)
CR3	C3bi (also LPS, *β*-glucan, fibrinogen, factor X, lipophosphoglycan and gp63 of *Leishmania*, filamentous hemagglutinin of *B. pertussis*, ICAM-1
CR4	C3bi (also lipophosphoglycan)
Lectin-like receptors	
Mannosyl fucosyl	Mannose, fucose, GlcNAc
β-Glucan	Glucan
Advanced glycosylation end product (AGE)	Glucose-modified protein
Miscellaneous	
Fibronectin receptor	Fibronectin
LFA-1	ICAM-1
Vitronectin receptor	Vitronectin
CD14	LPS; LPS-binding protein

[a]For references, see text.

surface adhesion (KISHIMOTO et al. 1989). CR3 and CR4 are also expressed on natural killer cells and certain cytotoxic T cells (KISHIMOTO et al. 1989; LANIER et al. 1985), cells that play a role in the cellular immune response to mycobacterial antigens (HANCOCK et al. 1989; OTA et al. 1990). CR3 and CR4 contain several extracellular binding domains (PETTY and TODD 1993). The third member of the β-2 integrins, LFA-1 (CD11a/CD18), shares an identical β chain with CR3 and CR4 but has an α chain that is less homologous (KISHIMOTO et al. 1989; LARSEN et al. 1989) and does not bind C3. It has been found to mediate uptake of the intracellular pathogen, *H. capsulatum* (BULLOCK and WRIGHT 1987).

Phagocytosis of a diverse group of intracellular pathogens is mediated by CR, a family of receptors that has been postulated to play an important role in the early survival of intracellular parasites because ligation of these receptors by C3-coated particles does not uniformly stimulate the release of oxygen metabolites or inflammatory mediators (WILSON et al. 1980; WRIGHT and SILVERSTEIN 1983; ADEREM et al. 1985). In particular, CR3 is important in mediating uptake of *Legionella pneumophila* (PAYNE and HORWITZ 1987), *Histoplasma capsulatum* (BULLOCK and WRIGHT 1987; NEWMAN et al. 1990), *Leishmania sp.* (BLACKWELL et al. 1985; WILSON and PEARSON 1988; MOSSER and EDELSON 1985; WOZENCRAFT et al. 1986; MOSSER et al. 1992; ROBLEDO et al. 1994), and *Listeria monocytogenes* (DREVETS and CAMPBELL 1991). CR1 mediates uptake of *L. pneumophila* (PAYNE and HORWITZ 1987) and *L. major* (DA SILVA et al. 1989).

The expression of CR changes during monocyte maturation. Compared to monocytes, CR1 expression increases 1.5-fold, CR3 expression increases five fold, and CR4 expression increases eight-fold on MDMS (MYONES et al. 1988). CR4 is expressed minimally on monocytes and is expressed abundantly on alveolar macrophages (MYONES et al. 1988; HOGG et al. 1986; SCHWARTING et al. 1985).

The expression of the MR is also increased during monocyte maturation. It is expressed on MDMS and tissue macrophages, including alveolar macrophages, but not monocytes SPEERT and SILVERSTEIN 1985; SHEPHERD et al. 1982). It binds glycoconjugates containing mannose and fucose units with highest avidity but also binds *N*-acetylglucosamine (GlcNAc) (reviewed in Stahl 1992). It is a C-type lectin since its binding function is calcium-dependent. The extracellular portion of the receptor contains a linear array of eight carbohydrate recognition domains (TAYLOR et al. 1990). Binding to the MR is highly pH-dependent, with optimal binding at pH 7 and poor binding at pH 5. This characteristic is postulated to allow dissociation of ligands within the acidic phagosome. MRs on human macrophages have an internal pool and undergo continual recycling to the cell surface. The MR has been found to be an important receptor for mediating uptake of *Leishmania sps.* (BLACKWELL et al. 1985; WILSON and PEARSON 1986, 1988).

Human mononuclear phagocytes express three classes of Fc receptors (reviewed in UNKELESS 1989). Monocytes express FcRI (CD64) and FcRII (CD32). Macrophages express a third Fc receptor (CD16). Other phagocyte receptors felt to play a role in phagocytosis of carbohydrate-coated particles include the β-glucan receptor (CZOP and KAY 1991) and the receptor for advanced glycosylation end products (VLASSARA et al. 1981). The latter receptor has been found to mediate uptake of *Leishmania sps.* (MOSSER et al. 1987). Still, other phagocyte receptors have been found to promote adhesion of microbes and microbial products. These include the fibronectin receptor, vitronectin receptor and CD14 (WRIGHT 1992). Whether these receptors directly mediate ingestion of particles is less clear.

4 The Mechanism of Entry of *Mycobacterium tuberculosis* into Phagocytes

Our electron microscopy studies have shown that entry of *M. tuberculosis* into human mononuclear phagocytes resembles conventional receptor-mediated phagocytosis in which phagocyte pseudopods move circumferentially around the bacterium and fuse at their distal tip, leaving the bacterium in a membrane-bound vacuole or phagosome (SCHLESINGER et al. 1990). Evidence for a *M. tuberculosis* invasion factor is limited to a recent report in which a cloned fragment of *M. tuberculosis* DNA in *E. coli* conferred enhanced invasiveness of this bacterium into mammalian cells in vitro (ARRUDA et al. 1993). We have observed that the levels of phagocytosis in fresh serum of live and heat-killed virulent Erdman *M. tuberculosis* by human MDMS are equivalent (unpublished studies). Thus, viability of the bacterium does not appear to be a requirement for the phagocytic process itself, although it plays an important role in subsequent events (ARMSTRONG and HART 1971).

5 Monocyte and Macrophage Receptors That Mediate Phagocytosis of *Mycobacterium tuberculosis*

5.1 Complement Receptors

Few early studies focused on the phagocytosis of *M. tuberculosis* by non-human mononuclear phagocytes (MAXWELL and MARCUS 1968). More recently, studies, including those from our laboratory, have demonstrated that CRS play a major role in the phagocytosis of *M. tuberculosis* by human monocytes and macrophages (SCHLESINGER et al. 1990; SCHLESINGER 1993; HIRSCH et al. 1994).

Our studies have determined that CR1 and CR3 mediate phagocytosis of the virulent Erdman strain of *M. tuberculosis* by human monocytes (SCHLESINGER et al. 1990). In these studies a combination of two monoclonal antibodies against distinct epitopes on the α chain of CR3, but not subtype control monoclonal antibodies, inhibited phagocytosis of *M. tuberculosis* to 20% of the control value. Monoclonal antibodies against the C3bi-binding epitope of CR3 as well as those against other ligand-binding epitopes inhibited phagocytosis. Thus, although steric hindrance may explain this observation, it is possible that more than one epitope is important in binding *M. tuberculosis*. In contrast, we determined that Fc receptors (in the presence of non-immune serum) and the β-glucan receptor play little or no role in mediating phagocytosis of *M. tuberculosis* by these cells.

Also in these studies, we determined that phagocytosis of *M. tuberculosis* was maximally enhanced in 2.5% fresh autologous non-immune serum (Schlesinger et al. 1990). This result indicated that serum components serve as opsonins to enhance phagocytosis of the bacterium. In subsequent experiments, we identified complement protein C3 in serum as an important bacterium-bound opsonin. To summarize these experiments, we determined that the serum factor was heat-labile, preopsonization of bacteria with fresh serum significantly enhanced phagocytosis, phagocytosis was reduced in C3-depleted serum, fab fragments of anti-C3 IgG markedly inhibited the phagocytosis of preopsonized bacteria and finally, C3 protein was fixed to the surface of serum-opsonized bacteria.

In our studies on phagocytosis of *M. tuberculosis* by 5 day old MDMS, a direct comparison among the virulent strains Erdman and H37Rv and the attenuated strain H37Ra was made (Schlesinger 1993). Adherence of the three strains to MDMS was enhanced greater than three fold in the presence of 2.5% fresh non-immune serum. Heat-labile components of serum were critical in this enhancement. Adherence of the three strains to MDMS was comparable both in terms of the mean number of bacteria per cell and the percentage of macrophages containing bacteria. Electron microscopy studies demonstrated comparable ingestion of all three strains and the bacteria were found within phagosomes. These studies demonstrated that phagocytosis of the H37Ra strain by human macrophages is comparable to that of the virulent strains and, thus, reduced phagocytosis is not a likely explanation for the reduced virulence of the H37Ra strain. These studies also demonstrated that *M. tuberculosis* is ingested by macrophages both in the presence and absence of serum, although the level of ingestion in the presence of serum is enhanced. In our model, there is a stepwise increase in the number of ingested bacteria both in the presence and absence of serum as the bacteria to MDM ratio is increased.

In our studies blockade of macrophage CR1, CR3 and CR4 by a combination of monoclonal antibodies against these receptors inhibited phagocytosis of all three strains to nearly 20% of the control value in the presence of serum, demonstrating a major role for CR in mediating phagocytosis of *M. tuberculosis* by macrophages (Schlesinger 1993). Monoclonal antibodies against CR also significantly inhibited the low level of phagocytosis of bacteria in the absence of serum. This result has raised two possibilities. The first is that there is a direct interaction between *M. tuberculosis* surface components and CR and the second is that functional C3 protein secreted by the macrophage during the assay binds to the bacterium and serves as the ligand for CR. That C3 secreted by the macrophage can function in this manner has been shown with other particles, including microbes (WOZENCRAFT et al. 1986; EZEKOWITZ et al. 1983). If the first possibility was true, then the binding sites for C3-opsonized bacteria and unopsonized bacteria on the extracellular domains of CR may differ.

Studies with human alveolar macrophages have also demonstrated an important role for CR in the phagocytosis of *M. tuberculosis*. In studies from our laboratory using the virulent Erdman strain (SCHLESINGER 1992), there was a significant reduction in phagocytosis in the presence of heat-inactivated serum or in the absence of serum compared to the fresh serum condition, although the level of inhibition (40%–50%) was not as striking as that observed with MDMS. The combination of monoclonal antibodies against CR1, CR3 and CR4 significantly inhibited bacterial adherence to human alveolar macrophages by approximately 50% (Fig. 2), a value comparable to the reduction seen with heat-inactivated serum but less than that observed with MDMS.

Two recently published studies support the importance of CR in the phagocytosis of *M. tuberculosis* by mononuclear phagocytes. In the first study

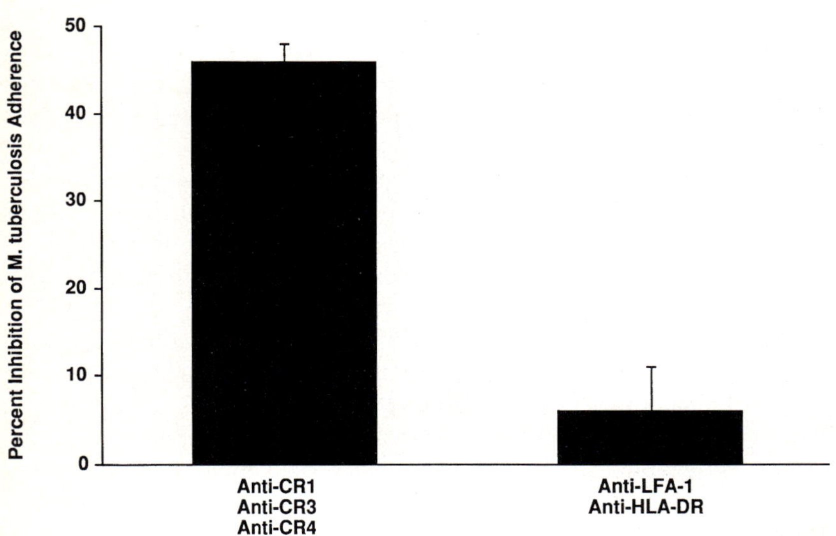

Fig. 2. Monoclonal antibodies against complement receptor 1 (CR1), CR3, and CR4 inhibit Erdman *M. tuberculosis* adherence to human alveolar macrophages in the presence of fresh serum. Freshly isolated human alveolar macrophages in monolayer culture were incubated with Erdman *M. tuberculosis* (1.25×10^6) in medium containing 2.5% fresh autologous non-immune serum in the presence of the Monoclonal antibodies (mAbs) indicated (final concentration 20 μg/ml each mAb) or in the absence of mAb (control). The mean number of bacteria per macrophage was determined by fluorescence microscopy. Data are the mean (± SD) of triplicate coverslips from one representative experiment. Percent inhibition on the vertical axis is relative to control. Anti-LFA-1 and Anti-HLA-DR mAbs served as subtype control mAbs

(HIRSCH et al. 1994), phagocytosis of the H37Ra strain of *M. tuberculosis* by human alveolar macrophages was greater than that by monocytes for a given multiplicity of infection. Phagocytosis was enhanced in the presence of fresh exogenous serum and CR1 and CR3 were found to be major receptors for phagocytosis by monocytes. In contrast, monoclonal antibodies against CR4 markedly decreased the phagocytosis by alveolar macrophages, indicating that CR4 is important for phagocytosis of the H37Ra strain by alveolar macrophages.

The other recent study examined the binding of *M. tuberculosis* to murine macrophages in a serum-free system (STOKES et al. 1993). Binding of *M. tuberculosis* was highly dependent upon the phenotype of the macrophage and was mediated by CR3. The binding site for *M. tuberculosis* on CR3 was distinct from the C3bi-binding site based on monoclonal antibody blocking experiments. Elicited and activated macrophages bound mycobacteria poorly despite expressing CR. Thus, in this model, binding of *M. tuberculosis* did not

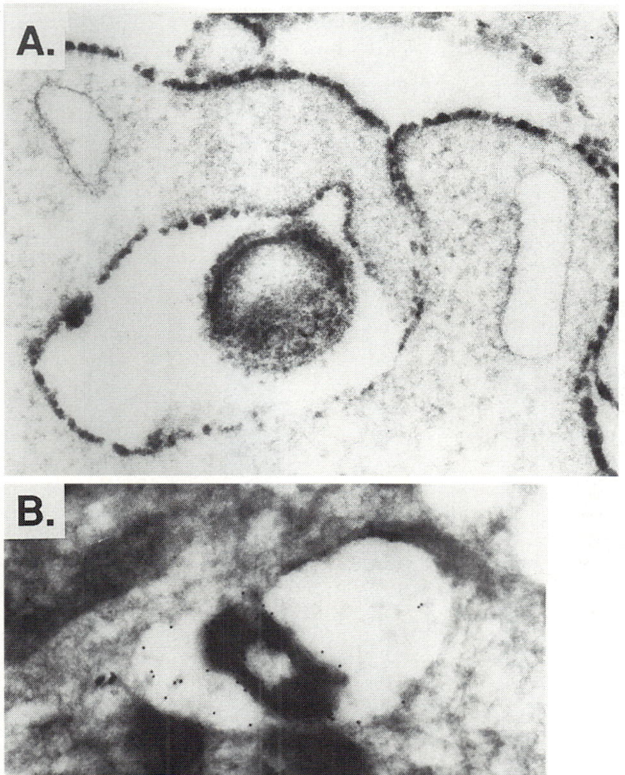

Fig. 3. Preembedding immunoperoxidase staining (**A** × 80 000) and cryosection immunogold staining (**B** × 76 000) of complement receptor (CR3) during phagocytosis of Erdman *M. tuberculosis* by *monocyte-derived macrophages* (MDMS), MDMS were fixed 6 min after incubation with bacteria. Complement receptor (CR3) is seen along the cell surface and inner layer of the developing phagosome (**A**) and within the newly formed phagosome (**B**)

correlate with expression of CR3, a finding the authors hypothesized was related to the functional state of the binding site on CR3.

In immunoelectron microscopy studies performed in our laboratory to localize CR3 during the phagocytosis of *M. tuberculosis* by human macrophages, CR3 was found primarily on the surface of MDMS and, as the bacteria are phagocytized, CR3 was seen on the inner leaflet of the developing phagosome (Fig. 3). In these studies, there was significant accumulation of CR3 visualized in some phagosomes containing *M. tuberculosis*.

5.2 Mannose Receptors

In contrast to CR, we have recently determined that the macrophage MR plays an important role in mediating adherence of the virulent strains Erdman and H37Rv but not the attenuated strain H37Ra (SCHLESINGER 1993). In these studies, preincubation of macrophages with soluble mannan or mannose BSA, multivalent ligands for the MR and therefore competitive inhibitors of this receptor, significantly inhibited adherence of only the virulent strains (Fig. 4). Selective down-modulation of the macrophage MR on a substrate of mannan and pre-

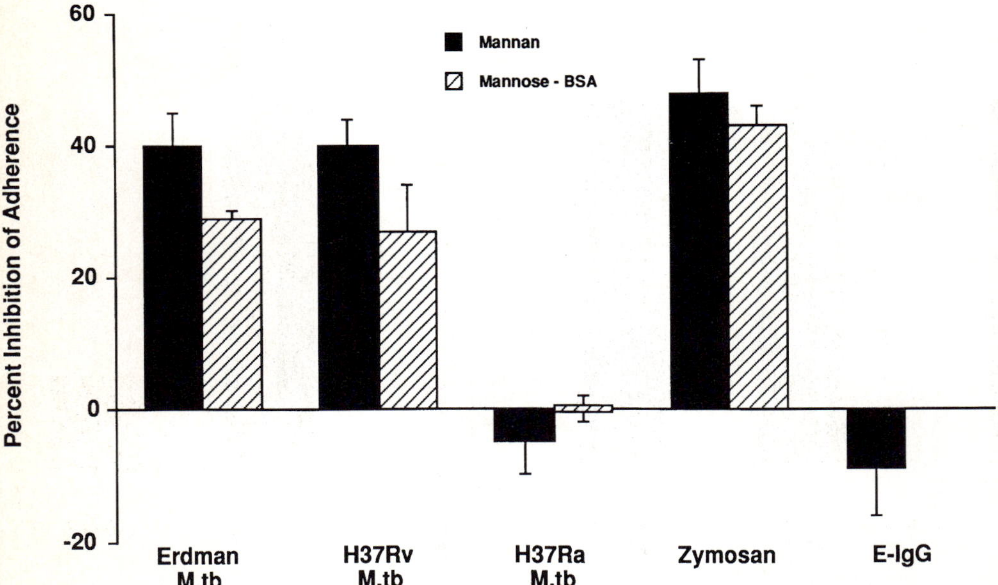

Fig. 4. Mannan and mannose-bovine serum albumin (BSA) in solution inhibit adherence of Erdman and H37Rv *M.tuberculosis* but not H37Ra *M. tuberculosis* to monocyte-derived macrophages (MDMS) in the absence of serum. MDMS in monolayer culture were incubated under serum-free conditions with *M. tuberculosis* (1.25×10^6), zymosan (3.0×10^6), or E-IgG (5.0×10^6, negative control) in the presence of mannan (1-4 mg/ml), mannose-BSA (0.02-2 mg/ml), or medium alone. Data are the mean ± SE from up to ten independent experiments. Percent inhibition in mean number of test particles per MDM from level of particle adherence in the presence of medium alone is shown on the vertical axis

incubation of macrophages with a polyclonal antibody to MR also inhibited adherence of the virulent strains selectively (Figs. 5, 6). These studies were initially performed in the absence of serum. In subsequent experiments, we determined that selective down-modulation of the macrophage MR in the presence of fresh serum also inhibited adherence of the Erdman and H37Rv strains. Simultaneous blockade of the MR and CR did not further inhibit the adherence of the virulent *M. tuberculosis* strains over that seen with blocking CR alone (80% inhibition). Thus, CR and MR may not be completely independent in their ability to mediate phagocytosis.

6 Host-Derived and Microbial Ligands in Phagocytosis of *Mycobacterium tuberculosis*

Fragments of C3 (Payne and Horwitz 1987; Mosser and Edelson 1985), fibronectin (Wyler et al. 1985; Wirth and Kierszenbaum 1984) and microbial surface molecules (Russell and Talamas-Rohana 1989; Van Strijp et al. 1993) are

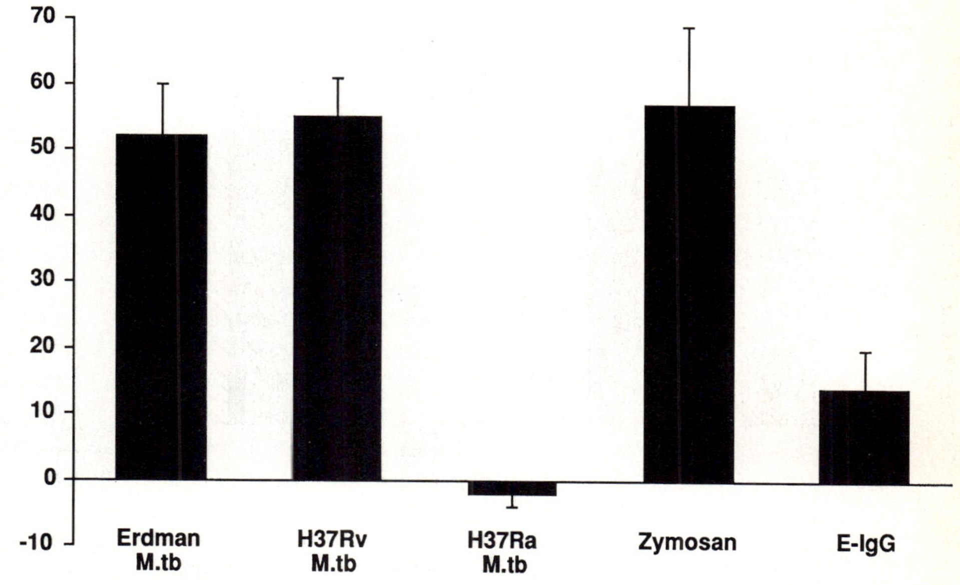

Fig. 5. Modulation of mannose receptors on a mannan substrate inhibits adherence of Erdman and H37Rv *M. tuberculosis* but not H37Ra *M. tuberculosis* to monocyte-derived macrophages (MDMS). MDMS were plated on glass coverslips coated with mannan or human serum albumin (HSA, control) for 2 h. MDMS in monolayer culture were then washed, incubated with *M. tuberculosis* (1.25×10^6), zymosan (3.0×10^6), or E-IgG (5.0×10^6, negative control) in the absence of serum, and adherence was measured by fluorescence microscopy. Data are the mean ± SE from up to four independent experiments. Percent inhibition in mean number of test particles per MDM from level of particle adherence with MDM plated on a substrate of HSA is shown on the vertical axis

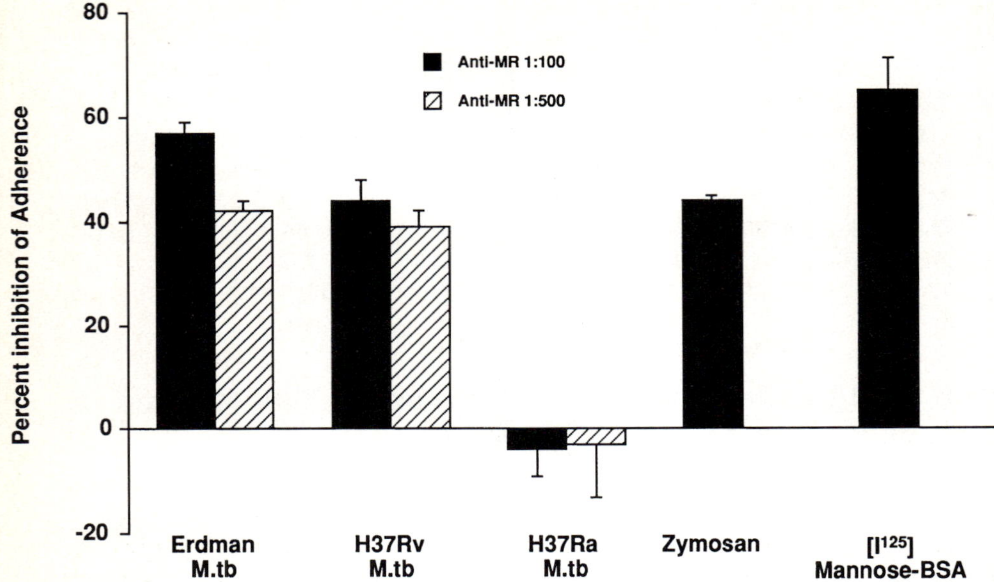

Fig. 6. Anti-mannose receptor (MR) inhibits adherence of Erdman and H37Rv *M. tuberculosis* but not H37Ra *M. tuberculosis* to monocyte-derived macrophages (MDMS) in the absence of serum. MDMS in monolayer culture were incubated with *M. tuberculosis* (1.25×10^6) under serum-free conditions in the presence or absence (control) of the indicated dilutions of rabbit anti-MR. Data are the mean (± SEM) of two independent experiments and represent the percent inhibition in the mean number of bacteria per MDM compared to controls. Inhibition of adherence of zymosan and [I^{125}]-labeled mannose-BSA (bovine serum albumin) in the presence of anti-MR served as positive controls

examples of molecules that serve to mediate the binding of intracellular pathogens to mononuclear phagocytes. The relative contribution of host-derived and microbial ligands and the phagocyte receptors they recognize is dependent upon whether serum is present during the assay. The importance of complement component C3 as an opsonin for *M. tuberculosis* is described above. Of particular relevance to *M. tuberculosis*, functional complement proteins are present in the alveolar fluid of lungs and are secreted by alveolar macrophages and thus may serve as bacterium-bound ligands in this milieu (FELS and COHN 1986). Although C3-acceptor molecules on the surface of other intracellular pathogens have been described (HORWITZ 1988), the major C3-acceptor molecule for *M. tuberculosis* is currently not known. One potential acceptor molecule for C3 on *M. tuberculosis* is the surface glycolipid trehalose dimycolate (cord factor) since this molecule has been shown to be a potent activator of the alternative complement pathway (RAMANATHAN et al. 1980). A recent study indicates that heat-stable opsonins other than antibody can mediate binding of mycobacteria to host cells and influence the host cell response (BLAIR et al. 1993).

The outer surface of *M. tuberculosis* contains large quantities of complex acylated carbohydrates and glycolipids. Because they are expressed on the bacterial surface, they are in a position to mediate initial interactions between

M. tuberculosis and phagocytes. Two major surface lipoglycans that have been studied in detail are lipoarabinomannan (LAM) and its smaller arabinose-free counterpart lipomannan (LM) (CHATTERJEE et al. 1992a). LAM regulates several macrophage effector functions (MORENO et al.1989; CHATTERJEE et al. 1992c; ROACH et al. 1993, 1994; BARNES et al. 1992; CHAN et al. 1989, 1991; SIBLEY et al. 1988) and thus has been postulated to be an important mycobacterial virulence factor. Immunoelectron microscopy studies demonstrate that LAM resembles a capsule surrounding individual bacteria (HUNTER and BRENNAN 1990). LAM from the Erdman strain (ManLAM) and LAM from a rapidly growing mycobacterial strain (AraLAM) contain core mannan units linked to long arabinose branches (CHATTERJEE et al. 1992b). Recently, CHATTERJEE et al. (1992b, 1993) have demonstrated that ManLAM differs from AraLAM in that it contains linear mannose oligosaccharide "caps" on the arabinose saccharides at the terminal portions of the molecule. These terminal mannosyl units are ideally situated to interact with the MR on macrophages.

In studies in our laboratory to examine the ability of LAM to serve as an adherence ligand for macrophages (SCHLESINGER et al. 1994), polystyrene microspheres were coated with known amounts of ManLAM, AraLAM, LM or buffer and incubated with macrophages in the absence of serum. Microspheres coated with ManLAM demonstrated a more than threefold increase in adherence to macrophages compared to the buffer microspheres. Microspheres coated with AraLAM or LM did not demonstrate enhanced adherence compared to controls. Selective down-modulation of the macrophage MR on a mannan substrate abrogated the enhanced adherence of microspheres mediated by ManLAM. Selective removal of the terminal mannosyl units of ManLAM by exomannosidase treatment abolished the enhanced adherence of ManLAM-coated microspheres to macrophages. Preincubation of Erdman *M. tuberculosis* with CS40, a monoclonal antibody directed against LAM, significantly inhibited adherence of bacteria to macrophages (up to 49% inhibition) compared to controls, confirming a role for ManLAM in adherence of intact bacteria to macrophages. Thus, the binding of the MR on macrophages to the linear mannosyl units at the terminal end of ManLAM on the bacterium is one receptor-ligand interaction in the phagocytosis of *M. tuberculosis*. These studies have led us to speculate that the linear array of the terminal mannosyl units of ManLAM may be important in enhancing the binding affinity of this ligand for the MR. Since mycobacteria that vary in their virulence properties have been found to contain LAM species with mannosyl caps (PRINZIS et al. 1993), it appears that the precise array, length and carbohydrate composition of the caps as well as the presence of specific exposed epitopes on the intact molecule may be important in determining the receptor to which LAM binds. A recent study indicates that AraLAM may interact with CD14 (ZHANG et al. 1993).

Terminal mannosylation of major microbial surface molecules may be an important common determinant of the phagocytic process leading to early survival and growth of intracellular pathogens in macrophages. *Leishmania sps.*, another group of intracellular pathogens, possess lipophosphoglycan and the glycoprotein gp63, two major surface molecules that play important roles in the interaction with macrophages, including phagocytosis (HALL and JOINER 1991). These molecules contain terminal α-linked mannosyl units, effectively coating these parasites with mannose (THOMAS et al. 1992; MCCONVILLE et al. 1990; OLAFSON et al. 1990). Because a reduction in gp63 may correlate with decreased virulence of *Leishmania* (WILSON and HARDIN 1988), we speculate that differences in the nature and amount of LAM or its terminal mannosyl units may correlate with the relative virulence of *M. tuberculosis* strains.

7 Receptors and Ligands Mediating Phagocytosis of Other Mycobacteria

Studies of several other pathogenic mycobacteria have demonstrated a major role for CR and the MR in phagocytosis by monocytes and macrophages. We and others have found that CR1 and CR3 are important receptors in mediating phagocytosis of *M. leprae* by monocytes (SCHLESINGER and HORWITZ 1990; LAUNOIS et al. 1992), and CR4, along with CR1 and CR3, mediates phagocytosis of *M. leprae* by MDMS (SCHLESINGER and HORWITZ 1991a). Although a role for CR in the presence and absence of serum was found for *M. leprae* on monocytes, monoclonal antibodies against CR did not further reduce the low level of adherence of *M. leprae* to MDMS in the absence of serum (SCHLESINGER and HORWITZ 1991a). C3 bound to phenolic glycolipid-1, a major glycolipid on the surface of *M. leprae*, is an important host-derived ligand for phagocytosis of this bacterium (SCHLESINGER and HORWITZ 1991b). A role for the MR in phagocytosis of *M. leprae* has not been found. Phagocytosis of *M. avium* by monocytes and macrophages (including alveolar macrophages) is mediated by CR3 and MR in the presence and absence of serum (BERMUDEZ et al. 1991; ROECKLEIN et al. 1992). *M. avium* adherence to monocytes is enhanced in fresh serum (SWARTZ et al. 1988) and preincubating serum with antibodies to C3 reduces *M. avium* adherence to macrophages (BERMUDEZ et al. 1991). These studies indicate that C3 is an important bacterium-bound opsonin for *M. avium*. Adherence of *M. bovis* BCG to human monocytes is mediated by CR1 and CR3 (LAUNOIS et al. 1992).

Several other receptors and ligands mediate uptake of *M. avium* by mononuclear phagocytes. A role for fibronectin has been implicated because preincubation of serum with anti-fibronectin antibody reduces *M. avium* binding (BERMUDEZ et al. 1991). The binding molecule for fibronectin on the surface of *M. avium* has been reported to be a β1 integrin (RAO et al. 1992). Other mycobacteria, including *M. tuberculosis*, bind fibronectin (ASLANZADEH et al. 1989; RAO et al. 1992; ESPITIA et al. 1992; RATLIFF et al. 1988; ABOU-ZEID et al. 1988). Whether bacterium-bound fibronectin serves as an opsonin for phagocyte fibronectin receptors in the phagocytosis of *M. tuberculosis* is unknown. Fibronectin has been found to enhance the binding to phagocytes of other intracellular pathogens (WYLER et al. 1985; WIRTH and KIERSZENBAUM 1984).

In addition to CR and MR, adherence of *M. avium* involves the phagocyte vitronectin receptor (CATANZARO and WRIGHT 1990; RAO et al. 1993) and a 68 Da microbial ligand (RAO et al. 1994). The involvement of other integrins in the binding of *M. avium* to host cells may be important in determining the cellular tropism of this pathogen and other mycobacterial pathogens (BYRD et al. 1993). In one study, uptake of *M. avium* by monocytes and human alveolar macrophages was found to involve Fc receptors, the transferring receptor and the β-glucan receptor (ROECKLEIN et al. 1992).

8 Regulation of Phagocytosis of *Mycobacterium tuberculosis* and Other Mycobacteria

Several host molecules regulate receptor-ligand interactions involving CR and the MR and thus may influence the phagocytosis of *M. tuberculosis* and other mycobacteria. These include pulmonary surfactant proteins, fibronectin, interferon-γ (IFNγ) and antibody (TENNER et al. 1989; WRIGHT et al. 1986; FIRESTEIN and ZVAIFLER 1987; ESPARZA et al. 1986; BEVILACQUE et al. 1981; BOHNSACK et al. 1985; POMMIER et al. 1983) (Table 2).

Human pulmonary surfactant protein-A (SP-A), a major surfactant-associated protein, belongs to a small group of proteins called collectins that contain homologous collagen-like sequence regions (HAAGSMAN 1994). The other members of this group include surfactant protein D, the complement protein Clq, mannose-binding protein, core-specific lectin, acetylcholinesterase, and conglutinin. In addition to regulating the level of lung surfactant, SP-A has been found to enhance Fc receptor and CR function (TENNER et al. 1989). In studies in progress, we have determined that human and rat SP-A proteins, either in so-

Table 2. Molecules that regulate phagocytosis of pathogenic mycobacteria[a]

Host molecule	Mechanism
Up-regulation	
Complement protein C3	Opsonin
Natural antibody	Enhances C3 binding to bacteria
Fibronectin	Up-regulates CR pathway, ? ligand
Pulmonary surfactant Protein-A	?
Down-regulation	
Interferon-γ	Decreases expression and/or function of CR and the MR
HIV envelope glycoprotein gp120	Effects mediated through binding of gp120 to CD4

CR, complement receptor; MR, mannose receptor.
[a]For references and further explanation, see text.

lution or on a matrix, enhance the phagocytosis of *M. tuberculosis* by MDMS and alveolar macrophages by a direct interaction between SP-A and the macrophage (GAYNOR and SCHLESINGER 1994). These studies indicate that SP-A can regulate *M. tuberculosis* phagocytosis in the alveoli of the lung. This action of SP-A may be particularly important in influencing *M. tuberculosis* phagocytosis in disease states such as alveolar proteinosis of HIV infection in which SP-A levels are increased (REYES and PUTONG 1980; PHELPS and ROSE 1991). Clq can also enhance phagocytic function (BOBAK et al. 1987). This protein has been found on the surface of *M. leprae* and its major surface glycolipid (SCHLESINGER and HORWITZ 1994).

In addition to serving as an opsonin, fibronectin up-regulates the activity of Fc receptors and CR on monocytes and macrophages (BOHNSACK et al. 1985; POMMIER et al. 1983). In the presence of fibronectin, such phagocytes exhibit enhanced adherence of E-IgG and complement-coated erythrocytes. We have found that *M. leprae* phagocytosis is enhanced by fibronectin in this fashion (SCHLESINGER and HORWITZ 1991a).

The lymphokine interferon-γ (IFNγ) plays a key role in host defense against a variety of intracellular pathogens in part by enhancing the antimicrobial capacity of mononuclear phagocytes. Among its effects, IFNγ down-regulates the expression and/or function of phagocyte CR and the MR (WRIGHT et al. 1986; FIRESTEIN and ZVAIFLER 1987; ESPARZA et al. 1986; MOKOENA and GORDON 1987; SCHREIBER et al. 1993). Macrophages activated with IFNγ exhibit decreased adherence of *M. leprae* (SCHLESINGER and HORWITZ 1991a), *M. avium* (TOBA et al. 1989) and *Leishmania sps.* (MOSSER and HANDMAN 1992). In one study, IFNγ-activation of macrophages led to a small decrease in adherence but enhanced intracellular multiplication of *M. tuberculosis* (DOUVAS et al. 1985). Although IFNγ impairs CR function, it does not lead to decreased surface expression of CR3 and CR4 (WRIGHT et al. 1986; ESPARZA et al. 1986; SCHLESINGER and HORWITZ 1991a). Inasmuch as several cytokines and inflammatory mediators, such as prostaglandins, influence surface expression of phagocyte receptors and receptor-mediated phagocytosis, it is likely that these mediators will be important in regulating the phagocytosis of *M. tuberculosis* as well as the early host cell response to phagocytosis. In this regard, interleukin(IL)-4 and prostaglandin (PG)E_2, two mediators that modulate the host response to mycobacterial infection, markedly up-regulate CR and/or MR expression and function and down-regulate phagocyte effector functions (SCHREIBER et al. 1993; STEIN et al. 1992; ABRAMSON and GALLIN 1990; SAMPSON et al. 1991; SCALES et al. 1989). In the study described earlier, in which CR4 was found to play a major role in phagocytosis of H37Ra *M. tuberculosis* by alveolar macrophages, the alveolar macrophages were found to limit the intracellular growth of *M. tuberculosis* more effectively than monocytes, in part through increased secretion of tumor necrosis factor (TNF)-α (HIRSCH et al. 1994). It is not known whether ligation of CR4 influences (TNF)-α secretion directly.

Natural antibody to several species of mycobacteria has been found in healthy PPD-negative persons (BARDANA et al. 1973). The roles of natural antibody and complement activation pathways in C3 fixation to *M. leprae* have been explored in our laboratory (SCHLESINGER and HORWITZ 1994). In 50% non-immune serum, the alternative complement pathway is the dominant pathway leading to C3 deposition on *M. leprae*. In contrast, both the alternative and classical complement pathways play important roles in mediating C3 deposition on the bacteria in 2.5% serum. Regardless of the pathway of complement activation, natural antibody in serum (very low levels of IgG or IgM) is required for C3 deposition to the *M. leprae* surface since C3 binding is markedly reduced in agammaglobulinemic serum. These studies indicate that natural antibody plays a role in disease pathogenesis for this intracellular pathogen by enhancing C3 deposition onto the bacillus and hence, phagocytosis by mononuclear phagocytes.

Later in infection with *M. tuberculosis* or during active disease, lysis of bacterial-laden macrophages leads to the phagocytosis of *M. tuberculosis* by neighboring macrophages in the presence of high titer immune antibody. In one previous study, high titer rabbit anti-mycobacterial immunoglobulin facilitated the multiplication of BCG in the spleens of mice (FORGET et al. 1976). In an in vitro study, rabbit anti–mycobacterial immunoglobulin had no effect on phagocytosis of H37Rv by mouse peritoneal macrophages (ARMSTRONG and HART 1975). The role of human immune antibody in influencing receptor-mediated phagocytosis and intracellular survival of *M. tuberculosis* by human cells has not been defined. In the murine model, specific antibody was found to enhance phagosome-lysosome fusion (ARMSTRONG and HART 1975).

Resistance to mycobacterial infection in the murine model is strongly dependent upon expression of the Bcg gene (DENIS et al. 1986). Although this gene appears in part to regulate phagosome-lysosome fusion, the levels of phagocytosis of *M. avium* by Bcg^s and Bcg^r macrophages are the same (DE CHASTELLIER et al. 1993).

Phagocytosis of *M. tuberculosis* and other mycobacteria by cells of the monocyte lineage can influence and be influenced by other microbes. In coinfection models, phagocytosis of *M. tuberculosis* can enhance directly HIV transcription in monocytic cells (SHATTOCK et al. 1994), a phenomenon that may not be true for more differentiated cells (NICOLACAKIS et al. 1994). Preincubation of human monocytes with the HIV envelope glycoprotein gp120 inhibits phagocytosis but enhances intracellular growth of *M. avium* (SHIRATSUCHI et al. 1994). It is of interest that there is enhanced expression of CR on monocytes from patients with HIV (PALMER and HAMBLIN 1993) and other disease states (SHAKOOR and HAMBLIN 1992), but the precise relationship between this enhanced expression of CR and the development of mycobacterial disease is unknown.

9 Consequences of Phagocytosis of *Mycobacterium tuberculosis*

Entry of *M. tuberculosis* via the CR pathway may provide the bacterium safe passage into mononuclear phagocytes by allowing it to avoid the toxic consequences of the oxidative burst (WRIGHT and SILVERSTEIN 1983; WILSON et al. 1980; YAMAMOTO and JOHNSTON 1984). In this regard, pathogenic mycobacteria as well as some other intracellular pathogens have been found to elicit little or no detectable oxidative burst during phagocytosis (HOLZER et al. 1986; WILSON et al. 1980; LAUNOIS et al. 1992; GANGADHARAM and EDWARDS 1984; DOUVAS et al. 1986; MCCABE and MULLINS 1990; JACOBS et al. 1984; EISSENBERG and GOLDMAN 1987). Mycobacteria, including *M. tuberculosis*, also have surface glycolipids that scavenge toxic oxygen radicals (LAUNOIS et al. 1989; NEILL and KLEBANOFF 1988; CHAN et al. 1989, 1991; BROZNA et al. 1991; PABST et al. 1988; VACHULA et al. 1989), providing another mechanism for enhancing the survival of these bacteria during phagocytosis.

It is perhaps not surprising that a diverse group of host-adapted intracellular pathogens has evolved to utilize two major classes of phagocyte receptors, CR and the MR, in host cell entry. These receptors are expressed abundantly on professional phagocytes and are competent in mediating phagocytosis. Differences in the host cell response to these microbes may relate rather to more subtle differences in the receptor binding epitopes utilized, the unique microbial ligands for these receptors or the specific host cell studied. The following three observations support these hypotheses. First, multiple extracellular binding domains on CR3 have been described (WRIGHT 1992). Ligation of the β-glucan binding domain on this receptor triggers a strong oxidative burst (ROSS et al. 1987; CAIN et al. 1987), whereas ligation of the C3bi binding domain does not. Recent studies using particles that bind to mononuclear phagocytes via β-glucan moieties in concert with other receptors have demonstrated that phagocyte secretory responses relate to the receptor sites that bind β-glucan (ELSTAD et al. 1994; HOFFMAN et al. 1993). Second, despite the use of CR1 and CR3 in the phagocytosis of *M. bovis* BCG and *M. leprae* by monocytes, only binding of the low virulent *M. bovis* BCG was found to generate a chemiluminescence response in the presence of fresh serum (LAUNOIS et al. 1992). Third, phagocytosis of live or heat-killed H37Rv *M. tuberculosis* by human neutrophils, cells that also express CR, is accompanied by detectable release of superoxide (MAY and SPAGNUOLO 1987), whereas phagocytosis of Erdman *M. tuberculosis* by human monocytes and macrophages is not (DOUVAS et al. 1986). Although strain differences may account for this last set of observations, subtle differences in binding to CR epitopes, the functional state of CR or the use of the MR on macrophages may also be a factor.

Tissue macrophages, including those containing intracellular mycobacteria, are long-lived cells that are in constant communication with extracellular matrix proteins and other cell types involved in the host cellular immune response.

Therefore, changes in the expression and/or function of macrophage receptors as a result of phagocytized *M. tuberculosis* could potentially serve to regulate the cellular immune response. Along these lines, *M. tuberculosis* and *M. leprae*-infected mononuclear phagocytes demonstrate altered Fc receptor function (BIRDI et al. 1983; MARIANO et al. 1977). In a recent study, phagocytosis of *M. tuberculosis* by a myeloid cell line was shown to increase expression of the adhesion molecule ICAM-1 on these host cells (LÓPEZ RAMIREZ et al. 1994).

10 Summary

Our working model for phagocytosis of *M. tuberculosis* includes several receptor-ligand interactions (Fig. 7). The bacteria bind C3 fragments that are produced by activation of the complement protein cascade. These fixed C3 fragments are, in turn, bound by CR on the surface of the phagocyte. Alter-

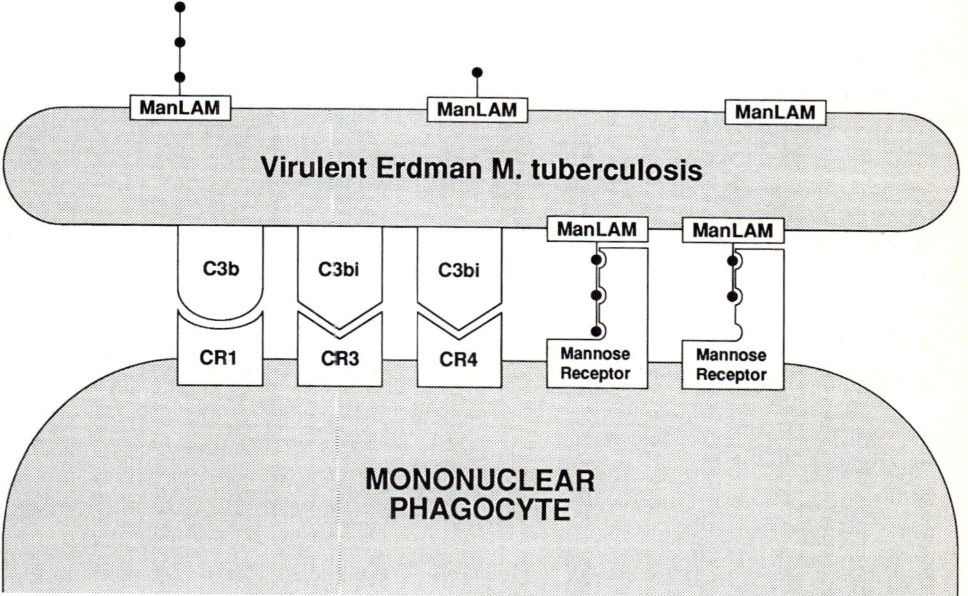

Fig. 7. Receptor-ligand interactions that mediate phagocytosis of Erdman *M. tuberculosis* by mononuclear phagocytes. See text for detailed explanation

natively, surface molecules on *M. tuberculosis* may bind to phagocyte CR directly, particularly at tissue sites where functional complement proteins are less abundant. Independent of these interactions, but perhaps facilitating them, virulent strains of *M. tuberculosis* bind to the phagocyte MR. This interaction plays an important role in mediating phagocytosis of bacteria by macrophages rather than monocytes since macrophages express the MR in abundance. In the case of the virulent Erdman strain of *M. tuberculosis*, the nonreducing end mannosides of ManLAM serve as a ligand for the macrophage MR. The end result of these receptor-ligand interactions is that the bacteria are phagocytized by macrophages to gain access to their host intracellular niche.

Thus, *M. tuberculosis*, a highly host-adapted intracellular pathogen, has evolved to utilize several different dominant host cell receptors to mediate its entry into professional phagocytes. The receptor-ligand interactions involved in entry are complex and are dependent on several variables including the tissue site, the degree of cell differentiation, the presence of inflammatory mediators and the animal model chosen. Since there are several extracellular binding domains on each of these receptors, particularly in the case of CR3, involvement of specific domains and cooperativity between different receptors will likely be important in defining the early host cell response during phagocytosis of *M. tuberculosis*.

With the development of highly purified *M. tuberculosis* surface molecules and in vitro models that allow for the study of individual receptors in phagocytosis, it will be possible to study the precise role of each receptor type during phagocytosis of *M. tuberculosis* and to determine whether the receptor-ligand pathway taken dictates the intracellular fate of this pathogen.

Acknowledgements. I acknowledge research support from the NIH (AI33004 and HL51990) and the Department of Veterans Affairs. I am grateful for the assistance and support of all of my colleagues (in particular Marcus Horwitz and Patrick Brennan) who have contributed to certain aspects of the work. I also thank Deb Nollen for her editorial assistance and William Nauseef and Bradley Britigan for their critical review of the manuscript.

References

Abou-Zeid C, Ratliff TL, Wiker HG, Harboe M, Bennedsen J, Rook GAW (1988) Characterization of fibronectin-binding antigens released by *Mycobacterium tuberculosis* and *Mycobacterium bovis BCG*. Infect Immun 56: 3046–3051

Abramson SL, Gallin JI (1990) IL-4 inhibits superoxide production by human mononuclear phagocytes. J Immunol 144: 625–630

Aderem AA, Wright SD, Silverstein SC, Cohn ZA (1985) Ligated complement receptors do not activate the arachidonic acid cascade in resident peritoneal macrophages. J Exp Med 161: 617–622

Armstrong JA, Hart PD (1971) Response of cultured macrophages to *Mycobacterium tuberculosis*, with observations on fusion of lysosomes with phagosomes. J Exp Med 134: 713–740

Armstrong JA, Hart PD (1975) Phagosome-lysosome interactions in cultured macrophages infected with virulent tubercle bacilli: reversal of the usual nonfusion pattern and observations on bacterial survial. J Exp Med 142: 1–16

Arnaout MA (1990) Structure and function of the leukocyte adhesion molecules CD11/CD18. Blood 75: 1037–1050

Arruda S, Bomfim G, Knights R, Huima-Byron T, Riley LW (1993) Cloning of an *M. tuberculosis* DNA fragment associated with entry and survival inside cells. Science 261: 1454–1457

Aslanzadeh J, Brown EJ, Quillin SP, Ritchey JK, Ratliff TL (1989) Characterization of soluble fibronectin binding to Bacille Calmette-Gurin. J Gen Microbiol 135: 2735–2741

Bardana EJ Jr, McClatchy JK, Farr RS, Minden P (1973) Universal occurrence of antibodies to tubercle bacilli in sera from non-tuberculous and tuberculous individuals. Clin Exp Immunol 13: 65–77

Barnes PF, Chatterjee D, Abrams JS, Lu S, Wang E, Yamamura M, Brennan PJ, Modlin RL (1992) Cytokine production induced by *Mycobacterium tuberculosis* lipoarabinomannan: Relationship to chemical structure. J Immunol 149: 541–547

Bermudez LE, Young LS, Enkel H (1991) Interaction of *Mycobacterium avium complex* with human macrophages: roles of membrane receptors and serum proteins. Infect Immun 59: 1697–1702

Bevilacque MP, Amrani D, Musesson MW, Bianco C (1981) Receptors for cold-insoluble globulin (plasma fibronectin) on human monocytes. J Exp Med 153: 42–60

Birdi TJ, Mistry NF, Mahadevan PR, Antia NH (1983) Alterations in the membrane of macrophages from leprosy patients. Infect Immun 41: 121–127

Blackwell J, Ezekowitz RAB, Roberts MB, Channon JY, Sim RB, Gordon S (1985) Macrophage complement and lectin-like receptors bind *Leishmania* in the absence of serum. J Exp Med 162: 324–331

Blair AL, Cree IA, Beck JS, Grange JM, Kardjito T (1993) Heat-stable opsonins in tuberculosis and leprosy. FEMS Immunol Med Microbiol 7: 197–204

Bobak DA, Gaither TA, Frank MM, Tenner AJ (1987) Modulation of FcR function by complement: subunit Clq enhances the phagocytosis of IgG-opsonized targets by human monocytes and culture-derived macrophages. J Immunol 138: 1150–1156

Bohnsack JF, O'shea JJ, Takahashi T, Brown EJ (1985) Fibronectin-enhanced phagocytosis of an alternative pathway activator of human culture-derived macrophages is mediated by the C4b/C3b complement receptor (CR1). J Immunol 135: 2680–2686

Brozna JP, Horan M, Rademacher JM, Pabst KA, Pabst MJ (1991) Monocyte responses to sulfatide from *Mycobacterium tuberculosis*: inhibition of priming for enhanced release of superoxide, associated with increased secretion of interleukin-1 and tumor necrosis factor alpha, and altered protein phosphorylation. Infect Immun 59: 2542–2548

Bullock WE, Wright SD (1987) Role of the adherence-promoting receptors, CR3, LFA-1, and p150, 95, in binding of *Histoplasma capsulatum* by human macrophages. J Exp Med 165: 195–210

Byrd SR, Gelber R, Bermudez LE (1993) Roles of soluble fibronectin and β_1 integrin receptors in the binding of *Mycobacterium leprae* to nasal epithelial cells. Clin Immunol Immunopathol 69: 266–271

Cain JA, Newman SL, Ross GD (1987) Role of complement receptor type three and serum opsonins in the neutorphil response to yeast. Complement 4: 75–86

Catanzaro A, Wright SD (1990) Binding of *Mycobacterium avium-intracellulare* to human leukocytes. Infect Immun 58: 2951–2956

Chan J, Fujiwara T, Brennan P, McNeil M, Turco SJ, Sibille J-C, Snapper M, Aisen P, Bloom BR (1989) Microbial glycolipids: possible virulence factors that scavenge oxygen radicals. Proc Natl Acad Sci USA 86: 2453–2457

Chan J, Fan X, Hunter SW, Brennan PJ, Bloom BR (1991) Lipoarabinomannan, a possible virulence factor involved in persistence of Mycobacterium tuberculosis within macrophages. Infect Immun 59: 1755–1761

Chatterjee D, Hunter SW, McNeil M, Brennan PJ (1992a) Lipoarabinomannan. Multiglycosylated form of the mycobacterial mannosylphosphatidylinositols. J Biol Chem 267: 6228–6233

Chatterjee D, Lowell K, Rivoire B, McNeil MR, Brennan PJ (1992b) Lipoarabinomannan of Mycobacterium tuberculosis. Capping with mannosyl residues in some strains. J Biol Chem 267: 6234–6239

Chatterjee D, Roberts AD, Lowell K, Brennan PJ, Orme IM (1992c) Structural basis of capacity of lipoarabinomannan to induce secretion of tumor necrosis factor. Infect Immun 60: 1249–1253

Chatterjee D, Khoo K-H, McNeil MR, Dell A, Morris HR, Brennan PJ (1993) Structural definition of the non-reducing termini of mannose-capped LAM from *Mycobacterium tuberculosis* through selective enzymatic degradation and fast atom bombardment-mass spectrometry. Glycobiology 3: 497–506

Czop JK, Kay J (1991) Isolation and characterization of β-glucan receptors on human mononuclear phagocytes. J Exp Med 173: 1511–1520

Da Silva RP, Hall BF, Joiner KA, Sacks DL (1989) CR1, the C3b receptor, mediates binding of infective *Leishmania major* metacylic promastigotes to human macrophages. J Immunol 143: 617–622
De Chastellier C, Frehel C, Offredo C, Skamene E (1993) Implication of phagosome-lysosome fusion in restriction of *Mycobacterium avium* growth in bone marrow macrophages from genetically resistant mice. Infect Immun 61: 3775–3784
Denis M, Forget A, Pelletier M, Turcotte R, Skamene E (1986) Control of the Bcg gene on early resistance in mice to infections with BCG substrains and atypical mycobacteria. Clin Exp Immunol 63: 517–525
Douvas GS, Looker DL, Vatter AE, Crowle AJ (1985) Gamma interferon activates human macrophages to become tumoricidal and leishmanicidal but enhances replication of macrophage-associated mycobacteria. Infect Immun 50: 1–8
Douvas GS, Berger EM, Repine JE, Crowle AJ (1986) Natural mycobacteriostatic activity in human monocyte-derived adherent cells. Am Rev Respir Dis 134: 44–48
Drevets DA, Campbell PA (1991) Roles of complement and complement receptor type 3 in phagocytosis of *Listeria monocytogenes* by inflammatory mouse peritoneal macrophages. Infect Immun 59: 2645–2652
Drevets DA, Canono BP, Campbell PA (1992) Listericidal and nonlistericidal mouse macrophages differ in complement receptor type 3-mediated phagocytosis of L. monocytogenes and in preventing escape of the bacteria into the cytoplasm. J Leukoc Biol 52: 70–79
Eissenberg LG, Goldman WE (1987) *Histoplasma capsulatum* fails to trigger release of superoxide from macrophages. Infect Immun 55: 29–34
Elstad MR, Parker CJ, Cowley FS, Wilcox LA, McIntyre TM, Prescott SM, Zimmerman GA (1994) CD11b/CD18 integrin and a β-glucan receptor act in concert to induce the synthesis of platelet-activating factor by monocytes. J Immunol 152: 220–230
Esparza I, Fox RI, Schreiber RD (1986) Interferon-γ-dependent modulation of C3b receptors (CR1) on human peripheral blood monocytes. J Immunol 136: 1360–1365
Espitia C, Sciutto E, Bottasso O, Gonzlez-Amaro R, Hernndez-Pando R, Mancilla R (1992) High antibody levels to the mycobacterial fibronectin-binding antigen of 30–31 kD in tuberculosis and lepromatous leprosy. Clin Exp Immunol 87: 362–367
Ezekowitz RAB, Sim RB, Hill M, Gordon S (1983) Local opsonization by secreted macrophage complement components: role of preceptors for complement in uptake of zymosan. J Exp Med 159: 244–260
Fels A, Cohn ZA (1986) the alveolar macrophage. J Appl Physiol 60: 353–369
Firestein GS, Zvaifler NJ (1987) Down regulation of human monocyte differentiation antigens by interferon-γ. Cell Immunol 104: 343–354
Forget A, Benoit JC, Turcotte R, Gusew-Chartrand N (1976) Enhanced activity of anti-mycobacterial sera in experimental Mycobacterium bovis (BCG) infection in mice. Infect Immun 13: 1301–1306
Gangadharam PRJ, Edwards CK, III (1984) Release of superoxide anion from resident and activated mouse peritoneal macrophages infected with Mycobacterium intracellulare. Am Rev Respir Dis 130: 834–838
Gaynor CD, Schlesinger LS (1994) Pulmonary surfactant protein A enhances adherence of *Mycobacterium tuberculosis* by human macrophages: evidence for a direct SP-A macrophage interaction. Clin Res 42: 301A (abstract)
Haagsman HP (1994) surfactant proteins A and D. Biochem Soc Trans 22: 100–106
Hall BF, Joiner KA (1991) Strategies of obligate intracellular parasites for evading host defences. Parasitol Today 7: A22–A27
Hancock GE, Cohn ZA, Kaplan G (1989) the generation of antigen-specific, major histocompatibility complex-restricted cytotoxic T lymphocytes of the CD4+ phenotype: Enhancement by the cutaneous administration of interleukin 2. J Exp Med 169: 909–919
Hieny S, Da Silva RP, Sher A (1992) Complement enhances the survival of metacyclic trypomastigotes of Trypanosoma cruzi in mouse peritoneal macrophages. FASEB J 2: A678 (abstract)
Hirsch CS, Ellner JJ, Russell DG, Rich EA (1994) complement receptor-mediated uptake and tumor necrosis factor-α-mediated growth inhibition of *Mycobacterium tuberculosis* by human alveolar macrophages. J Immunol 152: 743–753
Hoepelman AIM, Tuomanen EI (1992) consequences of microbial attachment: Directing host cell functions with adhesins. Infect Immun 60: 1729–1733
Hoffman OA, Standing JE, Limper AH (1993) Pneumocystis carinii stimulates tumor necrosis factor-α release from alveolar macrophages through a β-glucanmediated mechanism. J Immunol 150: 3932–3940

Hogg N, Takacs L, Palmer DG, Selvendran Y, Allen C (1986) The p150,95 molecule is a marker of human mononuclear phagocytes: comparison with expression of class II molecules. Eur J Immunol 16: 240–248

Holzer TJ, Nelson KE, Schauf V, Crispen RG, Anderson BR (1986) *Mycobacterium leprae* fails to stimulate phagocytic cell superoxide anion generation. Infect Immun 51: 514–520

Horwitz MA (1988) Intracellular parasitism. Curr Opin Immunol 1: 41–46

Hunter SW, Brennan PJ (1990) Evidence for the presence of a phosphatidylinositol anchor on the lipoarabinomannan and lipomannan of *Mycobacterium tuberculosis*. J Biol Chem 265: 9272–9279

Jacobs RF, Locksley RM, Wilson CB, Hass JE, Klebanoff SJ (1984) Interaction of primate alveolar macrophages and *Legionella pneumophila*. J Clin Invest 73: 1515–1520

Johnson WD Jr, Mei B, Cohn ZA (1977) The separation, long-term cultivation, and maturation of the humman monocyte. J Exp Med 146: 1613–1626

Joiner KA, Fuhrman SA, Miettinen HM, Kasper LH, Mellman I (1990) Toxoplasma gondii: fusion competencee of parasitophorous vacuoles in Fc receptor-transfected fibroblasts. Science 249: 641–646

Kishimoto TK, Larsen RS, Corbi AL, Dustin ML, Staunton DE, Springer TA (1989) The leukocyte integrins: LFA-1, Mac-1, and p150,95. Adv Immunol 46: 149–182

Lanier LL, Arnaout MA, Schwarting R, Warner NL, Ross GD (1985) p150,95, third member of the LFA-1/CR3 polypeptide family identified by anti-Leu-M5 monoclonal antibody. Eur J Immunol 15: 713–718

Larsen RS, Corbi AL, Berman L, Springer TA (1989) Primary structure of the LFA-1α subunit: an integrin with an embedded domain defining a protein superfamily. J Cell Biol 108: 703–712

Launois P, Blum L, Dieye A, Millan J, Sarthou JL, Bach MA (1989) Phenolic glycoplid-1 from *M. leprae* inhibits oxygen free radical production by human mononuclear cells. Res Immunol 140: 847–855

Launois P, Niang M, Dieye A, Sarthou J-L, Rivier F, Millan J (1992) Human phagocyte respiratory burst by Mycobacterium bovis BCG and *M. leprae*: Functional activation by BCG is mediated by complement and its receptors on monocytes. Int J Lepr Other Mycobact Dis 60: 225–233

López Ramírez GM, Rom WN, Ciotoli C, Talbot A, Martiniuk F, Cronstein B, Reibman J (1994) *Mycobacterium tuberculosis* alters expression of adhesion molecules on monocytic cells. Infect Immun 62: 2515–2520

Mariano M, Nikitin T, Malucelli BE (1977) Phagocytic potential of macrophages from within delayed hypersensitivity-mediated granulomata. J Pathol 123: 27–33

Maxwell KW, Marcus S (1968) Phagocytosis and intracellular fate of *Mycobacterium tuberculosis*: in vitro studies with guinea pig peritoneal and alveolar mononuclear phagocytes. J Immunol 101: 176–182

May ME, Spagnuolo PJ (1987) Evidence for activation of a respiratory burst in the interaction of human neutrophils with *Mycobacterium tuberculosis*. Infect Immun 55: 2304–2307

McCabe RE, Mullins BT (1990) Failure of Trypanosoma cruzi to trigger the respiratory burst of activated macrophages. Mechanism for immune evasion and importance of oxygen-independent killing. J Immunol 144: 2384–2388

McConville MJ, Thomas-Oates JE, Ferguson MAJ, Homans SW (1990) Structure of the lipophosphoglycan from *Leishmania major*. J Biol Chem 265: 19611–19623

Mokoena T, Gordon S (1987) Human macrophage activation: modulation of mannosyl, fucosyl receptor activity in vitro by lymphokines, γ and α interferons, and dexamethasone. J Clin Invest 75: 624–631

Moreno C, Taverne J, Mehlert A, Bate CAW Brealey RJ, Meager A, Rook GAW, Playfair JHL (1989) Lipoarabinomannan from *Mycobacterium tuberculosis* induces the production of tumour necrosis factor from human and murine macrophages. Clin Exp Immunol 76: 240–245

Mosser DM, Edelson PJ (1985) The mouse macrophage receptor for C3bi (CR3) is a major mechanism in the phagocytosis of *Leishmania* promastigotes. J Immunol 135: 2785–2789

Mosser DM, Edelson PJ (1987) The third component of complement (C3) is responsible for the intracellular survival of *Leishmania major*. Nature 327: 329–331

Mosser DM, Handman E (1992) Treatment of murine macrophages with interferon-γ inhibits their ability to bind *Leishmania* promastigotes. J Leukocyte Biol 52: 369–376

Mosser DM, Vlassara H, Edelson PJ, Cerami A (1987) *Leishmania* promastigotes are recognized by the macrophage receptor for advanced glycosylation endproducts. J Exp Med 165: 140–145

Mosser DM, Springer TA, Diamond MS (1992) *Leishmania* promastigotes require opsonic complement to bind to the human leukocyte integrin Mac-1 (CD11b/CD18). J Cell Biol 116: 511–520

Myones BL, Dalzell JG, Hogg N, Ross GD, (1988) Neutrophil and monocyte cell surface p150,955 has iC3b-receptor (CR4) activity resembling CR3. J Clin Invest 82: 640–651

Nakagawara A, Nathan CF, Cohn ZA (1981) Hydrogen peroxide metabolism in human monocytes during differentiation in vitro. J Clin Invest 68: 1243–1252

Neill MA, Klebanoff SJ (1988) The effect of phenolic glycolipid 1 from *Mycobacterium leprae* on the antimicrobial activity of human macrophages. J Exp Med 167: 30–42

Newman SL, Musson RA, Henson PM (1980) Development of functional complement receptors during in vitro maturation of human monocytes into macrophages. J Immunol 125: 2236–2244

Newman SL, Bocher C, Rhodes J, Bullock WE (1990) Phagocytosis of *Histoplasma capsulatum* yeasts and microconidia by human cultured macrophages and alveolar macrophages. Cellular cytoskeleton requirement for attachment and ingestion. J Clin Invest 85: 223–230

Nicolacakis K, Toossi Z, Rich EA (1994) Activation of human immunodeficiency virus-1 in alveolar macrophages from patients with acquired immunodeficiency syndrome by *Mycobacterium tuberculosis*. Clin Res 42: 157A (abstract)

Olafson RW, Thomas JR, Ferguson MAJ, Dwek RA, Chaudhuri M, Chang K-P, Rademacher TW (1990) Structures of the N-linked oiligosaccharides of gp63, the major surface glycoprotein, from Leishmania mexicana amazonensis. J Biol Chem 265: 12240–12247

Ota T, Okubo Y, Sekiguchi M (1990) Analysis of immunologic mechanisms of high natural killer cell activity in tuberculosis pleural effusions. Am Rev Respir Dis 142: 29–33

Pabst JJ, Gross JM, Prozna JP, Goren MB (1988) Inhibition of macrophage priming by sulfatide from *Mycobacterium tuberculosis*. J Immunol 140: 634–640

Palmer S, Hamblin AS (1993) Increased CD11/CD18 expression on the peripheral blood leucocytes of patients with HIV disease: relationship to disease severity. Clin Exp Immunol 93: 344–349

Payne N, Horwitz MA (1987) Phagocytosis of *Legionella pneumophila* is mediated by human monocyte complement receptors. J Exp Med 166: 1377–1389

Petty HR, Todd RF, III (1993) Receptor-receptor interactions of complement receptor type 3 in neutrophil membranes. J Leukoc Biol 54: 492–494

Phelps DS, Rose RM (1991) Increased recovery of surfactant protein A in AIDS-related pneumonia. Am Rev Respir Dis 143: 1072–1075

Pommier CG, Inada S, Fries LF, Takahashi T, Frank MM, Brown EJ (1983) Plasma fibronectin enhances phagocytosis of opsonized particles by human peripheral blood monocytes. J Exp Med 157: 1844–1854

Prinzis S, Chatterjee D, Brennan PJ (1993) Structure and antigenicity of lipoarabinomannan from *Mycobacterium bovis BCG*. J Gen Microbiol 139: 2649–2658

Ramanathan VD, Curtis J, Turk JL (1980) Activation of the alternative pathway of complement by mycobacteria and cord factor. Infect Immun 29: 30–35

Rao SP, Gehlsen KR, Catanzaro A (1992) Identification of aβ_1 integrin on *Mycobacterium avium-Mycobacterium intracellulare*. Infect Immun 60: 3652–3657

Rao SP, Ogata K, Catanzaro A (1993) *Mycobacterium avium-M. intracellulare* binds to the integrin receptor $\alpha_v\beta_3$ On human monocytes and monocyte-derived macrophages. Infect Immun 61: 663–670

Rao SP, Ogata K, Morris SL, Catanzaro A (1994) Identification of a 68 kd surface antigen of *Mycobacterium avium* that binds to human macrophages. J Lab Clin Med 123: 526–535

Ratliff TL, McGarr JA, Abou-Zeid C, Rook GAW, Stanford JL, Aslanzadeh J, Brown EJ (1988) Attachment of mycobacteria to fibronectin-coated surfaces. J Gen Microbiol 134: 1307–1313

Reyes JM, Putong PB (1980) Association of pulmonary alveolar lipoproteinosis with mycobacterial infection. Am J Clin Pathol 74: 478–485

Roach TIA, Barton CH, Chatterjee D, Blackwell JM (1993) Macrophage activation: Lipoarabinomannan from avirulent and virulent strains of *Mycrobacterium tuberculosis* differentially induces the early genes c-fos, KC, JE and tumor necrosis factor-α. J Immunol 150: 1886–1896

Roach TIA, Chatterjee D, Blackwell JM (1994) Induction of early-response genes KC and JE by mycobacterial lipoarabinomannans: regulation of KC expression in murine macrophages by *Lsh/Ity/Bcg* (candidate *Nramp*). Infect Immun 62: 1176–1184

Robledo S, Wozencraft A, Valencia AZ, Saravia N (1994) Human monocyte infection by *Leishmania* (*Viannia*) *panamensis:* role of complement receptors and correlation of susceptibility in vitro with clinical phenotype. J Immunol 152: 1265–1276

Roecklein JA, Swartz RP, Yeager H Jr (1992) Nonopsonic uptake of *Mycobacterium avium complex* by human monocytes and alveolar macrophages. J Lab Clin Med 119: 772–781

Ross GD (1986) Opsonization and membrane complement receptors. In: Ross GD (ed) Immunobiology of the complement system. Academic, Orlando, pp 87–114
Ross GD, Cain JA, Myones BL, Newman SL, Lachmann PJ (1987) Specificity of membrane complement receptor type three (CR3) for B-glucans. Complement 4: 61–74
Russell DG, Talamas-Rohana P (1989) *Leishmania* and the macrophage: a marriage of inconvenience. Immunol Today 10: 328–333
Sampson LL, Heuser J, Brown EJ (1991) Cytokine regulation of complement receptor-mediated ingestion by mouse peritoneal macrophages. J Immunol 146: 1005-1013
Scales WE, Chensue SW, Otterness I, Kunkel SL (1989) Regulation of monokine gene expression: prostaglandin E2 suppresses tumor necrosis factor but not interleukin-1α or β mRNA and cell-associated bioactivity. J Leukoc Biol 45: 416–421
Schlesinger LS (1992) Roles of complement receptors, the mannosyl-fucosyl receptor, and pulmonary surfactant protein, SP-A in phagocytosis of *Mycobacterium tuberculosis* by human macrophages. Clin Res 40: 383A (Abstract)
Schlesinger LS (1993) Macrophage phagocytosis of virulent but not attenuated strains of *Mycobacterium tuberculosis* is mediated by mannose receptors in addition to complement receptors. J Immunol 150: 2920–2930
Schlesinger LS, Horwitz MA (1990) Phagocytosis of leprosy bacilli is mediated by complement receptors CR1 and CR3 on human monocytes and complement component C3 in serum. J Clin Invest 85: 1304–1314
Schlesinger LS, Horwitz MA (1991a) Phagocytosis of *Mycobacterium leprae* by human monocyte-derived macrophages is mediated by complement receptors CR1(CD35), CR3(CD11b/CD18), and CR4(CD11c/CD18) and interferonγ activation inhibits complement receptor function and phagocytosis of the bacterium. J Immunol 147: 1983–1994
Schlesinger LS, Horwitz MA (1991b) Phenolic glycolipid-1 of *Mycobacterium leprae* binds complement component C3 in serum and mediates phagocytosis by human monocytes. J Exp Med 174: 1031–1038
Schlesinger LS, Horwitz MA (1994) A role for natural antibody in the pathogeneiss of leprosy: antibody in nonimmune serum mediates C3 fixation to the *Mycobacterium leprae* surface and hence phagocytosis by human mononuclear phagocytes. Infect Immun 62: 280–289
Schlesinger LS, Bellinger-Kawahara CG, Payne NR, Horwitz MA (1990) Phagocytosis of *Mycobacterium tuberculosis* is mediated by human monocyte complement receptors and complement component C3. J Immunol 144: 2771–2780
Schlesinger LS, Hull SR, Kaufman TM (1994) Binding of the terminal mannosyl units of lipoarabinomannan from a virulent strain of *Mycobacterium tuberculosis* to human macrophages. J Immunol 152: 4070–4079
Schreiber S, Perkins SL, Teitelbaum SL, Chappel J, Stahl PD, Blum JS (1993) Regulation of mouse bone marrow macrophage mannose receptor expression and activation by prostaglandin E and IFN-γ. J Immunol 151: 4973–4981
Schwarting R, Stein H, Wang CY (1985) The monoclonal antibodies anti-S-HCL 1 (anti-Leu-14) and anti-S-HCL 3 (anti-Leu-M5) allow the diagnosis of hairy cell leukemia. Blood 65: 974–983
Shakoor Z, Hamblin AS (1992) Increased CD11/CD18 expression on peripheral blood leucocytes of patients with sarcoidosis. Clin Exp Immunol 90: 99–105
Shattock RJ, Friedland JS, Griffin GE (1994) Phagocytosis of *Mycobacterium tuberculosis* modulates human immunodeficiency virus replication in human monocytic cells. J Gen Virol 75: 849–856
Shepherd VL, Campbell EJ, Senior RM, Stahl PD (1982) Characterization of mannose/fucose receptor on human mononuclear phagocytes. J Reticul Soc 32: 423–431
Shiratsuchi H, Johnson JL, Toossi Z, Ellner JJ (1994) Modulation of the effector function of human monocytes for *Mycobacterium avium* by human immunodeficiency virus-1 envelope glycoprotein gp120. J Clin Invest 93: 885–891
Sibley LD, Hunter SW, Brennan PJ, Krahenbuhl JL (1988) Mycobacterial lipoarabinomannan inhibitsγ interferon-mediated activation of macrophages. Infect Immun 56: 1232–1236
Speert DP, Silverstein SC (1985) Phagocytosis of unopsonized zymosan by human monocyte-derived macrophages: maturation and inhibition by mannan. J Leukoc Biol 38: 655–658
Stahl PD (1992) The mannose receptor and other macrophage lectins. Curr Opin Immunol 4: 49–52
Stein M, Keshav S, Harris N, Gordon S (1992) Interleukin 4 potently enhances murine macrophage mannose receptor activity: a marker of alternative immunologic macrophage activation. J Exp Med 176: 287–292

Stokes RW, Haidl ID, Jefferies WA, Speert DP (1993) Mycobacteria-macrophage interactions: macrophage phenotype determines the nonopsonic binding of *Mycobacterium tuberculosis* to murine macrophages. J Immunol 151: 7067–7076

Swartz RP, Naai D, Vogel CW, Yeager H Jr (1988) Differences in uptake of mycobacteria by human monocytes: a role for complement. Infect Immun 56: 2223–2227

Taylor ME, Conary JT, Lennartz MR, Stahl PD, Drickamer K (1990) Primary structure of the mannose receptor contains multiple motifs resembling carbohydrate-recognition domains. J Biol Chem 265: 12156–12162

Tenner AJ, Robinson SL, Borchelt J, Wright JR (1989) Human pulmonary surfactant protein (SP-A), a protein structurally homologous to C1q, can enhance FcR- and CR1-mediated phagocytosis. J Biol Chem 264: 13923–13928

Thomas JR, McConville MJ, Thomas-Oates JE, Homans SW, Ferguson MAJ, Gorin PAJ, Greis KD, Turco SJ (1992) Refined structure of the lipophosphoglycan of *Leishmania donovani*. J Biol Chem 267: 6829–6833

Toba H, Crowford JT, Ellner JJ (1989) Pathogenicity of *Mycobacterium avium* for human monocytes: absence of macrophage activating factor activity of γ interferon. Infect Immun 57: 239–244

Unkeless JC (1989) Function and heterogeneity of human Fc receptors for immunoglobulin G. J Clin Invest 83: 355–361

Vachula M, Holzer TJ, Anderson BR (1989) Suppression of monocyte oxidative response by phenolic glycolipid I of *Mycobacterium leprae*. J Immunol 142: 1696–1701

Van Strijp JAG, Russell DG, Tuomanen E, Brown EJ, Wright SD (1993) Ligand specificity of purified complement receptor type three (CD11b/CE18, $\alpha_m\beta_2$, Mac-1): Indirect effects of an ARG-GLY-ASP (RGD) sequence. J Immunol 151: 3324–3336

Vlassara H, Brownlee M, Cerami A (1981) High affinity receptor-mediated uptake and degradation of glucose-modified proteins: a potential mechanism for the removal of senescent macromolecules. Proc Natl Acad Sci USA 78: 5190–5195

Wilson CB, Tsai V, Remington JS (1980) Failure to trigger the oxidative metabolic burst by normal macrophages: possible mechanism for survival of intracellular pathogens. J Exp Med 151: 328–346

Wilson ME, Hardin KK (1988) The major concanavalin A-binding surface glycoprotein of *Leishmania donovani chagasi* promastigotes is involved in attachment to human macrophages. J Immunol 141: 265–272

Wilson ME, Pearson RD (1986) Evidence that *Leishmania donovani* utilizes a mannose receptor on human mononuclear phagocytes to establish intracellular parasitism. J Immunol 136: 4681–4688

Wilson ME, Pearson RD (1988) Roles of CR3 and mannose receptors in the attachment and ingestion of *Leishmania donovani* by human mononuclear phagocytes. Infect Immun 56: 363–369

Wirth JJ, Kierszenbaum F (1984) Fibronectin enhances macrophage association with invasive forms of *Trypanosoma cruzi*. J Immunol 133: 460–464

Wozencraft AD, Sayers G, Blackwell JM (1986) Macrophage type 3 complement receptors mediate serum-independent binding of *Leishmania donovani:* detection of macrophage-derived complement on the parasite surface by immunoelectron microscopy. J Exp Med 164: 1332–1337

Wright SD (1992) Receptors for complement and the biology of phagocytosis. In: Gallin JI, Goldstein IM, Snyderman R (eds) Inflammation: basic principles and clinical correlates, 2nd edn. Raven, New York, pp 477–495

Wright SD, Silverstein SC (1982) Tumor-promoting phorbol esters stimulate C3b and C3b' receptor-mediated phagocytosis in cultured human monocytes. J Exp Med 156: 1149–1164

Wright SD, Silverstein SC (1983) Receptors for C3b and C3bi promote phagocytosis but not the release of toxic oxygen from human phagocytes. J Exp Med 158: 2016–2023

Wright SD, Detmers PA, Jong MTC, Meyer BC (1986) Inteferon-γ depresses binding of ligand by C3b and C3bi receptors on cultured human monocytes, an effect reversed by fibronectin. J Exp Med 163: 1245–1259

Wyler DJ, Sypek JP, McDonald JA (1985) In vitro parasite-monocyte interactions in human *Leishmania*sis: possible role of fibronectin in parasite attachment. Infect Immun 49: 305–311

Yamamoto K, Johnston RB Jr (1984) Dissociation of phagocytosis from stimulation of the oxidative metabolic burst in macrophages. J Exp Med 159: 405–416

Youmans GP (1979) Tuberculosis. Saunders, Philadelphia

Zhang Y, Doerfler M, Lee TC, Guillemin B, Rom WN (1993) Mechanisms of stimulation of interleukin-1β and tumor necrosis factor-α by *Mycobacterium tuberculosis* components. J Clin Invest 91: 2076–2083

In Vitro Interaction of *Mycobacterium tuberculosis* and Macrophages: Activation of Anti-mycobacterial Activity of Macrophages and Mechanisms of Anti-mycobacterial Activity

L. O'Brien, B. Roberts, and P.W. Andrew

1 Introduction

Mycobacterium tuberculosis is an example of a facultative intracellular pathogen which will survive and multiply within macrophages from non-immune individuals. It is a central tenet of cell-mediated immunity to intracellular pathogens that macrophages are the principal effector cells and that, after exposure to products of specifically sensitized T cells, they acquire the ability to kill these pathogens (Mackaness 1968). Much of the basic data on which the idea of cell-mediated immunity is founded came from studies on immunity to mycobacterial infections (Lurie 1942; Suter 1953; Mackaness 1964, 1968).

It is believed that macrophages can be activated by lymphocyte products to kill *M. tuberculosis*. Therefore, the identification of the substances that activate macrophages to kill *M. tuberculosis* is desirable. Likewise, identification of the

Department of Microbiology and Immunology, University of Leicester, Leicester LE1 9HN, UK mechanisms that the activated macrophages use to kill or at least inhibit the replication of *M. tuberculosis* would be welcomed. These problems have attracted a considerable amount of experimental effort because the answer to either could have major implications for anti-tubercular therapy. However, even to the noncasual observer the state of the literature in these areas can be very confusing.

The literature records work on macrophages collected from different anatomical sites and from different animal species. The answer one gets seems to depend on the source of the macrophage. The bottom line for human macrophages is that we have no idea what the mechanisms of activation are or how they kill *M. tuberculosis*. In truth, we do not know if human macrophages can kill *M. tuberculosis*. Fundamentally, this is due to an inability to convincingly demonstrate an antitubercular activity by these macrophages in vitro. For macrophages from other species the clouds of ignorance have cleared a little, but we still have a far from complete picture of mechanisms of activation or of antitubercular activity.

In addition, confusion might arise because there is a parallel body of work done with *Mycobacterium avium* to identify activating factors and killing mechanisms and some progress has been made with human macrophages in this area. There may be a tendency to extrapolate from work on one mycobacterial species to another. However, *M. avium* is not *M. tuberculosis*. Not least, they differ markedly in virulence. Further confusion can arise because different laboratories use different strains of *M. tuberculosis* and *M. avium* and there is evidence that strains of both species differ in susceptibility to potential killing systems.

1.1 Technical Problems Associated with Study of *Mycobacterium tuberculosis*-Macrophage Interaction In Vitro

Mycobacterium tuberculosis is not the easiest microorganism to work with. Its serious pathogenic nature demands that special high containment measures are taken. In an attempt to avoid this problem some workers have chosen to work with *Mycobacterium bovis* BCG (Flesch and Kauffman 1987) or *Mycobacterium microti* (Walker and Lowrie 1981). Both are members of the *M. tuberculosis* complex (Wayne and Kubica 1986). However, when considering data on their interaction with human macrophages, it should be borne in mind that both are nonpathogenic for humans and both have been used as live vaccines (MRC 1972). BCG also is nonpathogenic for the mouse. However, *M. microti* is a pathogen of mice and therefore its use with murine macrophages presents less problems in interpretation.

It is a number of other technical factors associated with experimental design that present problems in showing intracellular killing. Such technical problems associated with study of *M. tuberculosis*-macrophage interaction in vitro were the subject of discussion by a number of authors recently (COLLINS 1990; CROWLE 1990; ELLNER 1990; ROOK 1990a). However, they deserve reemphasis here

No antibiotics should be included in cultures. Data derived from experiments in which antibiotics are included in the culture medium should be treated with a great deal of caution, because antibiotics can readily enter all intracellular compartments in macrophages (LOWRIE et al. 1979, 1982; DREVETS et al. 1994). Furthermore, the intracellular activity of the antibiotic may be influenced by the activation state of the macrophage (ALTES et al. 1985).

Mycobacteria, including *M. tuberculosis*, are hydrophobic and are very prone to forming clumps. Failure to avoid clumps, and not to use a unicellular suspension of bacteria, will cause serious problems. Treatments, such as use of detergent and/or mild sonication (O'BRIEN et al. 1991), must be included in the experiment. Ingestion of clumps will give problems in determining the rate of infection of macrophages, while intracellular formation of clumps would appear as intracellular killing. Clumping is clearly most acutely problematic during assays based on counting colony-forming units (CFUs).

Conceptually, counting colonies is the best method of assaying killing of microbes since one is directly measuring numbers of viable bacteria. Theoretically, intracellular *M. tuberculosis* could be induced to enter a state of long-term dormancy or they may be injured but not killed so that they will not grow immediately on agar. Such organisms could be mistakenly recorded as being killed. We have seen such an effect after exposure of *M. tuberculosis* to some cell-free killing systems (B. Roberts and P.W. Andrew, unpublished data). Agar plates, therefore, should be kept beyond the usual 4–6 weeks to minimize this problem.

Because of the problems of clumping, together with the fact that it can take weeks for colonies to become visible on solid media, some workers prefer to use assays based on measurement of metabolic activity of the bacteria. By far the most popular method of this type is that developed by ROOK and colleagues (ROOK and RAINBOW 1981; ROOK et al. 1985) and based on the uptake of radiolabelled uracil, which is incorporated mainly into RNA (ROOK et al. 1985). However, a major disadvantage of this technique is that it cannot distinguish stasis from killing of mycobacteria (ROOK et al. 1985). Thus, for rapid demonstration of an undefined antitubercular effect, metabolic assays are fine but colony counting should be used to show killing of *M. tuberculosis*.

A complete audit should be kept of the numbers of viable bacteria present at each stage of the experiment. The percentage viability of mycobacterial cultures can be surprisingly low. Viability of the culture should be confirmed to ensure that the desired ratio of bacteria to cells is obtained. At the end of the period of macrophage infection, numbers of extracellular bacteria should be reduced to a minimum by extensive, careful washing. The efficiency of this process should be monitored. The presence of extracellular bacteria may mask intracellular killing especially in situations in which phagocytosis is not efficient. After these washings, no further washing should be done. Most importantly, the numbers of bacteria present in the culture supernatant, in addition to cell-associated bacteria, should be determined at each time point. This final point is frequently overlooked, yet if bacteria associated with macrophages that detach from the monolayer during culture are not counted, an erroneous conclusion of bacterial killing may be drawn.

As pointed out by COLLINS (1990), many workers fail to monitor the health of the monolayer once the infection period is complete. The numbers of cells and their viability should be monitored throughout the experiment. As discussed above, loss of infected macrophages from monolayers into the supernatant can create artifacts. However, even if a full bacterial audit is kept, lysis of infected macrophages still can create artifacts. If the mycobacteria are released into fetal calf serum, they will multiply (ROOK 1990a) and this may mask killing. In other circumstances, lysis of macrophages may result in a picture of apparent bacterial killing. If macrophages lyse and mycobacteria are released into human serum, the bacteria do not multiply (ROOK 1990a). If the monolayer integrity is not monitored, the effect of human serum on mycobacterial growth could be interpreted as an antimycobacterial effect of the macrophages. In fact, mycobacteria may fare far less well in tissue culture medium containing human serum than inside macrophages (ROOK et al. 1986a; ROBERTSON and ANDREW 1991). It is worth bearing in mind that macrophage types are not equally proficient in remaining as monolayers. Human mononuclear phagocytes are more fragile than mouse peritoneal macrophages and are more likely to detach. Similarly, alveolar macrophages have a greater tendency to detach.

Failure to take account of the factors discussed above can lead to erroneous conclusions about the intracellular fate of *M. tuberculosis*.

2 In Vitro Activation of Macrophage Antitubercular Activity

It is now clear that macrophage function, including antimicrobial activity, can be modulated by cytokines, autocrines, hormones, vitamins and combinations thereof. Success, however, in identifying the cocktail to activate mononuclear phagocytes' antitubercular activity in vitro has been variable. It is relatively straightforward to activate murine macrophages in vitro to have some activity in restricting the growth of *M. tuberculosis*. We will therefore review the literature on these cells before dealing with human cells.

2.1 Murine Macrophages

Several groups have been able to activate murine peritoneal or pleural macrophages to slow the growth of *M. tuberculosis* by incubation either with sensitized lymphocytes or in culture supernatants from these cells (PATTERSON and YOUMANS 1970; KLUN and YOUMANS 1973; CAHALL and YOUMANS 1975; TURCOTTE et al. 1976; ZLOTNIK and CROWLE 1981; ROOK et al. 1985; VARESIO et al. 1990; SYPEK et al. 1993). The degree of inhibition of replication reported has been variable although in some cases complete stasis seemed to be achieved (ROOK et al. 1985). However, it has proved more difficult to show killing of virulent *M. tuberculosis*. It was not until the work of WALKER and LOWRIE (1981) that it was shown conclusively that murine macrophages could be activated by a cytokine mixture to kill a member of the *M. tuberculosis* complex. Murine macrophages cultured for 3 days in medium supplemented with culture supernatants from PPD-stimulated lymphocytes were able to kill >90% of phagocytosed *M. microti* in 24 h.

The observation of WALKER and LOWRIE (1981), combined with the availability of pure, recombinant cytokines, provided an impetus to identify the active components in lymphocyte supernatants. Although, in some circumstances, it has been possible, subsequently, to demonstrate in vitro activation of tuberculocidal mechanisms (KHOR et al. 1986; CHAN et al. 1992), many experiments with murine cells and purified cytokines have shown stasis and not killing of *M. tuberculosis* (ROOK et al. 1986a; FLESCH and KAUFMANN 1987; SYPEK et al. 1993). ROOK (1988) has suggested that the inability to demonstrate killing means that it is difficult to relate these in vitro findings to the in vivo situation. However, LOWRIE (1990) suggested that mycobacteriostasis might represent a good immune response in mice, where tuberculosis is characterised by the long-term persistence of a static population of bacteria (REES and HART 1961).

Interferon-γ (IFN-γ) is now established as a key factor for activating antitubercular activity in murine cells in vitro. The importance of IFN-γ to murine antitubercular immunity recently was reinforced strongly by infection of mice that have a targeted disruption of the gene for IFN-γ (COOPER et al. 1993; FLYNN et al. 1993) or the IFN-γ receptor (KAMIJO et al. 1993). Such mice were unable to restrict the growth of a normally nonlethal inoculum of *M. tuberculosis* (COOPER et al. 1993; FLYNN et al. 1993) or *M. bovis* BCG (KAMIJO et al. 1993).

IFN-γ given alone activates murine peritoneal macrophages to inhibit intracellular multiplication of *M. tuberculosis* (ROOK et al. 1986a; DENIS 1991b) and *M. bovis* BCG (SYPEK et al. 1993) and murine bone marrow-derived macrophages to inhibit *M. bovis* BCG and *M. tuberculosis* (FLESCH and KAUFMANN 1987). Interestingly, the susceptibility of *M. tuberculosis* to IFN-activated bone marrow cells varied according to strain (FLESCH and KAUFMANN 1987). Growth of *M. tuberculosis* H37Rv was significantly inhibited whereas *M. tuberculosis* Middleberg was resistant to inhibition. This observation confirmed an earlier report (LOWRIE et al. 1985), that the susceptibility of *M. tuberculosis* to the antimicrobial mechanisms of activated macrophages varied from strain to strain. Here, alveolar macrophages taken from the lungs of *M. bovis* BCG-vaccinated guinea pigs were able to kill *M. tuberculosis* during 24 h in vitro. However, the extent of killing varied with the strain. Obviously such variations in susceptibility of strains may explain variations in results between laboratories.

The timing of addition of IFN-γ to murine peritoneal macrophages appears to influence its effects. KHOR and colleagues (1986) found that addition of IFN-γ prior to infection caused macrophages to be cidal to *M. microti*. This cidal action mainly took place during the first 15 min following infection and had subsided by 30 min. However, if IFN-γ was added after the infection period macrophages were bacteriostatic. These observations have important consequences when considering mechanisms of antibacterial activity. They suggest the presence of at least two antimycobacterial mechanisms. Later, FLESCH and KAUFMANN (1990b) and SYPEK and colleagues (1993) also found that the timing of IFN-γ addition was important. They reported that IFN-γ added prior to infection, inhibited growth of *M. bovis* BCG but that it was less effective (FLESCH and KAUFMANN 1990b) or ineffective (SYPEK et al. 1993) if added after infection.

Over recent years there has been an explosion in the characterization and availability of cytokines. Attention has been directed to investigating if any can substitute for, or synergise with, IFN-γ. To date there have been no reports of a mediator able to mimic the effect of IFN-γ on the antitubercular effect of murine macrophages. ROOK (1990b) reported that tumor necrosis factor-α (TNF-α) and 1,25-(dihydroxy) vitamin D_3 (calcitriol) had no effect on the growth of *M. tuberculosis* inside murine peritoneal macrophages. Commenting on these observations, ROOK (1990a) suggested that the absence of effect of calcitriol in this system was reasonable because IFN-γ does not activate the key enzyme, 25-hydroxy-vitamin D_3-1-hydroxylase, necessary for the formation of calcitriol. Furthermore, treatment of tuberculous mice with calcitriol failed to alter the course of disease. The lack of activation of 25-hydroxy-vitamin D_3-1-hydroxylase in murine cells is particularly relevant when considering the observation of RASTOGI (1990), that 4 μg calcitriol/ml completely inhibited the replication of *M. tuberculosis* in the J774 murine macrophage line. This concentration is 700-fold higher than physiological and the attainment of this concentration locally seems unlikely given the absence of 1-hydroxylase activation (ROOK 1990a).

ROOK (1990b) considered that the absence of an effect with TNF-α was more noteworthy, since in vivo experiments with BCG-infected mice had shown that TNF-α was protective (KINDLER et al. 1989). Administration of anti-TNF-α antibody to these mice resulted in poor granuloma formation and disseminated infection. Of course, there is no proof that TNF-α was acting at the level of the macrophage in this model. Analysis of the interaction of TNF-α, macrophages and mycobacteria is complicated by the fact that TNF-α will always be present in macrophage-*M. tuberculosis* cultures because *M. tuberculosis* lipoarabinomannan (LAM) is a powerful trigger of TNF-α production by macrophages (MORENO et al. 1989). Other mycobacterial antigens may also trigger TNF-α from macrophages (VALONE et al. 1988). Thus optimal concentrations of TNF-α may exist so that addition of extra TNF-α has no effect. Alternatively, may be cell surface TNF-α is more effective as an autocrine. In this regard it is interesting that SYPEK and colleagues (1993) also found that exogenous TNF-α had no effect on growth of *M. bovis* BCG but that anti-TNF-α blocked the activation of infected cells by sensitized lymphocytes.

FLESCH and KAUFMANN (1990a) also found that TNF-α given alone had no effect on *M.bovis* BCG subsequently given to bone marrow-derived macrophages. However, they found that it synergized with IFN-γ in inducing stasis.

CHAN and colleagues (1992) also showed that TNF-α had a synergistic effect in inhibiting uracil incorporation by intracellular *M. tuberculosis* when added to IFN-γ-primed macrophages. They also showed that lipopolsaccharide (LPS) added to IFN-γ-primed macrophages rendered them tuberculocidal.

What of other cytokines? Interleukin (IL)-2 (ROOK 1990b), IL-4 (FLESCH and KAUFMANN 1990a; ROOK 1990b) and IL-6 (FLESCH and KAUFMANN 1990b) have been reported to have no effect on the antitubercular activity of macrophages when added before infection. However, as with IFN-γ, timing of addition may be important. Addition of IL-4 or IL-6 to bone marrow-derived macrophages infected with *M. bovis* BCG induced a small inhibition of uracil incorporation (FLESCH and KAUFMANN 1990a,b). It is interesting that *M. tuberculosis* stimulates the release of IL-6 from THP-1 cells, a human myelomonocytic cell line (FRIEDLAND 1993).

There is no direct evidence for the involvement of other cytokines in the antitubercular activity of murine macrophages but some circumstantial evidence suggests they may play a role. *Mycobacterium tuberculosis* is a potent stimulus of IL-8 release from THP-1 cells in vitro (FRIEDLAND et al. 1992). *Mycobacterium avium* induced production of IL-10 in mice; IL-10 suppressed in vitro anti-*M.avium* activity of murine peritoneal macrophages; and anti-IL-10 antibody reduced the extent of *M. avium* infection (BERMUDEZ and CHAMPSI 1993). IL-12 also is produced by mononuclear phagocytes (D'ANDREA et al. 1992) and may play a role in resistance to tuberculosis in mice (FLYNN and BLOOM referenced in CHAN and KAUFMANN 1994). It remains to be seen if these agents influence the antitubercular activity of murine macrophages and whether the in vivo effects recorded are due to direct or indirect effects on macrophages. However, it is already clear that for the murine macrophage modulation of antitubercular activity may be a complex, dynamic system subject to up- and down-regulation by cytokines, autocrines and mycobacterial products and differentially modulated by the timing of exposure to these agents.

As often seems the case in biology, in trying to unravel the molecular mechanisms for activation of antitubercular activity of murine macrophages, we have moved from a simple explanation for activation by IFN-γ alone to a complex, rather unclear picture apparently involving several cytokines. Nevertheless, there is a feeling that progress is being made. However, it is very hard to say the same when it comes to understanding the situation with human cells.

2.2 Human Macrophages

In our view, no one has been able to establish convincingly the conditions required to activate human mononuclear phagocytes in vitro to kill *M. tuberculosis*. The best that has been achieved is a slowing of the rate of replication, and even this is an inconsistent phenomenon. The ability of human mononuclear phagocytes to inhibit the replication of *M. tuberculosis* in vitro was first reported in 1981 (CROWLE and MAY 1981). Supernatants from cultures of lymphocytes from tuberculin-positive individuals stimulated with mycobacterial antigen activated syngeneic monocyte-derived macrophages to inhibit the replication of *M. tuberculosis*. Unfortunately, attempts to identify the molecules required, even for the induction of stasis, have been unsuccessful. Perhaps this is because the requirements are more complex than was suggested by the early experiments with murine cells. Sadly, to date it appears that we cannot directly apply the conclusions made with murine cells to human cells. Nowhere is this better illustrated than with IFN-γ.

In direct contrast to the mouse, the consensus is that IFN-γ has little or no effect on human cells and even may be detrimental. Certainly this is true for IFN acting alone, but its role in conjunction with other molecules is still an open question. These views have not changed since we last reviewed the role of IFN in immunity to mycobacteria (Andrew and Lowrie 1987).

Rook and colleagues, in a series of publications, reported their findings on the effect of IFN-γ on antitubercular activity of human cells. The overall inhibition of uracil uptake by *M. tuberculosis* in IFN-γ-treated human alveolar macrophages was a nonsignificant 18% compared to 90% by murine peritoneal macrophages (Steele et al. 1986). Cells from two individuals gave significant inhibition, about 34%. These persons were smokers so it is possible that their cells represented inflammatory macrophages primed to respond to IFN and/or mycobacteria. However, the observation was not a general effect of smoking, since cells from other smokers had no significant effect on uracil uptake.

If cultured human monocytes were given IFN-γ either before or concomitant with infection, there was a trend to a small inhibition of uptake of uracil by *M. tuberculosis* (Rook et al. 1986b). However, it was a variable finding and even the largest effects were much less than those seen with murine macrophages. Others also have found that IFN-γ given to monocyte-derived macrophages before (Robertson and Andrew 1991) or at the time of (Denis et al. 1990) infection did not change the rate of growth of *M. tuberculosis*.

With freshly explanted monocytes there was an even more variable effect (Rook et al. 1986b). Significant inhibition of *M. tuberculosis* was seen with cells from some individuals, though again less than with murine macrophages. However, with cells from other individuals, IFN-γ treatment caused a significant increase in growth of *M. tuberculosis*. Douvas and colleagues (1985) also found that IFN-γ treatment of human macrophages enhanced the rate of replication of *M. tuberculosis* in the macrophages even though IFN-γ pretreatment activated the macrophages to kill *Leishmania donovani* and murine tumor cells.

IFN-γ alone also appears unable to activate human cells to alter the growth of *M. avium* (Bermudez and Young 1988; Denis 1991a; Zerlauth et al. 1991). Perhaps even more surprisingly, IFN-γ alone could not activate human macrophages against the nonpathogenic *Mycobacterium phlei* (Robertson and Andrew 1991), even though activity against the protozoan *Toxoplasma gondii* was activated. This observation, and that of Douvas and coworkers (1985) and Zerlauth and coworkers (1991), means that the concept that activation of macrophages during a cell-mediated immune response is nonspecific (Mackaness 1964, 1968) may be simplistic and implies that macrophages may require a particular cocktail of activation signals in order to deal with an individual type of pathogen. We still have not found what this cocktail is for *M. tuberculosis*.

Calcitriol, the active form of vitamin D_3, was known to cause differentiation of murine and human cells in culture (ABE et al. 1981; AMENTO et al. 1984; MANGELSDORF et al. 1984). ROOK and colleagues (1986b) therefore speculated that it might alter the response of monocytes to IFN-γ or their ability to control *M. tuberculosis*. They found that monocytes cultured for 3 days in $10^{-7} M$ calcitriol were able to inhibit consistently the growth of *M. tuberculosis*. Other vitamin D_3 metabolites at this concentration were not active. When calcitriol and IFN-γ were included together, the effects were additive and of similar magnitude, causing greater than 60% inhibition of uptake of uracil by *M. tuberculosis*. CROWLE and colleagues (1987) also found that calcitriol could activate monocyte-derived macrophages to slow the replication of *M. tuberculosis*, but they found no additive effect of coincubation of calcitriol and IFN-γ. Furthermore, calcitriol concentrations of 4 μg/ml (approximately $10^{-5} M$) were required, much higher than the normal concentration in the circulation (SCHWARZTMAN and FRANK 1987). However, CROWLE and colleagues (1987) argued that higher concentrations might be achieved in the confines of a granuloma. This seems plausible given that human macrophages can synthesize calcitriol (KOEFFLER et al. 1985) and synthesis is stimulated by IFN-γ (ROOK et al. 1986b).

Another immunologically active metabolite of a vitamin can activate human monocyte-derived macrophages. CROWLE and ROSS (1989) reported that physiological levels of retinoic acid, a metabolite of vitamin A, added before infection resulted in a substantial slowing of *M. tuberculosis* growth. ROOK et al. (1986b) and CROWLE and ROSS (1989) suggested that the in vitro effects of vitamins A and D may explain why cod liver oil was useful in the treatment of tuberculosis before the age of chemotherapy. However, this view is not supported by the observation that the plasma concentration of 25-hydroxy-vitamin D_3 did not influence the outcome of tuberculosis in the guinea pig (HERNANDEZ-FRONTERA and MCMURRAY 1993). Unfortunately, however, levels of calcitriol were not measured in the guinea pigs in this study.

Even though the effects of calcitriol appear encouraging, ROOK (1988, 1990b) has argued that the level of inhibition of replication of *M. tuberculosis* is too low to represent a protective pathway, representing only a decrease in the rate of replication from three to two generations over 4 days. Rather, ROOK (1988, 1990b) has proposed that, because calcitriol and IFN-γ prime macrophages for LAM-induced release of TNF-α, calcitriol represents part of a pathological mechanism rather than a protective one. TNF-α, it is argued (ROOK 1988, 1990b), is responsible for much of the pathology seen in tuberculosis. In addition, ROOK (1990b) suggested that TNF-α could not activate antitubercular activity in human macrophages.

Others (FAZAL et al. 1992) also have suggested that TNF-α is unlikely to have a role in controlling the growth of tubercle bacilli in human macrophages. Furthermore, ROOK and colleagues have reported that human cells, including a monocytic cell line, are much more susceptible to the toxic effects of TNF-α when infected with *M. tuberculosis* (FILLEY and ROOK 1991; FILLEY et al. 1992).

However, the exclusion of TNF-α and calcitriol from a protective pathway may be an extreme view. Maybe given the correct, though as yet undefined, concentrations, timing of exposure or some other condition, they will indeed have a role in protection and activation of antitubercular activity in human macrophages. Admittedly, most of the in vitro evidence for this alternative view is indirect. Mycobacterial-induced release of TNF-α from human macrophages varies according to strain both with *M. tuberculosis* (BARNES et al. 1992) and *M. avium* (SHIRATSUCHI et al. 1993). Also, TNF-α appears to play a role in the inhibition of growth of *M. avium* by human macrophages (BERMUDEZ et al. 1990; DENIS 1991a; ZERLAUTH et al. 1991). HIRSCH and colleagues (1994) reported that human alveolar macrophages were less permissive for growth of *M. tuberculosis* than monocytes when cultured for 7 days. They suggested the slower growth was mediated by TNF-α since alveolar cells produced more TNF-α than monocytes and intracellular growth was enhanced by anti-TNF-α antibody. However, it was DENIS (1991c) who provided direct evidence that TNF-α could induce antitubercular activity in human macrophages in vitro.

DENIS (1991c) reported that a combination of IFN-γ, TNF-α and calcitriol activated human monocyte-derived macrophages not simply to inhibit replication but to kill *M. tuberculosis*. This was an exciting development, which, if confirmed, would represent a significant advance. Unfortunately, it appears that the optimism that this report has generated (FRIEDLAND 1993; CHAN and KAUFMANN 1994) is misplaced. Independent experiments (WARWICK-DAVIES et al. 1994) failed to repeat the findings of DENIS (1991c). Indeed, in some cases, it was found that the combination of IFN-γ, TNF-α and calcitriol sensitized the cells to a subsequent cytotoxic effect of *M. tuberculosis*. However, even in those monolayers showing no evidence of cytotoxicity, there was no evidence that the modulators influenced the fate of *M. tuberculosis*. It was suggested (WARWICK-DAVIES et al. 1994) that part of the reason for the difference in the reports was due to differences in the accounting procedures for viable bacteria.

Other cytokines have been tested, but IL-2, IL-4 and IFN-α given alone were all ineffective in influencing the growth of *M. tuberculosis* inside human macrophages (DENIS et al. 1990; ROOK 1990b).

2.3 Reasons for the Discrepancy Between Experience with Human and Murine Macrophages

The failure to show convincingly that human macrophages could kill or inhibit the replication of *M. tuberculosis* has led some to suggest that the time has come when we should question the dogma that the basis of immunity to this organism lies within the macrophage. According to this line of thought, we must seek other effector systems (ELLNER 1990; ROOK 1988, 1990a). Neutrophils and cytotoxic CD4 and CD8 cells are often suggested (ROOK 1988; KAUFMANN 1993). To these might be added the general environment within the granuloma.

Although it is perfectly legitimate and desirable to continually challenge dogma, it seems to us that it is far too early to abandon the human monocyte/macrophage system. Is it fair to compare the activity of a bone marrow-derived macrophage isolated from a specific pathogen-free, inbred mouse with a peripheral blood monocyte-derived macrophage isolated from a city-living, outbred human?

There are a number of reasons why we conclude that the antitubercular potential of human cells has not been fully explored. Some of these reasons were nicely outlined in 1990 (Ellner 1990) and they are still legitimate today. Three reasons for the apparent difference in human and murine macrophages can be put forward. They are: variation in the human cells used and their handling following isolation; use of the wrong human cell; use of the wrong regimen for activation or a combination of these reasons.

When working with human cells we have to deal with variation between races, between individual donors and within an individual donor.

In comparing data from human and murine macrophages, we are not comparing like with like. Difference in the animal source may explain the discrepancy between observations. Susceptibility to tuberculosis differs between animal species (Brown 1983). Furthermore, there are clear biochemical differences between human and murine macrophages. The difference in 25-hydroxyvitamin D_3-1-hydroxylase activity (Rook 1990a) is an example that we have discussed. Another is in the metabolism of arginine (see below).

There is variation between donors of macrophages. Significant activity against *M. tuberculosis* has been reported in some cases when using cells from particular individuals. One of the main problems seems to be a lack of consistency in observations with cells from different subjects. There was a great deal of donor-to-donor variation (Crowle and May 1981; Rook et al. 1986a). Experience shows that there is a great deal of variability built into experiments with outbred animals, yet we don't seem to take this into account when looking at data from humans. Furthermore, there is evidence that macrophages from different races of people have inherent differences in the ability to handle *M. tuberculosis* (Crowle and Elkins 1990). To overcome variation between individuals in animals, we resort to inbred animals, often kept in special environments. Even though inbred strains of mice vary in innate resistance to mycobacteria (Gros et al. 1983), single inbred strains are used to remove variation between individuals. Essentially all of the experiments with murine cells described above were done with inbred animals. Obviously we cannot do this with humans, but we should at least recognize the problem.

There also is variation at the level of the individual donor. Bermudez and Young (1990) have pointed out that CD14+/CD16+ cells have a reduced ability to respond to activating agents compared to CD14+/CD16- cells. Furthermore, they reported that heavily infected macrophages are impaired in their response to activating signals. Anyone working with intracellular pathogens and macrophages has observed that individual macrophages within a population appear to vary widely in their phagocytic activity. Maybe we should be devising strategies to look at the antitubercular activity of individual macrophages.

An obvious reason why we cannot get human cells to exert a consistent antitubercular activity in vitro is that we are using the wrong cell. Apart from one report using human alveolar macrophages (STEELE et al. 1986), all of the reports discussed in the previous section were with blood monocytes or, more usually, macrophages derived from monocytes during a period in culture. Obviously there are alternatives. It is clear that cells in different tissues vary in their ability to kill *M. tuberculosis*. Some tissues, such as the liver and spleen, can clear mycobacteria efficiently while others, such as the lung and lymph nodes, cannot (COLLINS 1990; NORTH and IZZO 1993). Perhaps we should investigate the antitubercular activity of Kupffer cells or splenic macrophages. Working with splenic macrophages is more possible, logistically, because splenectomy is a relatively common surgical procedure. Given the observations with murine peritoneal macrophages, why do we not investigate human peritoneal macrophages. Patients undergoing laparoscopy are a source of "normal" macrophages and these cells do have activity against intracellular pathogens (ISRAELSKI et al. 1990).

An observation with murine peritoneal macrophages is that elicited cells have a greater microbicidal activity and response to cytokines than resting cells (MURRAY and COHN 1980). There is evidence that human peritoneal macrophages from patients undergoing chronic peritoneal dialysis are examples of elicited cells (PETERSON et al. 1985). They have an enhanced antimicrobial activity (PETERSON et al. 1985). Maybe cells entering granuloma should be considered as elicited macrophages.

An alternative approach to using elicited or partially activated macrophages is to begin with a virgin cell population which may be more homogeneous. FLESCH and KAUFMANN followed this rationale in using murine bone marrow macrophages (FLESCH and KAUFMANN 1987, 1988, 1991), and perhaps we should try to emulate these studies with human cells. Mononuclear cells derived in culture from human bone marrow cells appear to have a homogeneous phenotype (KELLER et al. 1993).

A commonly stated explanation for the inability to activate antitubercular activity in human macrophages is that we have not used the correct combination of cytokines. This is plausible but it may not be as simple as this because, as discussed above, the sequence and timing of addition of cytokines can govern the resulting phenotype. Furthermore, it is also possible that soluble cytokines alone cannot do the job and that macrophage-lymphocyte contact is required (CHENG et al. 1988, 1993; SYPEK et al. 1993).

Culture and the conditions of culture may influence antitubercular activity. It has been noted before that conditions of culture and infection seem to favor the growth of the mycobacterium rather than the antimycobacterial activity of the macrophage (COLLINS 1990; CROWLE 1990; ELLNER 1990). Maybe we should try to tilt the balance the other way. COLLINS (1990) suggested that we should explore opsonization of mycobacteria as a solution, while ELLNER (1990) has questioned the use of plastic culture-ware. Perhaps we could also consider the amount of oxygen we provide. Within a granuloma oxygen availability is likely to be limited and the availability of oxygen can influence the development of tuberculosis

(CHANDLER et al. 1965). Furthermore, *M. tuberculosis* has reduced intracellular growth rates in human macrophages when cultured at physiological oxygen pressure compared to cultures in ambient oxygen concentration (MEYLAN et al. 1992). The ratio of macrophages to bacteria also needs to be considered, since this can influence antimycobacterial activity (BERMUDEZ and YOUNG 1990; CHAN et al. 1992).

Typically, human macrophages are allowed a period in culture of anything from 3 to 10 days before addition of activating agents and/or mycobacteria. Yet the antimicrobial activity of mononuclear phagocytes changes with time in culture (DOUVAS et al. 1985; MARTIN and EDWARDS 1993). In addition, differences in the conditions of culture may govern antimycobacterial phenotype. Variation in serum is likely to be particularly problematic here (FLESCH and KAUFMANN 1987; CROWLE and ELKINS 1990; ROOK 1990a). The use of serum-free culture (FLESCH and KAUFMANN 1987) could overcome this problem.

Finally, we need to think about the choice of strain of *M. tuberculosis* to be used in studies with human cells. Work with murine cells has shown a difference in sensitivity to macrophages among strains of *M. tuberculosis* and *M. avium*. Strains of these organisms also vary in sensitivity to killing in cell-free experiments (DOI et al. 1993; O'BRIEN et al. 1994).

There is clearly still a lot to do before we can exclude human marcophages as antitubercular cells.

3 Mechanisms of Macrophage Activity Against *Mycobacterium tuberculosis*

We do not know exact details of how macrophages kill or inhibit the replication of *M. tuberculosis*. However, there are data for and against the involvement of several antimicrobial mechanisms. We shall summarize these data, but before doing so we will make some general points.

Macrophages have a large repertoire of antimicrobial mechanisms. The antimicrobial mechanisms of macrophages often are classified into two groups on the basis of oxygen requirement: they are said to be oxygen-dependent or oxygen-independent (ANDREW et al. 1985; LOWRIE and ANDREW 1988). In spite of the recognition that macrophages have such a variety of means of killing microorganisms, there seems to be a tendency in the literature to try to define one all-powerful mechanism that explains all of the antimicrobial activities of macrophages. For practical reasons it is probably inevitable that we study each mechanism in isolation, but it should be remembered that within the cell they will interact with each other. Furthermore, in analysis of the data pertinent to a particular antimicrobial mechanism, we should consider the possibility that the overall antimicrobial system of phagocytes has a great deal of redundancy, as a means of coping with short-term changes in microbial phenotype. Thus, data which excludes a particular mechanism from being the predominant or sole killing agent, cannot exclude that mechanism from having some antimicrobial role.

The identity of the all-powerful agent changes with time and, as each new champion emerges, the evidence that supported the old contenders is pushed into the background. In the 1950s it was a series of toxic, oxygen-independent agents; then lysosomes became important. By the late 1970s, these had been discarded in favor of oxygen-dependent systems. However, by the mid-1980s, oxygen-independent systems were making a comeback. Now, without doubt, the favorite antimicrobial agents are the reactive nitrogen intermediates.

A more profitable way of looking at macrophages' antimicrobial activities is that they are not arranged as an autocracy, but as an interacting and interdependent community of agents.

There are some other general points worth making. The particular group of macrophage mechanisms that are effective will vary from microbe to microbe. Likewise, the effective mechanisms may change between types of macrophages. It also seems reasonable to suppose that mechanisms responsible for alteration in the rate of growth may be different, in quantity if not quality, from those mechanisms that kill. Thus, when trying to understand how an activated macrophage kills a particular microbe, one can draw lessons from work with other microbes and/or macrophage type, but one cannot extrapolate completely from one to another.

Understanding the mechanisms active against *M. tuberculosis* has been hindered because of a lack of suitable in vitro models. Thus, while work with other intracellular pathogens proceeded apace, evidence with *M. tuberculosis* mainly was indirect, from in vivo work and from work with cell-free systems. As outlined above, it was not until the mid-1980s that suitable in vitro models of antitubercular activity of macrophages became available. Since that time evidence has been derived from cells from two species: murine peritoneal and bone marrow macrophages and guinea pig alveolar macrophages.

As explained above for murine cells, models mainly were of tuberculostasis. For this and other reasons, we have used an alternative model, the guinea pig alveolar macrophage. We have found that by using a protocol in which guinea pigs were given three doses of *M. bovis* BCG over a 5 week period we obtained alveolar macrophages able to kill *M. tuberculosis* in vitro (O'Brien et al. 1991). The guinea pig also is a good model of human tuberculosis because the pathology of the disease in this animal mirrors that in humans, which is not true for the mouse (Smith and Wiegeshaus 1989; McMurray 1994).

Clearly, since we cannot demonstrate in vitro antitubercular activity with human cells, evidence of how they might do it is absent! However, we can draw inferences on the role of those metabolic processes that can be activated in human macrophages, in the absence of killing of tubercle bacilli.

3.1 Oxygen-Dependent Mechanisms

Phagocytosis by polymorphonuclear and mononuclear phagocytes is accompanied by a cyanide-insensitive increase in the consumption of oxygen. This is the respiratory burst (STAHELIN et al. 1956; SBARRA and KARNOVSKY 1959). During the respiratory burst there is a one-electron reduction of molecular oxygen to superoxide, two molecules of which react together to form hydrogen peroxide. Further interaction between these species can lead to the formation of other reactive forms of oxygen; hydroxyl radicals and singlet oxygen (HALLIWELL and GUTTERRIDGE 1984). The electrons are derived from NADPH and formation of superoxide is catalyzed by a multicomponent redox system, the NADPH oxidase. Information on the structure of the oxidase and its role in antimicrobial activity of phagocytes has been reviewed recently (DINAUER and ORKIN 1992; SEGAL and ABO 1993).

Each of the reactive species has been implicated in killing of microbes. Futhermore, the efficiency of hydrogen peroxide as a microbicidal agent can be dramatically increased by the presence of halide ions and peroxidase, the so-called Klebanoff system (KLEBANOFF and HAMON 1972). It also has been suggested that, in addition to the directly toxic nature of its products, the NADPH oxidase has a microbicidal activity by its involvement in modulation of the pH within the phagocytic vacuole (HENDERSON et al. 1988; SEGAL and ABO 1993).

The evidence for the antimicrobial and antitubercular activity of each species of reactive oxygen and of the Klebanoff system was comprehensively reviewed before (ANDREW et al. 1985; LOWRIE et al. 1985). Around that time, such evidence that existed was compelling that macrophages relied on products of the respiratory burst to kill *M. tuberculosis*. Now the data are less persuasive.

Broadly, the evidence for the involvement of peroxide in defense against *M. tuberculosis* was of three kinds (ANDREW et al. 1985; LOWRIE et al. 1985). There is a correlation between low virulence in the guinea pig and susceptibility to peroxide (MITCHISON et al. 1963). Secondly, peroxide-sensitive mutants of *M. tuberculosis* fared far less well in the lungs of guinea pigs than the peroxide-resistant parent, and in vaccinated animals the effect of peroxide sensitivity was far greater as immunity became effective. What was more, the ability of alveolar macrophages to undergo a respiratory burst in response to phagocytosis of *M. tuberculosis* in vitro was correlated with the immune status of the guinea pig (JACKET et al. 1981a,b). Finally, there was the observation that addition of catalase abolished the killing of *M. microti* by murine macrophages activated with a lymphocyte culture supernatant (WALKER and LOWRIE 1981).

In humans, the evidence is all circumstantial. Individuals whose macrophages are defective in a respiratory burst can develop a disseminated infection when vaccinated with *M. bovis* BCG (MACKAY et al. 1980). Furthermore, it has been suggested that peroxide-susceptible strains of *M. tuberculosis* may be less virulent in humans (TRIPATHY et al. 1969). However, the fact that macrophages can be activated to mount an enhanced respiratory burst, yet not for enhanced antitubercular activity (ROOK et al. 1986a; ROBERTSON and ANDREW 1991), does not support a role for oxygen-dependent mechanisms.

With the advent of the in vitro systems described above, efforts were made to corroborate the conclusion from in vivo observations that the respiratory burst was an important antitubercular mechanism. Evidence was sought in four areas: Was intracellular fate correlated with susceptibility to peroxide? Was the occurrence and/or the magnitude of the respiratory burst of the macrophage correlated with intracellular fate of *M. tuberculosis*? Were intracellular *M. tuberculosis* protected by scavengers of the respiratory burst products? What is the fate of *M. tuberculosis* in cells incapable of mounting a respiratory burst? These experiments, with murine and guinea pig macrophages, have provided evidence that macrophages do not require the products of the respiratory burst to kill or inhibit the replication of *M. tuberculosis* (FLESCH and KAUFMANN 1988; O'BRIEN and ANDREW 1991; O'BRIEN et al.; CHAN et al. 1992).

Peroxide-susceptible strains of *M. tuberculosis* were not killed by guinea pig macrophages more readily than peroxide-resistant strains (LOWRIE et al. 1985; O'BRIEN and ANDREW 1991). An explanation for this result could have been that the peroxide-susceptible strains happened to be less effective in triggering a respiratory burst. However, in a survey of eight strains of *M. tuberculosis* covering a range of peroxide susceptibilities, there was no correlation between killing by macrophages and triggering of the respiratory burst (O'BRIEN and ANDREW 1991). All strains were equally effective in stimulating a respiratory burst by normal and activated guinea pig alveolar macrophages.

It has been suggested (SCHLESINGER et al. 1990; CHAN and KAUFMANN 1994) that *M. tuberculosis* may avoid triggering the respiratory burst because they enter cells via complement receptors, a pathway which does not trigger a respiratory burst in nonactivated human macrophages (WRIGHT and SILVERSTEIN 1983). However, it is clear that under the correct circumstances tubercle bacilli can trigger a respiratory burst even in nonactivated macrophages (O'BRIEN and ANDREW 1991).

One explanation may be that CR3 receptors on normal guinea pig alveolar macrophages are in a different activation state from those on resting cells from other species. Alternatively, for activated and nonactivated guinea pig macrophages, uptake of tubercle bacilli is more dependent on mannose receptors (SCHLESINGER 1993) which do mediate a respiratory burst (BERTON and GORDON 1983). SCHLESINGER (1993) reported that adherence to human monocytes of virulent but not avirulent strains of *M. tuberculosis* is mediated by mannose receptors. Interestingly, we found that association of *M. tuberculosis* with activated guinea pig macrophages inversely correlated with virulence (O'BRIEN and ANDREW 1991). No such correlation was found with normal macrophages. Either way, the mechanism of adherence of tubercle bacilli to guinea pig macrophages and how these receptors change with activation seems worthy of further study.

The effects of scavengers of toxic oxygen products on the intracellular fate of tubercle bacilli in activated macrophages has been studied in three different models (Flesch and Kaufmann 1988; O'Brien and Andrew 1991; Chan et al. 1992) but in no case did a scavenger influence the fate of the bacteria.

Evidence against a role for the respiratory burst also has come from cells that cannot mount a respiratory burst. Chan and colleagues (1992) reported that a clone of the murine macrophage cell line J774 that was unable to produce a respiratory burst was as capable of killing *M. tuberculosis* as the respiratory burst-sufficient parental clone. We also found that IFN-γ treated murine fibroblasts could inhibit the replication of *M. tuberculosis*, even though they did not produce a detectable respiratory burst (S. O'Brien, L. O'Brien, B. Roberts and P.W. Andrew, unpublished data).

Thus the weight of in vitro evidence indicates that the respiratory burst products are not essential for antitubercular activity of activated macrophages. Nevertheless, the data do not exclude the oxygen-dependent mechanisms having a role; they merely exclude an exclusive role. Furthermore, we still have to explain how the pre-1985 data fit in with this conclusion, especially the strong correlation between virulence and peroxide suceptibility (Mitchison et al. 1963). One explanation is that in vivo the peroxide is not derived from macrophages. A second is that peroxide susceptibility is only a manifestation of a wider susceptibility. We will return to this point later. Finally, it is worth noting that virulence is determined by infecting nonvaccinated animals (Mitchison et al. 1963) and the state of activation that macrophages reach in these animals may be far different to that in the BCG-vaccinated animals. Peroxide may still have a central role for macrophages in a nonvaccinated host.

How can we explain the difference between the observations of Walker and Lowrie (1981) and Chan and colleagues (1992) on the role of oxygen-dependent killing mechanisms? One explanation is that the cocktail of mediators in the cytokine soup of Walker and Lowrie produced a different macrophage phenotype from that produced by IFN-γ plus LPS or TNF-α (Chan et al. 1992). A simpler explanation is that catalase not only scavengers hydrogen peroxide but influences the availability of another macrophage product – nitric oxide (Li et al. 1992).

3.2 Oxygen-Independent Mechanisms

The data described in the previous section, while failing to support a role for oxygen-dependent systems, simultaneously support the existence of oxygen-independent antitubercular mechanisms. It is now clear that oxygen-independent systems have a role in the antitubercular activity of macrophages. What is not clear is which oxygen-independent mechanisms are effective against *M. tuberculosis* in macrophages.

The list of substances making up the oxygen-independent antimicrobial mechanisms is far more extensive than those forming the oxygen-dependent systems. Various enzymes, peptides, organic acids and lipids present in macrophages and other cells can kill *M. tuberculosis*. Some of these have been described for some time, yet their role in antitubercular activity is still a mystery. For example, DUBOS had described how organic acids are lethal to tubercle bacilli in 1950 (DUBOS 1950). Among these was lactic acid, a metabolic product of macrophages (YOUNG and ZYGAS 1986).

We have reviewed the role of individual oxygen-independent mechanisms in the anti-tubercular activity of macrophages before (ANDREW et al. 1985; LOWRIE and ANDREW 1988). Since some have not been the subject of active research since that time, we will concentrate only on those areas where there have been some recent developments.

3.2.1 Lysosomal Mechanisms

In 1971, ARMSTRONG and HART observed (ARMSTRONG and HART 1971) that macrophage phagosomes containing living *M. tuberculosis* were inhibited in fusing the lysosomes, whereas fusion with phagosomes containing dead *M. tuberculosis* was not inhibited. More recent observations suggest that the interaction of phagosomes containing mycobacteria with lysosomes may be even more complex (MCDONOUGH et al. 1993; STURGILL-KOSZYCKI et al. 1994). However, it is not our intention to review the literature on this topic or on the possible mediators of inhibition of phagosome-lysosome fusion. Both are reviewed elsewhere in this volume.

The observation of ARMSTONG and HART (1971) concentrated attention on lysosomes as a potential site of antitubercular material. Macrophages possess cytoplasmic granules (lysosomes) containing a large variety of hydrolytic enzymes (ANDREW et al. 1985). Although the majority of these enzymes have a role in digestion rather than killing of microorganisms, lysosomes do contain material capable of killing microorganisms (ANDREW et al. 1985)

It has been known for some time that lysosomes contain antimycobacterial substances. These substances have been isolated from the lysosomes of spleens and lungs of tuberculous mice (KANAI and KONDO 1968, 1969). The granule fractions of the lungs of *M. bovis* BCG-vaccinated mice had more activity against *M. tuberculosis* than those from normal mice (KANAI and KONDO 1970). Part of this activity was found in a water-soluble peptide fraction (KANAI and KONDO 1970).

Material capable of killing *M. tuberculosis*, but not rapidly growing mycobacteria, was isolated from extracts of guinea pig glycogen-elicited peritoneal cells (KOTANI et al. 1962). The activity was very pH-dependent, with an optimum at pH 6.2. However, there was no evidence that antimycobacterial activity was increased by vaccination with *M. bovis* BCG. Later, SHARMA and MIDDLEBROOK (1977) described the partial purification of a water-soluble antimicrobial substance produced by caseinate-elicited peritioneal cells from *M. bovis* BCG-vaccinated guinea pig provided they had been incubated with PPD.

Recently, we found that the lysosomes from guinea pig alveolar macrophages contained acid-extractable material that killed *M. tuberculosis* (L. O'Brien, S. O'Brien and P.W. Andrew, unpublished data). This activity was very pH dependent, being active at pH 5.5 and inactive at pH 7.0. More interestingly, the activity was found only in lysosomes from macrophages that had been activated to kill *M. tuberculosis* by *M. bovis* BCG vaccination of the guinea pigs (L. O'Brien, S. O'Brien and P.W. Andrew, unpublished data). No activity was found in lysosomes of macrophages from nonimmune animals.

3.2.2 Antimicrobial Proteins and Peptides

The identity of the material responsible for the activities described above is still a matter for speculation. It is possible that they are part of a group of proteins and peptides which are attracting a growing amount of interest. They are often described as peptide antibiotics (Spitznagel 1990; Selsted et al. 1992; Lehrer et al. 1991, 1993). These proteins and peptides exhibit an antimicrobial activity that is not dependent on enzymatic activity. They have been described as part of the antimicrobial system of neutrophils, macrophages and intestinal epithelium (Spitznagel 1990; Selsted et al. 1992; Lehrer et al. 1991,1993).

The lysosomes of human neutrophils contain seven of these proteins (Spitznagel 1990): four human neutrophil peptides (HNP 1–4), also known as defensins; cathepsin G; a 37 kDa cationic antimicrobial protein (CAP37) also known as azurocidin; and bactericidal permeability increasing protein (BPI), also known as CAP57. These molecules are cationic in nature due to the relatively high proportion of lysine and/or arginine (Spitznagel 1990). In addition, three distinct peptides have been isolated from the granules of bovine neutrophils. These are the proline-and arginine-rich bactenecins (Frank et al. 1990), the tryptophan-rich indolicin (Delsal et al. 1992) and a cyclic dodecapeptide (Storici et al. 1992).

The antimycobacterial activity of these molecules is largely unreported but the defensins are the most promising candidates. These are active against a broad range of microorganisms, whereas only activity against gram-negative bacteria has been assigned to the others (Spitznagel 1990). Although their activity against *M. tuberculosis* is unknown (Lehrer et al. 1993), defensins are active against *M. avium* (Ogata et al. 1992) and *M. fortuitum* (Lehrer et al. 1993). The defensins HNP 1,2 and 3, from human neutrophils, are equally active in killing *M. avium* (Ogata et al. 1992). However, we should also test the activity of the antibiotic peptides to synergize with other agents. For example, CAP37 was shown to enhance the cidal activity of elastase or cathepsin G against the oral bacterium, *Capnocytophaga sputigena*, even though it was apparently not active alone (Miyasaki and Bodeau 1992).

The significance of the peptide antibiotics in macrophages is unresolved. The observations of Silva and colleagues (1989) suggests that in vivo, macrophages may obtain the peptides by phagocytosis of neutrophils. However, macrophages also may synthesise these molecules. Two defensins have been isolated from rabbit alveolar macrophages (Patterson-Delafield et al. 1980). These cells also contain cathepsin G (Andrew et al. 1983). To date, these represent the best described examples in macrophages. However, the antimicrobial agent found in guinea pig inflammatroy peritoneal cells was a cationic protein which absorbed to the surface of target bacteria (SHARMA and MIDDLEBROOK 1977). More recently, the isolation of three lysine-rich cationic proteins from the lysosomes of murine peritoneal macrophages was described (HIEMSTRA et al. 1993). The proteins were present in resting and IFN-γ-activated cells. They had a broad spectrum of antibacterial activity, including against *M. fortuitum*. The authors also reported that there was evidence for additional microbicidal proteins, including one whose synthesis may have been induced by IFN-γ. To date, defensins have not been found in human monocytes (OGATA et al. 1992).

3.2.3 Iron Sequestration

Most bacteria need iron for growth and it is known that iron availability can determine the rate of growth. Therefore, it is reasonable to speculate that it will determine the intracellular rate of growth of *M. tuberculosis*. Sadly, evidence with mycobacteria in this area is scanty. Most information on the role of iron in behavior of intracellular pathogens has been obtained with *Legionella pneumophila*. Sequestration of iron using chelators such as lactoferrin or transferrin will inhibit intracellular and extracellular growth of *L. pneumophila* (BYRD and HORWITZ 1989, 1991, 1993). Conversely, high iron can promote the intracellular growth of the intracellular pathogens *L. pneumophila* (BYRD and HORWITZ 1989) and *Listeria monocytogenes* (ALFORD et al. 1991). However, iron availability is not a straightforward story, since in some situations iron can enhance the antimicrobial activity of macrophages, possibly by enhancing oxygen-dependent killing (JIANG and BALDWIN 1993).

Macrophages may use neutrophil-derived lactoferrin to decrease levels of intracellular iron. It has been reported that during chronic mycobacterial infections in the mouse, macrophages ingest neutrophils and thus acquire their lactoferrin (SILVA et al. 1989). The availability of iron inside macrophages also may be decreased by reducing the uptake of iron chelated to serum transferrin. It has been suggested that inhibition of growth of *L. pneumophila* by IFN-γ-activated human monocytes is due to a decrease in the numbers of transferrin receptors (BYRD and HORWITZ 1989).

A similar mechanism may be effective against mycobacteria. It is known that transferrin can be bacteriostatic for *M. tuberculosis*, as can human serum unless iron is added (KOCHAN et al. 1963; KOCHAN 1973). ROOK and colleagues (1986a) reported that, when cultured in 20% human serum, the rate of growth of *M. tuberculosis* and *M. bovis* BCG inside normal human monocytes was enhanced when iron was added as ferric ammonium citrate. Human serum also is inhibitory for *M. avium* and this is due to the presence of transferrin (DOUVAS et al. 1993). Addition of ferrous iron enhanced growth of *M. avium* inside human macrophages. In contrast, addition of apotransferrin inhibited intracellular replication of *M. avium*. However, the influence of an iron chelator depends on the degree of its iron saturation (BYRD and HORWITZ 1991; DOUVAS et al. 1993). Holotransferrin at 500 μg/ml enhanced growth of *M. avium* inside human macrophages (Douvas et al. 1993).

3.2.4 Metabolism of Amino Acids by Macrophages

At various times, the metabolism of three amino acids has been proposed to have a role in antimicrobial activity. From the 1950s, metabolites of ornithine have been proposed as important. In the 1980s, degradation of tryptophan was highly topical. At the moment however, there is no doubt that it is the metabolism of arginine that is attracting most attention.

3.2.4.1 Metabolism of Ornithine – A Source of Polyamines

Metabolism of ornithine results in formation of antimycobacterial agents first identified over 40 years ago but essentially forgotten by mycobacteriologists since then.

Ornithine is the starting material for the biosynthesis of the polyamines spermine and spermidine (Morgan 1987). These polyamines are converted, in an oxidative deamination reaction catalyzed by polyamine oxidase, to an aminoaldehyde, ammonia and hydrogen peroxide (Morgan 1987). In 1952, Hirsch and Dubos (Hirsch and Dubos 1952) identified spermine as the tuberculostatic agent that they had previously extracted from animal tissues using acidic water and ethanol (Dubos 1951). Later, Hirsch (1953) showed that an enzyme present in various tissues and sera was required for spermine to inhibit the replication of *M. tuberculosis*. It is now known that this enzyme is polyamine oxidase (Morgan 1987).

The work of Hirsch was the first demonstration of the antimicrobial activity of spermine and polyamine oxidase. However, he did not exclude the possibility that the tuberculostasis was due to the hydrogen peroxide and his experiments could not distinguish stasis from killing. Since that time polyamines seem to have been largely forgotten by mycobacteriologists. In the meantime, however, they have attracted a great deal of attention from workers in other fields. It is now known that products of spermine and spermidine oxidation are toxic to viruses (Bachrach et al. 1965), bacteria (Bachrach and Persky 1964) and protozoa (Ferranate et al. 1986). Although also cytotoxic to mammalian cells (Morgan 1987), it has been shown that oxidation of polyamines will kill intracellular microorganisms without host cell death (Morgan et al. 1986). Furthermore, it has been shown that the antimicrobial activity is not dependent on hydrogen peroxide and that the aminoaldehyde product is the main antimicrobial product (Bachrach and Persky 1964; Ferrante et al. 1986).

In addition to a direct antimicrobial activity, spermine and spermidine play an important role in the initiation and regulation of cell proliferation, cell differentiation and the activation of differentiated cells (Heby 1986), including macrophages (Messina et al. 1992). A role in apoptosis also has been suggested (Chayen et al. 1990).

We have recently looked anew at the affect of polyamine oxidation products on mycobacteria (B. Roberts and P.W. Andrew, unpublished data). We found that both spermine and spermidine, in the presence of polyamine oxidase, were toxic to *M. tuberculosis*. However, we have gone further than Hirsch and shown that both of these reagents kill *M. tuberculosis*, rather than inhibiting its replication. We also showed that this toxicity was not dependent on the production of hydrogen peroxide. Furthermore, we found that resistance of strains of *M. tuberculosis* to spermine or spermidine plus polyamine oxidase tended to reflect their virulence in the guinea pig.

The first step in the formation of spermine and spermidine is the conversion of ornithine to putrescine, catalyzed by ornithine decarboxylase (ODC) (MORGAN et al. 1986). This is the rate limiting step in the formation of spermine and spermidine (MORGAN et al. 1986). Inhibition of ODC blocks the formation of spermine and spermidine and it will block cell activities, such as apoptosis, cell proliferation and macrophage activation (HEBY 1986; CHAYEN et al. 1990; MESSINA et al. 1992). ODC activity is enhanced in macrophages or macrophage cell lines treated with LPS or mycobacterial cell walls (NICHOLS and PROSSER 1980; TAFFET and HADDOX 1985). Treatment of the human monocyte cell line U937 with IFN-γ has been shown to increase ODC activity (L. O'Brien and P.W. Andrew, unpublished data) and the accumulation of ODC mRNA (MESSINA et al. 1990).

3.2.4.2 Degradation of Tryptophan – A Mechanism of Nutrient Depletion

The enzyme indoleamine 2,3-dioxygenase (IDO) is the first enzyme in the conversion of tryptophan to *N*-formylkynurenine (BYRNE 1987). The induction of this enzyme in human cells by IFN-γ is well documented (YOSHIDA et al. 1981; HYAISHI 1985). Several groups have shown that these events are associated with the inhibition of replication of several intracellular pathogens such as *Toxoplasma gondii* (PFEFFERKORN 1984), *Chlamydia psittaci* (BYRNE et al. 1986) and *Leishmania donovani* (MURRAY et al. 1989). This antimicrobial effect is not due to the production of toxic products but rather to depletion of an essential nutrient. Supplementing the culture medium with excess tryptophan reversed the inhibition of microbial growth (PFEFFERKORN 1984; BYRNE et al. 1986; MURRAY et al. 1989).

There is a second way that this pathway may influence macrophage antimicrobial function: it is a pathway of autocrine production. The final product of tryptophan metabolism is picolinic acid. Picolinic acid synergises with IFN-γ in the activation of macrophages (VARESIO et al. 1990; MELILLO et al. 1993).

Although this system has attracted a good deal of attention, it is by no means a universal antimicrobial system. Degradation of tryptophan appears not to be the explanation of IFN-γ-induced inhibition of *Rickettsia prowazeki* in human and murine cells (TURCO and WINKLER 1986) or *Chlamydia trachomatis* in murine cells (MAZA et al. 1985). Indeed there are conflicting data on whether an inducible indoleamine 2,3-dioxygenase (IDO) exists in the mouse: compare (MAZA et al. 1985; TURCO and WINKLER 1986; BYRNE 1987; MURRAY et al. 1989) and (YOSHIDA et al. 1981).

There are no published data on the role of tryptophan degradation in the antimycobacterial activity of macrophages. Therefore, as a long-shot, we investigated the possibility that it may have a role. Sadly, we have been disappointed! We found that addition of excess tryptophan did not reverse the killing of *M. tuberculosis* by guinea pig alveolar macrophages (B. Roberts and P.W. Andrew, unpublished data). Perhaps this is not surprising given that tryptophan is not an essential amino acid for mycobacteria and mycobacteria do not appear to use tryptophan as a sole carbon source (Grange 1976). However, in one in vitro situation it has been shown that tryptophan may be advantageous to *M. tuberculosis*. In a synthetic medium, when tryptophan was included in a mixture with other amino acids, the rate of growth was greater than with any mixture not including this amino acid (Sundaram and Venkitasubramanian 1978).

3.2.4.3 Arginine – A source of Reactive Nitrogen Oxides

Circumstatial evidence that mammals produced nitric oxide (·NO) has existed since 1916 (Mitchell et al. 1916), yet it was not until 1987 that it was shown that mammalian cells could synthesize. ·NO (Palmer et al. 1987). Subsequently, there has been an explosion in the number of publications on this topic (Knowles and Moncada 1994). ·NO is produced during the conversion of L-arginine to citrulline in mammalian cells. The reaction is catalyzed by the enzyme nitric oxide synthase (NOS). The mechanism of formation of ·NO, the nature of the enzymes and the biological consequences of ·NO formation currently are amongst the hottest issues in biology and have been the subject of several reviews (Moncada 1992; Stuehr and Griffith 1992; Marletta 1993; Prince and Gunson 1993; Knowles and Moncada 1994).

In mammals, ·NO is synthesized by a wide range of tissues and cells (Marletta 1993; Knowles and Moncada 1994). Synthesis was first demonstrated with cells from the central nervous system, vascular endothelium and macrophages (Knowles and Moncada 1994). cDNA cloning and functional experiments have shown that the nitric oxide synthases from these three sources are structurally and functionally distinct (Knowles and Moncada 1994). One of the features of the macrophage's enzyme, that distinguishes it from the other NOSs, is that its synthesis is induced following exposure of the macrophage to activating agents (Moncada 1992), whereas in other tissues the enzyme is constitutive. Thus the macrophage enzyme is often refered to as iNOS.

In 1987, Hibbs and colleagues suggested that ·NO formed from arginine was responsible for the antitumor activity of murine macrophages (Hibbs et al. 1987). Since then, ·NO and related nitrogen oxides (reactive nitrogen intermediates RNI), have been implicated in the antimicrobial activity of macrophages against many intracellular pathogens (see references in Murray and Teitelbaum 1992; Schneemann et al. 1993). Indeed, the casual observer could be forgiven for concluding that it is now the only antimicrobial mechanism!

A number of studies have concluded that RNIs are responsible for the activity of murine macrophages against *M. tuberculosis* (Denis 1991b; Chan et al.

1992), *M. bovis* BCG (FLESCH and KAUFMANN 1991), *M. leprae* (ADAMS et al. 1991) and *M. avium* (DOI et al. 1993). From these studies the evidence is that mycobacteria are sensitive to killing by RNI formed in acidic nitrite solution; RNI production by murine macrophages correlates with antimycobacterial activity; inhibition of RNI synthesis blocks antimycobacterial activity. Further, circumstantial evidence for a role for RNI came from the observation that in vivo production of RNI was greatly reduced in IFN-γ knock-out mice undergoing an unrestricted infection with *M. tuberculosis* (FLYNN et al. 1993).

An interesting by-product of these studies was that they called into question the validity of some of the conclusions drawn with scavengers of ROIs. Conclusions drawn from experiments with superoxide dismutase (SOD) may be difficult to interpret given that SOD may increase the amount of ·NO macrophages produce (CHAN et al. 1992). Likewise, catalase may not only affect the availability of hydrogen peroxide since it has been reported to inhibit nitric oxide synthesis also (LI et al. 1992). It has been suggested (CHAN and KAUFMANN 1994) that this may explain the effect of catalase seen by WALKER and LOWRIE (1981). However, in considering macrophage activity against intracellular pathogens in general, it seems too simplistic to dismiss evidence for ROI as an artifact. There are other ways we can reconcile the earlier evidence for an involvement of ROI with the newer evidence for RNI.

As discussed above, the superoxide-generating NADPH oxidase may be a mechanism for lowering the intraphagosomal pH. This would favor the conversion of nitrite to ·NO (Taylor et al. 1927). A second explanation is that the most potent antimicrobial products are formed from the interaction of ROI and ·NO. Interaction of superoxide anion and ·NO results in the formation of peroxynitrite anion ($OONO^-$) which decomposes in acidic conditions to form hydroxyl radical (·OH) and nitrogen dioxide radical ($\cdot NO_2$) (HALLIWELL and GUTTERIDGE 1992). Peroxynitrite is also formed by reaction of hydrogen peroxide and ·NO (CARMICHAEL et al. 1993; NORONHA-DUTRA et al. 1993). This reaction also generates singlet oxygen (NORONHA-DUTRA et al. 1993). All of these products are powerful oxidants. The antimicrobial activity of singlet oxygen and hydroxyl radical is well known (ANDREW et al. 1985) but peroxynitrite also is a potent antimicrobial agent (ZHU et al. 1992; DENICOLA et al. 1993), although apparently not to all microorganisms (ASSREUY et al. 1994). It has not been confirmed if peroxynitrite is cidal to *M. tuberculosis*, but exposure to SIN-1, a model for the simultaneous generation of superoxide and ·NO (HOGG et al. 1992), is cidal for *M. tuberculosis* (L. O'Brien and P.W. Andrew, unpublished data). We are currently investigating the active species in this system.

DOI and colleagues (1993) suggested that the differences in the susceptibilities of strains of *M. tuberculosis* and *M. avium* to the antimycobacterial effects of IFN-γ- treated murine macrophages (FLESCH and KAUFFMAN 1987; DOI et al. 1993) was due to differences in susceptibility to RNI. They went on to show that this was true for *M. avium* (DOI et al. 1993). It now appears that it is true also for *M. tuberculosis*. We have shown that strains of *M. tuberculosis* differ in susceptibility and furthermore that resistance correlates with virulence in the guinea pig (O'BRIEN et al. 1994). However, it is not clear how this relates to the antimycobacterial mechanisms of guinea pig macrophages, because we have been unable to detect nitrite production from guinea pig alveolar macrophages (L. O'BRIEN and P.W. ANDREW, unpublished data). However, because nitrite production is not a totally reliable measure of NOS activity (KNOWLES and MONCADA 1994), we are currently investigating citrulline production from these cells. It now appears that virulence of *M. tuberculosis* in the guinea pig correlates with susceptibility to three antimicrobial agents: hydrogen peroxide, aminoaldehydes and RNI. This could be evidence for the requirement of three distinct antimicrobial systems for full tuberculocidal activity. Alternatively, however, it may indicate a common target for each system (DNA?) which differs in sensitivity between strains and/or a common method of microbial resistance which differs in efficacy between strains; recA is a possible candidate here (CARLSSON and CARPENTER 1980). This possibility is doubly intriguing given the unusual nature of the recA of *M. tuberculosis* (DAVIS et al. 1994).

While there is a plethora of reports implicating RNI in the antimicrobial mechanism of murine macrophages, the role of RNI in the antimicrobial activity of human macrophages is much more controversial. There are some reports that human mononuclear phagocytes produce small amounts of ·NO (HUNT and GOLDING 1992; MARTIN and EDWARDS 1993).

Immunohistochemical data suggested that macrophages in human breast carcinoma tissue express iNOS and the tissue was shown to produce ·NO (THOMSEN et al. 1993). WEINBERG and colleagues (1993) used RT-PCR to investigate iNOS in human macrophages. They found that iNOS mRNA was inducible but found no evidence that these cells produced ·NO.

The increased generation of nitrate in response to injury and infection (OCHOA et al. 1991) or cytokine therapy (HIBBS et al. 1992) also was supportive of an iNOS in humans. However, recently genes for iNOS have been cloned from human hepatocytes (GELLER et al. 1993) and chondrocytes (CHARLES et al. 1993) and these could have been the source of nitrate.

There is a report that TNF-α- and granulocyte/macrophage colony stimulating factor (GM-CSF)-activated human macrophages produce RNI and that the activity of these cells against *M. avium* was dependent on RNI (DENIS 1991d). However, in complete contrast, Bermudez reported (BERMUDEZ 1993) that human macrophages activated with TNF-α, GM-CSF or IFN-γ did not produce RNI and that RNI was not required for the anti-*M. avium* activity of these cells. The reasons for the discrepancy between these reports is not obvious. However, several other investigators working with other pathogens have concluded that activated human macrophages do not produce RNI and iNOS is not part of the antimicrobial mechanisms of human macrophages (JAMES et al 1992; MURRAY and TEITELBAUM 1992; PADGETT and PRUETT 1992; SCHNEEMANN et al. 1993).

It has been suggested that the failure to detect RNI production from human macrophages was due to a failure to produce the essential cofactor tetrahydrobiopterin (CAMERON et al. 1990) but this is not the whole answer (SAKAI and MILSTEIN 1993; WEINBERG et al. 1993). Another suggestion is that it is due to a failure to use the correct mixture of activating agents (CHAN and KAUFMANN 1994). Indeed, unpublished data suggest that it is possible to construct a cocktail that will activate human macrophages to produce RNI (FY Liew, personal communication). The publication of the details of this experiment are eagerly anticipated. Maybe the conditions for activation for RNI production will also activate for killing of *M. tuberculosis*! However, it seems likely that skepticism about the role of RNI as a general antimicrobial mechanism of human macrophages will remain, given that specialized conditions of activation for RNI synthesis are not required for activation of activity against many pathogens (MURRAY and TEITELBAUM 1992).

4 Concluding Remarks

To conclude, we will restate some of the points we have made before. Much more work must be done with human mononuclear phagocytes to investigate if they are capable of antitubercular activity. Ultimately we must not be afraid to throw out the dogma on these cells, but at the moment it is too early for this action. We do not believe that one antimicrobial system will be shown to be exclusively effective against *M. tuberculosis* in all mononuclear phagocytes. We should consider that the phagocytes will have more than one means of killing a particular microorganism and that these systems will interact with each other. Finally, it is worth remembering that individual killing systems will have specific conditions for optimum operation. Systems that are effective in test tubes may not function intracellularly.

References

Abe E, Miyaura C, Sakagami H, Takeda M, Konno K, Yamazaki T, Yoshiki S, Suda T (1981) Differentiation of mouse myeloid leukemic cells induced by 1,25 dihydroxy vitamin D3. Proc Natl Acad Sci 78: 4990–4994

Adams LB, Franzblau SG, Vavrin Z, Hibbs JB, Krahenbuhl JL (1991) L-arginine-dependent macrophage effector functions inhibit metabolic activity of *Mycobacterium leprae*. J Immun 147: 1642–1646

Alford CE, King TE, Campbell PA (1991) Role of transferrin, transferrin receptors and iron in macrophage listericidal activity. J Exp Med 174: 459–466

Altes C, Steele J, Stanford JL, Rook GAW (1985) The effect of lymphokines on the ability of macrophages to protect mycobacteria from a bactericidal antibiotic. Tubercle 66: 261–266

Amento EP, Bhalla AK, Kurnick JT, Kradin RL, Clemens TL, Holick SA, Holick MF, Krane SM (1984) 1,25 dihydroxyvitamin D3 induces maturation of the human monocyte cell line U937 and in association with a factor from human T lymphocytes augments production of the monokine, mononuclear cell factor. J Clin Invest 73: 731–739

Andrew PW, Lowrie DB (1987) Role of interferon in immunity to mycobacteria. In: Byrne GI, Turco J (eds) Interferon and nonviral pathogents. Dekker,New York, pp 263–286

Andrew PW, Hughes KT, Davies M, Peters TJ (1983) Subcellular distribution of the neutral proteinases of rabbit alveolar macrophages. Cell Mol Biol 29: 315–321

Andrew PW, Jackett PS, Lowrie DB (1985) Killing and degradation of microorganisms by macrophages. In: Dean RT, Jessup W (eds) Mononuclear phagocytes: physiology and pathology. Elsevier, Amsterdam, pp 311–334

Armstrong JA, Hart PD (1971) Response of cultured macrophages to *Mycobacterium tuberculosis*, with observations on fusion of lysosomes with phagosomes. J Exp Med 134: 713–740

Assreuy J, Cunha FQ, Epperlein M, Noronha-Durta A, O'Donnell CA, Liew FY, Moncada S (1994) Production of nitric oxide and superoxide by activated macrophages and killing of *Leishmania major*. Eur J Immunol 24: 672–676

Bachrach U, Persky S (1964) Antibacterial action of oxidised spermine. J Gen Microbiol 37: 195–204

Bachrach U, Rabina S, Loebenstein G, Eilon G (1965) Antiviral action of oxidised spermine-inactivation of plant viruses. Nature 208: 1095–1096

Barnes PF, Chatterjee D, Abrams JS, Lu SH, Wang E, Yamamura M, Brennan PJ, Modlin RL (1992) Cytokine production induced by *Mycobacterium tuberculosis* lipoarabinomannan. Relationship to chemical structure. J Immunol 149: 541–547

Bermudez LE (1993) Differential mechanisms of intracellular killing of *Mycobacterium avium* and *Listeria monocytogenes* by activated human and murine macrophages. The role of nitric oxide. Clin Exp Immunol 91: 277–281

Bermudez LE, Champsi J (1993) Infection with *Mycobacterium avium* induces production of interleukin-10 (IL-10), and administration of anti-IL-10 antibody is associated with enhanced resistance to infection in mice. Infect Immun 61: 3093–3097

Bermudez LE, Young LS (1988) Recombinant tumor necrosis factor alone or in combination with interleukin-2 but not gamma-interferon is associated with killing of *M. avium* complex in AIDS patients. J Immunol 140: 3006–3013

Bermudez LE, Young LS (1990) Killing of *Mycobacterium avium*: insights provided by the use of recombinant cytokines. Res Microbiol Inst Pasteur 141: 241–243.

Bermudez LE, Young LS, Gupta S (1990) 1,25 dihydroxyvitamin D3-dependent inhibition of growth or killing of *Mycobacterium avium* complex in human macrophages is mediated by TNF and GM-CSF. Cell Immunol 127: 241–243

Berton G, Gordon S (1983) Modulation of macrophage mannosyl specific receptors by cultivation on immobilized zymosan – effects of superoxide anion release and phagocytosis. Immunology 49: 705–715

Brown IN (1983) Animal models and immune mechanisms in mycobacterial infection. In: Stanford J, Ratledge C (eds) Biology of mycobacteria. Academic, London, pp 311–334

Byrd T, Horwitz MA (1989) Interferon gamma-activated human monocytes down-regulate transferrin receptors and inhibit the intracellular multiplication of *Legionella pneumophila* by limiting the availability of iron. J Clin Invest 83: 1457–1465

Byrd T, Horwitz MA (1991) Lactoferrin inhibits or protects *Legionella* intracellular multiplication in non-activated and interferon gamma-activated human monocytes depending upon its degree of iron saturation. J Clin Invest 88: 1103–1112

Byrd T, Horwitz MA (1993) Regulation of transferrin receptor expression and ferritin content in human mononuclear phagocytes. Coordinate upregulation by iron transferrin and downregulation by interferon gamma. J Clin Invest 91: 969–976

Byrne GI (1987) Interferons, immunity and chlamydiae. In: Byrne GI (eds) Interferon and nonviral pathogens. Dekker, New York, pp 263–286

Byrne GI, Lehmann LK, Landry GJ (1986) Induction of tryptophan catabolism is the mechanism for gamma-interferon–mediated inhibition of intracellular *Chlamydia psittaci* replication in T24 cells. Infect Immun 53: 347–351

Cahall DL, Youmans GP (1975) Molecular weight and other characteristics of mycobacterial growth inhibitory factor produced by spleen cells obtained from mice immunised with viable attenuated mycobacterial cells. Infect Immun 12: 841–850

Cameron ML, Granger DL, Weinberg JB, Kozumbo WJ, Koren HS (1990) Human alveolar and peritoneal macrophages mediate fungistasis independently of L-arginine oxidation to nitrite or nitrate. Am J Respir Dis 142: 1313–1319

Carlsson J, Carpenter VS (1980) The recA+ gene product is more important than catalase and superoxide dismutase in protecting Escherichia coli against hydrogen peroxide. J Bacteriol 142: 319–321

Carmichael AJ, Steel-Goodwin L, Gray B, Arroyo CM (1993) Nitric oxide interaction with lactoferrin and its production by macrophage cells studied by EPR and spin trapping. Free Radical Res Commun 19: S201–S209

Chan J, Kaufmann SHE (1994) Immune mechanisms of protection In: Bloom BR (ed) Tuberculosis: pathogenesis, protection and control ASM, Washington, pp 389–415

Chan J, Xing Y, Magliozzo RS, Bloom BR (1992) Killing of virulent *Mycobacterium tuberculosis* by reactive nitrogen intermediates produced by activated murine macrophages. J Exp Med 175: 1111–1122

Chandler PJ, Allison KJ, Margolis G, Gerszten E (1965) The effects of intermittent hyperbaric oxygen therapy on the development of tuberculosis in rabbits. Am Rev Respir Dis 91: 855–860

Charles IG, Palmer RMJ, Hickery MS, Bayliss MT, Chubb AP, Hall VS, Moss DW, Moncada S (1993) Cloning, characterization and expression of a cDNA-encoding an inducible nitric oxide synthase from human chrondrocytes. Proc Natl Acad Sci USA 90: 11419–11423

Chayen J, Pitsillides AA, Bitensky L, Muir IH, Taylor PM, Askonas BA (1990) T-cell-mediated cytolysis: evidence for target-cell suicide. J Exp Pathol 71: 197–208

Cheng SH, Walker L, Poole J, Aber VR, Walker KB, Mitchison DA, Lowrie DB (1988) Demonstration of increased antimycobacterial activity in peripheral blood monocytes after BCG vaccination in British school children. Clin Exp Immun 74: 20–25

Cheng SH, Walker KB, Lowrie DB,Mitchison DA, Swamy R, Datta M, Prabhaker R (1993) Monocyte antimycobacterial activity before and after *Mycobacterium bovis* BCG vaccination in Chingleput, India and London, United Kingdom. Infect Immun 61: 4501–4503

Collins FM (1990) In vivo vs in vitro killing of virulent *Mycobacterium tuberculosis*. Res Microbiol Inst Pasteur 141: 191–270

Cooper AM, Dalton DK, Stewart TA, Griffin JP, Russell DG, Orme IM (1993) Disseminated tuberculosis in interferon -γ- gene-disrupted mice. J Exp Med 178: 2243–2247

Crowle AJ (1990) Intracellular killing of mycobacteria. Res Microbiol Inst Pasteur 141: 231–236

Crowle AJ, Elkins N (1990) Relative permissiveness of macrophages from black and white people for virulent tubercle bacilli. Infect Immun 58: 632–638

Crowle AJ, May M (1981) Preliminary demonstration of human tuberculoimmunity in vitro. Infect Immun 31: 453–464

Crowle AJ, Ross EJ (1989) Inhibition by retinoic acid of multiplication of virulent tubercle bacilli in cultured human macrophages. Infect Immun 57: 840–844

Crowle AJ, Ross EJ, May M (1987) Inhibition by 1,25(OH)2-vitamin D3 of the multiplication of virulent tubercle bacilli in cultured human macrophages. Infect Immun 55: 2945–2950

D'Andrea A, Rengaraju M, Valiente NM, Chehimi J, Kubin M, Aste M, Chan SH, Kobayashi M, Young D, Nickbarg E, Chizzonite R, Wolf SF, Trinchieri G (1992) Production of natural killer cell stimulatory factor (interleukin 12) by peripheral blood mononuclear cells. J Exp Med 176: 1387–1398

Davis EO, Thangaraj HS, Brooks PC, Colston MJ (1994) Evidence of selection for protein introns in the RecAs of pathogenic mycobacteria. EMBO J 13: 699–703

Delsal GD, Storici P, Schneider C, Romeo D, Zanetti M (1992) cDNA cloning of the neutrophil bactericidal peptide indolicidin. Biochem Biophys Res Commun 187: 467–472

Denicola A, Rubbo H, Rodriguez D, Radi R (1993) Peroxynitrite-mediated cytotoxicity to *Trypanosoma cruzi*. Arch Biochem Biophys 304: 279–286

Denis M (1991a) Growth of *Mycobacterium avium* in human monocytes: identification of cytokines which reduce and enhance intracellular microbial growth. Eur J Immun 21: 391–395

Denis M (1991b) Interferon-gamma-treated murine macrophages inhibit growth of tubercle bacilli via the generation of reactive nitrogen intermediates. Cell Immun 132: 150–157

Denis M (1991c) Killing of *Mycobacterium tuberculosis* within human monocytes: activation by cytokines and calcitrol. Clin Exp Immun 84: 200–206

Denis M (1991d) Tumor necrosis factor and granulocyte macrophage-colony stimulating factor stimulate human macrophages to restrict growth of virulent *Mycobacterium avium* and to kill avirulent *M. avium*: killing effector mechanism depends on the generation of reactive nitrogen intermediates. J Leukoc Biol 49: 380–387

Denis M, Gregg EO, Ghandirian E (1990) Cytokine modulation of *Mycobacterium tuberculosis* growth in human macrophages. Int J Immunopharmacol 12: 721–727

Dinauer MC, Orkin SH (1992) Chronic granulomatous disease. Annu Rev Med 43: 117–124

Doi T, Ando M, Akaike T, Suga M, Sato K, Maeda H (1993) Resistance to nitric oxide in *Mycobacterium avium* complex and its implication in pathogenesis. Infect Immun 61: 1980–1989

Douvas GS, Looker DL, Vatter AE, Crowle AJ (1985) Gamma interferon activates human macrophages to become tumoricidal and leishmanicidal but enhances replication of macrophage-associated mycobacteria. Infect Immun 50: 1–8

Douvas GS, May MH, Crowle AJ (1993) Transferrin, iron and serum lipids enhance or inhibit *Mycobacterium avium* replication in human macrophages. J Infect Dis 167: 857–864

Drevets DA, Canono BP, Leenen PJM, Campbell PA (1994) Gentamicin kills intracellular *Listeria monocytogenes*. Infect Immun 62: 2222–2228

Dubos RJ (1950) The effect of organic acids on mammalian tubercle bacilli. J Exp Med 92: 319–332

Dubos RJ (1951) A tuberculostatic agent present in animal tissues. Am Rev Tuberc 63: 119

Ellner JJ (1990) Sources of variability in assays of the interaction of mycobacteria with mononuclear phagocytes: of mice and men. Res Microbiol Inst Pasteur 141: 237–240

Fazal N, Lammas DA, Raykundalia C, Bartlett R, Kumararatne DS (1992) Effect of blocking TNF-α on intracellular BCG (Bacillus Calmette Guerin) growth in human monocyte-derived macrophages. FEMS Microbiol Immunol 105: 337–346

Ferrante A, Jungstrom IL, Rezepczyk CM, Morgan DML (1986) Differences in the sensitivity of *Schistosoma mansoni, Schistosoma dirofilaria-immitis* microfilariae, and *Nematospiroides dubius* 3rd stage larvae to damage by the polyamine oxidase-polyamine system. Infect Immun 53: 606–610

Filley EA, Rook GAW (1991) Effect of mycobacteria on sensitivity to the cytotoxic effects of tumor necrosis factor. Infect Immun 59: 2567–2572

Filley EA, Bull HA, Dowd PM, Rook GAW (1992) The effect of *Mycobacterium tuberculosis* on the susceptibility of human cells to the stimulatory and toxic effects of tumour necrosis factor. Immunology 77: 505–509

Flesch I, Kaufmann SHE (1987) Mycobacterial growth inhibition by interferon-γ-activated bone marrow macrophages and differential susceptibility among strains of *Mycobacterium tuberculosis*. J Immunol 138: 4408–4413

Flesch I, Kaufmann SHE (1988) Attempts to characterise the mechanisms involved in mycobacterial growth inhibition by gamma-interferon-activated bone marrow macrophages. Infect Immun 56: 1464–1469

Flesch IEA, Kaufmann SHE (1990a) Activation of tuberculostatic macrophage functions by gamma interferon, interferon-4, and tumor necrosis factor. Infect Immun 58: 2675–2677

Flesch IEA, Kaufmann SHE (1990b) Stimulation of antibacterial macrophage activities by B-cell stimulatory factor 2 (interleukin-6). Infect Immun 58: 269–271

Flesch I, Kaufmann SHE (1991) Mechanisms involved in mycobacterial growth inhibition by gamma-interferon-activated bone marrow macrophages: role of reactive nitrogen intermediates. Infect Immun 59: 3213–3218

Flynn JL, Chan J, Triebold KJ, Dalton DK, Stewart TA, Bloom BR (1993) An essential role for interferon γ in resistance to *Mycobacterium* tuberculosis infection. J Exp Med 178: 2249–2254

Frank RW, Gennaro R, Schneider K, Przybylski M, Romeo D (1990) Amino acid sequences of two proline-rich bactenecins. J Biol Chem 265: 18871–18874

Friedland JS, (1993) Cytokines, phagocytosis, and *Mycobacterium tuberculosis*. Lymphokine Cytokine Res 12: 127–133

Friedland JS Remick DG, Shattock R, Griffen GE (1992) Secretion of interleukin-8 following phagocytosis of *Mycobacterium tuberculosis* by human monocyte cell lines. Eur J Immunol 22: 1373–1378

Geller DA, Lowenstein CJ, Shapiro RA, Nussler AK, Silvo MD, Wang SC, Nakayama DK, Simmons RL, Snyder SH, Billiar TM (1993) Molecular cloning and expression of inducible nitric oxide synthase from human heptocytes. Proc Natl Acad Sci USA 90: 3491–3495

Grange JM (1976) Enzymic breakdown of amino acids and related compounds by suspensions of washed mycobacteria. J Appl Bacteriol 41: 425–431

Gros P, Skamene E, Forget A (1983) Cellular mechanisms of genetically controlled host resistance to *Mycobacterium bovis* (BCG). J Immunol 131: 1966–1972

Halliwell B, Gutterridge JMC (1984) Oxygen toxicity, oxygen radicals, transition-metals and disease. Biochem J 219: 1–14

Halliwell B, Gutteridge JMC (1992) Biologically relevant metal ion-dependent hydroxyl radical generation. FEBS Lett 307: 108–112

Hayaishi O (1985) Indoleamine 2,3-dioxygenase – with special reference to the mechanism of interferon action. Biken J 28: 39–49

Heby O (1986) Recent advances in the biochemistry of polyamines in eukaryotes. Biochem J 234: 249–262

Henderson LM, Chappell JB, Jones OTG (1988) Internal pH changes associated with the activity of NADPH oxidase of human neutrophils. Biochem J 251: 563–567

Hernandez-Frontera E, McMurray DN (1993) Dietary vitamin D affects cell-mediated hypersensitivity but not resistance to experimental pulmonary tuberculosis in guinea pigs. Infect Immun 61: 2116–2121

Hibbs JB, Vavrin Z, Taintor RR (1987) L-arginine is required for expression of the activated macrophage effector mechanism causing selective metabolic inhibition in target cells. J Immunol 138: 550–565

Hibbs JB, Westenfelder C, Taintor RR, Vavrin Z, Kablitz C, Baranowski RL, Ward JH, Menlove RL, McMurray MP, Kushner JP, Samlowski WE (1992) Evidence for cytokine-inducible nitric oxide synthesis from L-arginine in patients receiving interleukin-2 therapy. J Clin Invest 89: 867–877

Hiemstra PS, Eisenhauer PB, Harwig SSL, Barselaar MTVD, Furth RV, Lehrer RI (1993) Antimicrobial proteins of murine macrophages. Infect Immun 61: 3038–3046

Hirsch CS, Ellner JJ, Russell DG, Rich EA (1994) Complement receptor mediated uptake and tumor necrosis factor-α-mediated growth inhibition of *Mycobacterium tuberculosis* by human alveolar macrophages. J Immunol 152: 743–753

Hirsch JG (1953) The essential participation of an enzyme in the inhibition of growth of tubercle bacilli by Spermine J Exp Med 97: 327–343

Hirsch JG, Dubos RJ (1952) The effect of spemine on tubercle bacilli. J Exp Med 95: 191–208

Hogg N, Darley-Usmar VM, Wilson MT, Moncada S (1992) Production of hydroxyl radicals from the simultaneous generation of superoxide and nitric oxide. Biochem J 281: 419–424

Hunt NCA, Goldin RD (1992) Nitric oxide production by monocytes in alcoholic liver disease. J Hepatol 14: 146–150

Israelski DM, Araujo FG, Wachtel JS, Heinrichs L, Remington JS (1990) Differences in microbicidal activities of human macrophages against *Toxoplasma gondii* and *Typanosoma cruzi*. Infect Immun 58: 263–265

Jackett PS, Aber VR, Mitchison DA, Lowrie DB (1981a) The contribution of hydrogen peroxide resistance to virulence of *Mycobacterium tuberculosis* during the first six days after intravenous infection of normal and BCG vaccinated guinea pigs. Br J Exp Pathol 62: 34–40

Jackett PS, Andrew PW, Aber VR, Lowrie DB (1981b) Hydrogen peroxide and superoxide release by alveolar macrophages from normal and BCG vaccinated guinea pigs after intraveneous challenge with *Mycobacterium tuberculosis*. Br J Exp Pathol 62: 419–428

James SL, Cook KW, Lazdins JK (1992) Activation of human monocyte derived macrophages to kill schistosomula of *Schistosoma mansoni* in vitro. J Immunol 145: 2686–2690

Jiang X, Baldwin CL (1993) Iron augments macrophage-mediated killing of *Brucella abortus* alone and in conjunction with interferon-γ. Cell Immun 148: 397–407

Kamijo R, Le J, Shapiro D, Havell ED, Huang S, Aguet M, Bosland M, Vilcek J (1993) Mice that lack the interferon-γ receptor have profoundly altered responses to infection with bacillus calmette-guerin and subsequent challenge with lipopolysaccharide. J Exp Med 178: 1435–1440

Kanai K, Kondo E (1968) Studies on lysosomal response to tuberculous infection in mice. Jpn J Med Sci Biol 21: 415–422

Kanai K, Kondo E (1969) Lysosomes in tuberculous infection. Jpn J Med Sci Biol 22: 131

Kanai K, Kondo E (1970) A suggested role of the lysosomal membrane as a part of the defence mechanism against tuberculous infection. Jpn J Med Sci Biol 23: 295–302

Kaufmann SHE (1993) Immunity to intracellular bacteria. Annu Rev Immunol 11: 129–163

Keller R, Keist R, Joller P, Groscurth P (1993) Mononuclear phagocytes from human bone marrow progenitor cells; morphology, surface phenotype, and functional properties of resting and activated cells. Clin Exp Immunol 91: 176–182

Khor M, Lowrie DB, Mitchison DA (1986) Effects of recombinant interferon-gamma and chemotherapy with isoniazid and rifampicin on infections of mouse peritoneal macrophages with *Listeria monocytogenes* and *Mycobacterium microti* in vitro. Br J Exp Pathol 67: 707–717

Kindler V, Sappino AP, Gran GE, Piquet PF, Vassalli P (1989) The inducting role of tumor necrosis factor in the development of bactericidal granulomas during BCG infection. Cell 56: 731–740

Klebanoff SJ, Hamon CB (1972) Role of myeloperoxidase-mediated antimicrobial systems in intact leukocytes. J Reticulo Soc 12: 170–196

Klun CL, Youmans GP (1973) The effect of lymphocyte supernatant fluids on the intracellular growth of virulent tubercle bacilli. J Reticulo Soc 13: 263–274

Knowles RG, Moncada S (1994) Nitric oxide synthases in mammals. Biochem J 298: 249–258
Kochan I (1973) The role of iron in bacterial infections, with special consideration of the host-tubercle bacillus interaction. In: Capron A, Compans RW, Cooper M et al. (eds) Current Topics in Microbiology and Immunology, vol 60. Springer, Berlin Heidelberg New York, pp 1–30
Kochan I, Patton C, Ishak K (1963) Tuberculostatic activity of normal sera. J Immunol 90: 711-719
Koeffler HP, Reichel H, Bishop JE, Norman AW (1985) Interferon stimulates production of 1, 25-dihydroxyvitamin D3 by normal human macrophages. Biochem Biophys Res Commun 127: 596–603
Kotani S, Kitaura T, Inui S, Hashimoto S, Chimora M (1962) Antimycobacterial activity of extracts of mononuclear leukocytes from peritoneal exudates of guinea pigs. Biken J 5: 133–154
Lehrer RI, Ganz T, Selsted ME (1991) Defensins: endogenous antibiotic peptides of animal cells. Cell 64: 229–230
Lehrer RI, Lichtenstein AL, Ganz T (1993) Defensins: antimicrobial and cytotoxic peptides of mammalian cells. Annu Rev Immunol 11: 105–128
Li Y, Severn A, Rogers MV, Palmer RMJ, Moncada S, Liew FY (1992) Catalase inhibits nitric oxide synthesis and killing of intracellular *Leishmania major* in murine macrophages. Eur J Immunol 22: 441–446
Lowrie DB (1990) Is macrophage death on the field of battle essential to victory, or a tactical weakness in immunity against tuberculosis. Clin Exp Immunol 80: 301–302
Lowrie DB, Andrew PW (1988) Macrophage antimycobacterial mechanisms. Br Med J 44: 624–634
Lowrie DB, Aber VR, Carrol MEW (1979) Division and death rates of *Salmonella typhimurium* inside macrophages: use of penicillin as a probe. J Gen Microbiol 110: 409–419
Lowrie DB, Peters TJ, Scoging A (1982) Benzyl penicillin transport and subcellular distribution in mouse peritoneal macrophage monolayers. Biochem Pharmacol 31: 423–432
Lowrie DB, Jackett PS, Andrew PW (1985) Activation of macrophages for antimycobacterial activity. Immunol Lett 11: 195–203
Lurie MB (1942) The fate of tubercle bacilli ingested by mononuclear phagocytes derived from normal and immunized animals. J Exp Med 75: 247–270
Mackaness GB (1964) The immunological basis of acquired cellular resistance. J Exp Med 120: 105–120
Mackaness GB (1968) The immunology of anti-tuberculous immunity. Am Rev Respir Dis 97: 337–344
Mackay A, Alcorn MJ, Macleod IM, Stack BHR, Macleod T, Laidlaw M, Millar JS, White RG (1980) Fatal disseminated BCG infection in an 18-year-old boy. Lancet ii: 1332–1334
Mangelsdorf DJ, Keoffler HP, Donaldson CA, Pike JW, Haussler MR (1984) 1, 25 dihydroxyvitamin D3-induced differentiation in a human promyelocytic leukaemia cell line (HL60): receptor-mediated maturation to macrophage-like cells. J Cell Biol 98: 391–399
Marletta MA (1993) Nitric oxide synthase structure and mechanism. J Biol Chem 268: 12231–12234
Martin JHL, Edwards SW (1993) Changes in mechanisms of monocyte/macrophage-mediated cytotoxicity during culture. J Immunol 150: 3478–3486
Maza LMDL, Peterson EM, Fennie CW, Czarniecki CW (1985) The antichlamydial and anti-proliferative activities of recombinant murine interferon-γ are not dependent on typtophan concentrations. J Immunol 135: 4198–4200
McDonough KA, Kress Y, Bloom BR (1993) Pathogenesis of tuberculosis: interaction of *Mycobacterium tuberculosis* with macrophages. Infect Immun 61: 2763–2773
McMurray DN (1994) Guinea pig model of tuberculosis. In: Bloom BR (ed) Tuberculosis: pathogenesis, protection and control. ASM Washington, pp 135–147
Melillo G, Cox GW, Radzioch D, Varesio L (1993) Picolinic acid, a catabolite of L-tryptophan, is a costimulus for the induction of reactive nitrogen intermediate production in murine macrophages. J Immunol 150: 4031–4040
Messina L, Arcidiacono A, Spampinato G, Malaguarnera L, Berton G, Kaczmarek L, Messina A (1990) Accumulation of ornithine decarboxylase mRNA accompanies activation of human and mouse monocytes/macrophages. FEBS Lett 268: 32–34
Messina L, Spampinato G, Arcidiacono A, Malaguarnera L, Pagano M, Kaminska B, Messina A (1992) Polyamine involvement in functional activation of human macrophages. J Leuk Biol 52: 585–587
Meylan PRA, Richman DD, Kornbluth RS (1992) Reduced intracellular growth of mycobacteria in human macrophages cultivated at physiologic oxygen pressure. Am Rev Respir Dis 145: 947–953
Mitchel HH, Schonle HA, Grinolly HS (1916) The origin of the nitrates in the urine. J Biol Chem 24: 461–490

Mitchison DA, Selkon JB, Lloyd J (1963) Virulence in the guinea pig, susceptibility to hydrogen peroxide and catalase activity of isoniazid-sensitive tubercle bacilli from South Indian and British patients. J Pathol Bacteriol 86: 377–386

Miyasaki KT, Bodeau AL (1992) Human neutrophil azurocidin synergizes with leukocyte elastase and cathepsin G in the killing of *Capnocytopaga sputigena*. Infect Immun 60: 4973–4975

Moncada S (1992) The L-arginine: nitric oxide pathway. Acta Physiol Scand 145: 201–227

Moreno C, Taverne J, Mehlert A, Bate CA, Brealey RJ, Meager A, Rook GAW, Playfair JHL (1989) Lipoarabinomannan from *Mycobacterium tuberculosis* induces the production of tumor necrosis factor from human and murine macrophages. Clin Exp Immunol 76: 240–245

Morgan DML (1987) Polyamines. Essays Biochem 23: 82–115

Morgan DML, Bachrach U, Assaraf YG, Harari E, Golenser J (1986) The effect of purified aminoaldehydes produced by polyamine oxidation on the development in vitro of *Plasmodium falciparum* in normal and glucose-6-phosphate dehydrogenase deficient erythrocytes. Biochem J 236: 97–101

MRC (1972) BCG and vole bacillus vaccines in the prevention of tuberculosis in adolescence and early life. Bull WHO 46: 371–385

Murray HW, Cohn ZA (1980) Macrophage oxygen-dependent antimicrobial activity. Enhanced oxidative metabolism as an expression of macrophage activation. J Exp Med 152: 1596–1609

Murray HW, Teitelbaum RF (1992) L-arginine-dependent reactive nitrogen intermediates and the antimicrobial effect of activated human mononuclear phagocytes. J Infect Dis 165: 513–517

Murray HW, Szuro-Sudol A, Wellner D, Oca MJ, Granger AM, Libby DM, Rothermel CD, Rubin BY (1989) Role of tryptophan degradation in respiratory burst-independent antimicrobial activity of gamma interferon-stimulated human macrophages. Infect Immun 57: 845–849

Nichols WK, Prosser FH (1980) Induction of ornithine decarboxylase in macrophages by bacterial lipopolysaccharides (LPS) and mycobacterial cell walls. Life Sci 27: 913–920

Noronha-Dutra AA, Epperlein MM, Woolf N (1993) Reaction of nitric oxide with hydrogen peroxide to produce potentially cytotoxic singlet oxygen as a model for nitric oxide-mediated killing. FEBS Lett 321: 59–62

North RJ, Izzo AA (1993) Mycobacterial virulence. Virulent strains of *Mycobacteria tuberculosis* have faster in vivo doubling times and are better equipped to resist growth-inhibiting functions of macrophages in the presence and absence of specific immunity. J Exp Med 177: 1723–1733

O'Brien L, Carmichael J, Lowrie DB, Andrew PW (1994) Strains of *Mycobacterium tuberculosis* differ in susceptibility to reactive nitrogen intermediates in vitro. Infect Immun 62: 5187–5190

O'Brien S, Andrew PW (1991) Guinea pig alveolar macrophage killing of *Mycobacterium tuberculosis* in vitro does not require hydrogen peroxide or hydroxyl radical. Microbiol Pathol 11: 229–236

O'Brien S, Jackett PS, Lowrie DB, Andrew PW (1991) Guinea pig alveolar macrophages kill *Mycobacterium tuberculosis* in vitro but killing is independent of susceptibility to hydrogen peroxide or triggering of the respiratory burst. Microbiol Pathol 10: 199–207

Ochoa JB, Udekwu AO, Billiar TR, Curran RD, Cerra FB, Simmons RL, Peitzman AB (1991) Nitrogen oxide levels in patients after trauma and during sepsis. Ann Surg 214: 621–626

Ogata K, Linzer BA, Zuberi RI, Ganz T, Lehrer RI, Catanzaro A (1992) Activity of defensins from human neutrophilic granulocytes against *Mycobacterium avium-Mycobacterium intracellulare*. Infect Immun 60: 4720–4725

Padgett EL, Pruett SB (1992) Evaluation of nitrite production by human monocyte-derived macrophages. Biochem Biophys Res Commun 186: 775–781

Palmer RMJ, Ferrige AG, Moncada S (1987) Nitric oxide release accounts for the biological activity of endothelium derived relaxing factor. Nature 327: 524–526

Pattersonn RJ, Youmans GP (1970) Demonstration in tissue culture of lymphocyte-mediated immunity to tuberculosis. Infect Immun 1: 600–603

Patterson-Delafield J, Martinez RJ, Lehrer RI (1980) Microbicidal cationic proteins in rabbit alveolar macrophages: a potential host defense mechanism. Infect Immun 30: 180–192

Peterson PK, Gaziano E, Suh HJ, Devalon M, Peterson L, Keane WF (1985) Antimicrobial activities of dialysate-elicited and resident human peritoneal macrophages. Infect Immun 49: 212–218

Pfefferkorn ER (1984) Interferon blocks the growth of *Toxoplasma gondii* in human fibroblasts by inducing the host cells to degrade tryptophan. Proc Nat Acad Sci USA 81: 908–912

Prince RC, Gunson DE (1993) Rising interest in nitric oxide synthase. TIBS 18: 35–36

Rastogi N (1990) Killing intracellular mycobacteria in in vitro macrophage systems: what may be the role of known host microbicidal mechanisms? Res Microbiol Inst Pasteur 141: 217–230

Rees RJW, Hart PDA (1961) Analysis of the host-parasite equilibrium in chronic murine tuberculosis by total and viable bacillary counts. Br J Exp Pathol 42: 83–88

Robertson AK, Andrew PW (1991) Interferon gamma fails to activate human monocyte derived macrophages to kill or inhibit the replication of a nonpathogenic mycobacterial species. Microbiol Pathol 11: 283–288

Rook GAW (1988) Role of activated macrophages in the immunopathology of tuberculosis. Br Med J 44: 611–623

Rook GAW (1990a) Discussion 5th forum in microbiology. Res Microbiol Inst Pasteur 141: 267–268

Rook GAW (1990b) The role of activated macrophages in ptotection and immunopathology in tuberculosis. Res Microbiol Inst Pasteur 141: 253–256

Rook GAW, Rainbow S (1981) An isotope incorporation assay for the antimycobacterial effects of human monocytes. Ann Immun (Inst Pasteur) 132D: 281–289

Rook GAW, Champion BR, Steele J, Varley AM, Stanford JL (1985) I-A restricted activation by T cell lines of anti-tuberculosis activity in murine macrophages. Clin Exp Immunol 59: 414–420

Rook GAW, Steele J, Ainsworth M, Champion BR (1986a) Activation of macrophages to inhibit proliferation of *Mycobacterium tuberculosis*: comparison of the effects of recombinant gamma-interferon on human monocytes and murine peritoneal macrophages. Immunology 59: 333–338

Rook GAW, Steele J, Fraher L, Barker S, Karmali R, O'Riordan J, Stanford J (1986b) Vitamin D3, gamma interferon, and control of proliferation of *Mycobacterium tuberculosis* by human monocytes. Immunology 57 159–163

Sakai N, Milstein S (1993) Availability of tetrahydrobiopterin is not a factor in the inability to detect nitric oxide production by human macrophages. Biochem Biophys Res Commun 193: 378–383

Sbarra AJ, Karnovsky ML (1959) The biochemical basis of phagocytosis. Metabolic changes during the ingestion of particles by polymorphonuclear leukocytes. J Biol Chem 234: 1355–1362

Schlesinger LS (1993) Macrophage phagocytosis of virulent but not attenuated strains of *Mycobacterium tuberculosis* is mediated by mannose receptors in addition to complement receptors. J Immunol 150: 2920–2930

Schlesinger LS, Bellinger-Kawahara CG, Payne NR, Horwitz MA (1990) Phagocytosis of *Mycobacterium tuberculosis* is mediated by human monocyte complement receptors and complement component C3. J Immunol 144: 2771–2780

Schneemann M, Schoedon G, Hofer S, Blau N, Guerrero L, Schaffner A (1993) Nitric oxide synthase is not a constituent of the antimicrobial armature of human mononuclear phagocytes. J Infect Dis 167: 1358–1363

Schwarztman MS, Frank WA (1987) Vitamin D toxicity complicating the treatment of senile, postmenopausal, and glucocorticoid-induced osteoporosis. Am J Med 82: 224–230

Segal AW, Abo A (1993) The biochemical basis of the NADPH oxidase of phagocytes. TIBS 18: 43–47

Selsted ME, Miller SI, Henschen AH, Ouellette AJ (1992) Enteric defensins: antibiotic peptide components of intestinal host defense. J Cell Biol 118: 929–936

Sharma SD, Middlebrook G (1977) Partial purification and properties of an antibacterial product of peritoneal exudate cell cultures from BCG-infected guinea pigs. Infect Immun 15: 737–744

Shiratsuchi H, Toossi Z, Mattler MA, Ellner JJ (1993) Colonial morpho-type as a determinant of cytokine expression by human monocytes infected with *Mycobacterium avium*. J Immunol 150: 2945–2954

Silva MT, Silva MNT, Appelberg R (1989) Neutrophil-macrophage cooperation in the host defence against mycobacterial infections. Microbiol Pathol 6: 369–380

Smith DW, Wiegeshaus EH (1989) What animal models can teach us about the pathogenesis of tuberculosis in humans. Rev Infect Dis 11: S385–S393

Spitznagel JK (1990) Antibiotic proteins of human neutrophils. J Clin Invest 86: 1381–1386

Stahelin H, Suter E, Karnovsky ML (1956) Studies on the interaction between phagocytes and tubercle bacilli. 1. Observations on the metabolism of guinea pig leukocytes and the influence of phagocytosis. J Exp Med 104: 121–136

Steele J, Flint KC, Pozniak AL, Hudspith B, Johnson MM, Rook GAW (1986) Inhibition of virulent *Mycobacterium tuberculosis* by murine peritoneal macrophages and human alveolar lavage cells: the effects of lymphokines and recombinant gamma interferon. Tubercle 67: 289–294

Storici P, Delsal GD, Schneider C, Zanetti M (1992) cDNA sequence analysis of an antibiotic dodecapeptide from neutrophils FEBS Lett 314: 187–190

Stuehr DJ, Griffith OW (1992) Mammalian nitric oxide synthases. Adv Enzymol Relat Areas Mol Biol 65: 287–346

Sturgill-Koszycki S, Schlesinger PH, Chakroborty P, Haddix PL, Collins HL, Fok AK, Allen RD, Gluck SL, Heuser J, Russell DG (1994) Lack of acidification in mycobacterium phagosomes produced by exclusion of the vesicular proton-ATPase. Science 263: 678–681

Sundaram KS, Venkitasubramanian TA (1978) Tryptophan uptake by *Mycobacterium tuberculosis* H37Rv: effect of rifampicin and ethambutol. Antimicrol Agents Chemother 13: 726–730

Suter E (1953) Multiplication of tubercle bacilli within mononuclear phagocytes in tissue cultures derived from normal animals and animals vaccinated with BCG. J Exp Med 97: 235–247

Sypek JP, Jacobson S, Vorys A, Wyler DJ (1993) Comparison of gamma interferon, tumor necrosis factor, and direct cell contact in activation of antimycobacterial defense in murine macrophages. Infect Immun 61: 3901–3906

Taffet SM, Haddox MK (1985) Bacterial lipopolysaccharide induction of ornithine decarboxylase in the macrophage-like cell line RAW 264: requirement of an inducible factor. J Cell Physiol 122: 215–220

Taylor TWJ, Wignall EW, Cowley JF (1927) The decomposition of nitrous acid in aqueous solution. J Chem Soc 11: 1923–1930

Thomsen LL, Miles DW, Happerfield L, Riveros-Moreno V, Bobrow LG, Moncada S (1993) Nitric oxide synthase activity in human breast cancer tissue: localisation within macrophages and myoepithelial cells. Endothelium 1: s51

Tripathy SP, Menon NK, Mitchison DA, Narayana ASL, Somasundaram PA, Stott H, Velu S (1969) Response to treatment with isoniazid plus PAS of tuberculous patients with primary isoniazid resistance. Tubercle 50: 257–268

Turco J, Winkler HH (1986) Gamma-interferon-induced inhibition of the growth of *Rickettsia prowazeki* in fibroblasts cannot be explained by the degradation of tryptophan or other amino acids. Infect Immun 53: 38–46

Turcotte R, Desormeaux Y, Borduas AG (1976) Partial characterization of a factor extracted from sensitized lymphocytes that inhibits the growth of *Mycobacterium tuberculosis* within macrophages in vitro. Infect Immun 14: 337–344

Valone SE, Rich EA, Wallis RS, Ellner JJ (1988) Expression of tumor necrosis factor in vitro by human mononuclear phagocytes stimulated with whole *Mycobacterium bovis* BCG and mycobacterial antigens. Infect Immun 56: 3313–3335

Varesio L, Clayton M, Blasi E, Ruffman R, Radzioch D (1990) Picolinic acid, a catabolite of tryptophan, as the second signal in the activation of IFN-γ-primed macrophages. J Immunol 145: 4265–4271

Walker L, Lowrie DB (1981) Killing of *Mycobacterium microti* by immunologically activated macrophages. Nature 293: 69–70

Warwick-Davies J, Dhillon J, O'Brien L, Andrew PW, Lowrie DB (1994) Apparent killing of *Mycobacterium tuberculosis* by cytokine-activated human monocytes can be an artefact of a cytotoxic effect on the monocytes. Clin Exp Immun 96: 214–217

Wayne LG, Kubica GP (1986) *Mycobacterium*. In: Sneath PHA, Mair NS, Sharpe EM, Holt JG (eds) Bergeys Manual of Systematic Bacteriology. Williams and Wilkins, Baltimore, pp 1436–1457

Weinberg JB, Misukonis MA, Wood ER, Smith GL, Shami PL, Mason SN, Granger DL (1993) Human mononuclear phagocyte nitric oxide synthase (NOS): evidence for induction of NOS mRNA and protein without detectable capacity for nitric oxide production. Blood 82: 186a

Wright SD, Silverstein SC (1983) Receptors for C3b and C3bi promote phagocytosis but not the release of toxic oxygen from human phagocytes. J Exp Med 158: 2016–2023

Yoshida R, Imanishi J, Oku T, Kishida T, Hayashi O (1981) Induction of pulmonary indoleamine 2,3-dioxygenase by interferon. Proc Natl Acad Sci USA 78: 129–132

Young PR, Zygas AP (1986) Secretion of lactic acid peritoneal macrophages during extracellular phagocytosis. Possible role of local hyperacidity in inflammatory demyelination. J Neuroimmunol 15: 295–308

Zerlauth G, Eibl MM, Mannhalter JW (1991) Induction of anti-mycobacterial and anti-listerial activity of human monocytes requires different activation signals. Clin Exp Immunol 85: 90–97

Zhu L, Gunn C, Beckman JS (1992) Bactericidal activity of peroxynitrite. Arch Biochem Biophys 298: 452–457

Zlotnik A, Crowle AJ (1981) Lymphokine induced mycobacteriostatic activity in mouse pleural macrophages. Infect Immun 37: 786–793

Virulence Determinants of *Mycobacterium tuberculosis*

F.D. Quinn, G.W. Newman, and C.H. King

1 State of the Art

Microbial pathogenicity has been defined as "the biochemical mechanisms whereby microorganisms cause disease" (Smith 1968); however, the actual process is much more dramatic. The interaction between host and pathogen during disease is a dynamic confrontation where the microbe's strategies for

Division of AIDS, STD, and TB Laboratory Research, Centers for Disease Control and Prevention, Atlanta, GA 30333, USA survival meet face to face with the formidable defenses of the immune system. The tactics employed by both participants provide fascinating topics for researchers of many disciplines. Advances, fueled largely by the application of molecular biology, have been made in the biochemistry, immunology, and cell biology of the host–parasite interaction. Among the many new insights is the recognition that bacterial pathogens have evolved sophisticated signal transduction systems controlling the coordinate expression of virulence determinants.

Virulence is intrinsically coupled to disease and is, therefore, most readily understood in terms of morbidity and mortality. However, one must consider that the degree of host injury does not necessarily correlate with evolutionary success for a pathogen. Multiplication of the organism is clearly most important, while disease is simply a result of the complex interactions required to accomplish this goal within the environment of the host.

We favor a broad definition of virulence determinants that includes all those factors contributing to infection as well as to disease, with the exception of "housekeeping" functions required for efficient multiplication on artificial substrates. While some factors fit comfortably into such a scheme such as adhesion molecules, toxins, and resistance to host defenses, others are in a gray area that may more closely resemble housekeeping functions. Therefore, bacterial factors that aid in the acquisition of iron in the host are likely virulence factors, given the enormous effort the host makes to withhold this vital nutrient from pathogens (WEINBERG 1993). However, conditions can be set up in a laboratory in which the same bacterial factors are required for growth on iron-chelated laboratory media. In addition, the fact that mutations in genes encoding the enzymes of purine and pyramidine biosynthesis render some bacteria avirulent in animal models could qualify these enzymes as virulence factors. Interestingly, the coordinate regulation of housekeeping genes with a virulence gene can help support the housekeeping function as a virulence factor, such as the coregulation by iron of both catechol siderophore synthesis and cytotoxin production in *Escherichia coli*.

The complex nature of host-bacterial interactions indicates that more than one virulence factor is usually involved in the disease-causing process. This fact has been supported by several studies showing that specific virulence determinants such as adhesins and toxins (many times coordinately expressed) contribute to unique steps in microbial pathogenesis.

Mycobacterium tuberculosis has been recognized as the etiologic agent of tuberculosis since Robert Koch identified the organism in 1882. However, over the decades surprisingly little information has become available concerning the specific virulence mechanisms available to the bacterium. This fact is not due to lack of effort, and some very interesting and useful information was obtained before the technological revolution of the 1970s, 1980s and 1990s. Neverthless, due to technical difficulties, financial constraints, and lengthy incubation periods that are not involved in the study of other bacterial pathogens, several major issues remain to be addressed and elucidated.

It has become apparent from studies with other pathogenic organisms that several common themes are repeatedly used by microbes for infection, and

M. tuberculosis is no exception. It is the purpose of this review to summarize, not only our current knowledge of the known pathogenic mechanisms of *M. tuberculosis* and the known and suspected damage that they can cause, but also the approaches used by investigators to identify these factors (Table 1). For example, as with most pathogenic microorganisms, the model system used to identify and study the suspected virulence mechanisms should be one of the most important considerations. Unfortunately, as with all human pathogens, using the most appropriate alternative to the "intact human" becomes a matter for strong debate. We will examine the variety of animal and tissue culture models available and describe briefly the advantages and disadvantages of each. Also, molecular biology and bacterial genetics methodologies will be described;

Table 1. Known and suspectec virulence factors of *M. tuberculosis*

Putative factors	Proposed activity	Reference
Invasion protein	Invasion of nonprofessional phagocytes and intracellular survival	ARRUDA et al. 1993
Fibronectin binding proteins	Adhesion and invasion	SCHOREY et al. 1995
Complement and mannose receptor binding	Entry into professional phagocytes	SCHLESINGER et al. 1990; SCHLESINGER and HOROWITZ 1990; HIRSCH et al. 1994
Alternate phagosomal pathway (phagosomal trafficking)	Survival in the phagosome	CLEMONS and HOROWITZ 1995
Prevention of phagolysosome fusion	Survival in the phagosome	ARMSTRONG and HART 1971; ARMSTRONG and HART 1975; McDONOUGH et al. 1993
Escape into the cytoplasm and budding into a novel phagosome	Survival in the cytoplasm	MYRVIK et al. 1984; McDONOUGH et al. 1993
Hemolysin/phospholipase	Lysis of the phagosome; leakage of the phagosome	KING et al. 1993; LEAO et al. 1995
Cytotoxicity	Tissue destruction and enhanced TNF effects	GADEA et al. 1993; McDONOUGH and KRESS 1995
Lipoarabinomannans	Reduced TNF induction and macrophage effector function	ROACH et al. 1993; ADAMS et al. 1993; CHATTERJEE et al. 1992
Heat shock proteins	Induction of inflammatory cytokines	RETZLAFF et al. 1994
Cord factor	Tissue destruction, necrosis	MIDDLEBROOK et al. 1947; SILVA et al. 1985; BEHLING et al. 1993
Sulfolipid	Induction of inflammatory mediators	GOREN et al. 1974; ZHANG et al. 1988
Superoxide dismutase catalase	Inhibits reactive oxygen and nitrogen radicals within the phagosome	MIDDLEBROOK and COHN 1953; COHN et al. 1954. KOBAYASHI et al. 1988; ANDERSON et al. 1991; DESHPANDE et al. 1993; O'BREIN et al. 1994
Exochelins mycobactin	Acquisition of ferric iron	BARCLAY and RATLEDGE 1988; GOBIN et al. 1995

TNF, tumor necrosis factor

only recently has significant progress been made on addressing some of the technical problems in these fields particular to *M. tuberculosis*.

Additionally, studies on the interaction between *M. tuberculosis* and the immune system have demonstrated that the host possesses a formidable deterrent that must be breached or avoided by most pathogenic organisms. For those instances when infection does occur, the destructive effects of virulence factors on host tissues will be examined in detail.

2 Methods for the Identification of Virulence Factors

2.1 Virulence Models

Bacterial pathogens, including *M. tuberculosis*, are highly adapted microorganisms with survival strategies that require multiplication on or within another living organism. Additionally, tissue tropism likely plays a role in the development of a particular disease. Therefore, in order to identify all but the most basic constitutively expressed biochemical pathways, appropriate virulence model systems are required.

2.1.1 Animals

Since the initial isolation of the organism, various animal models have played key roles in elucidating many aspects of tuberculosis and the infectious process. These models primarily have included monkeys (Rock et al. 1995), mice (Dhillon and Mitchison 1994), genetic variant mice (North and Izzo 1993; Cooper et al. 1995), rats (Greenberg et al. 1995), guinea pigs (Balasubramanian et al. 1994), and rabbits (Malpani et al. 1992). Detailed aspects about these models are described fully elsewhere in this book.

While these models have been useful in answering many experimental questions, *M. tuberculosis* may be primarily a human pathogen. Thus, with the exception of primates, animal models may be of limited use in the study of human disease. Although some model systems do present with a lung infection, the immune and other physiological responses are different from those bacteria would encounter during human tuberculosis. Therefore, when identifying the bacterial genes required to perpetrate human disease, as well as to identify correct components of the human immune response, association with human cells is required.

2.1.2 Organ Culture

The nasopharyngeal organ culture system developed by Stephens et al.(1989) permits study of the interaction between bacteria and tissues of the upper respiratory tract. However, PEROSIO and FRANK (1993) and GILBERT et al. (1993) have examined the interaction between *M. tuberculosis* and human lung tissues. Though there is much potential for obtaining useful data, particularly microscopic, tissues for these types of studies are difficult to obtain becasue they have been severely traumatized during surgery, may be diseased, are variable from donor to donor, require the initial use of antibiotics, and have limited viability.

2.1.3 Primary and Transformed Tissue Culture Monolayers

Since the 1950s, tissue culture monolayers have played a crucial role in day-to-day pathogenesis studies, although animal models have remained the ultimate "gold standard" when definitive proof is required. Human tissue culture monolayers are simpler to work with, can be maintained under controlled conditions, and are relevant to human disease. Though many primary and transformed cell lines have been examined over the years, including human endometrial carcinoma cells (HeLa) (ARRUDA et al. 1993), the most common have been cells of macrophage/monocyte lineage and neutrophils. This makes the most sense since phagocytic cells are likely the first cells to be encountered during infection in the lungs. Using these cells of both human and animal origin, researchers have examined bacterial and host adhesion factors, mechanisms of intracellular survival, and the production of factors necessary for intracellular and extracellular survival in the host (ROCK and BLOOM 1994).

Recently, as the complexity of the cellular immune response to *M. tuberculosis* infections has begun to be understood, it has become evident that not only macrophages and neutrophils contribute to the infectious process. Cell types, such as lung pneumocyte epithelial cells and lung vascular endothelial cells, may be infected, or at least involved in production of inflammatory and immune cell recruiting cytokines and chemokines (STADNYK 1994; LUKACS et al. 1995).

2.1.4 Multiple Layer Tissue Culture Systems

When infecting a human host, the bacillus interacts with multiple layers of cells, and therefore, monolayer studies have severe limitations. A more relevant model incorporates the added complexity of cell-to-cell interactions associated with multiple layers. BIRKNESS et al. (1995) recently developed an artificial tissue system incorporating epithelial and endothelial monolayers on a microporous membrane as an approach to examine the process of attachment and passage that occurs as the bacterium makes its way from the mucosal surface through the epithelial cells and into the vascular system (Fig. 1). The layering of one cell type over another allows the cells to communicate and interact, essential components for the polarization, growth, and differentiation of cell types in vivo. Though the preliminary report by BIRKNESS et al.(1995) examined the interaction of Neisseria meningitidis with the bilayer system,a report soon to be published (K.A. Birkness, personal communication) examined the attachment and invasion

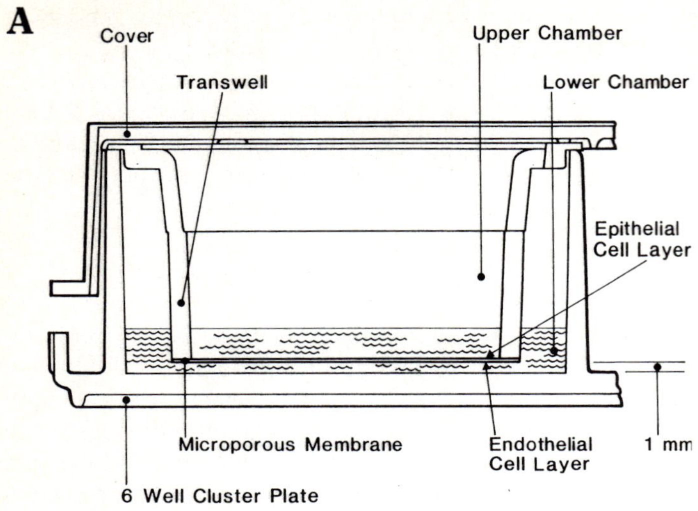

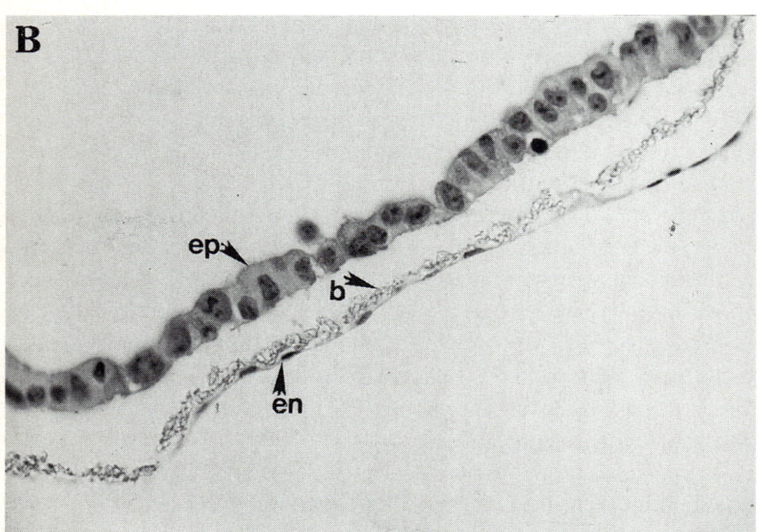

Fig. 1A,B. Tissue culture bilayer virulence model system. **A** Schematic diagram of a Transwell-COL chamber (Costar, Cambridge, MA) with layers of epithelial and endothelial cells in one well of a six-well tissue culture dish. **B** Uninfected bilayer with dierentiated columnar epithelial cells (*ep*), a 3.0 μm microporous membrane (*b*), and thin, flat human microvascular endothelial cells (*en*). X400

by *M. tuberculosis* of a bilayer composed of human pneumocyte epithelial cells (A549) (McDonough and Kress 1995) and human lung endothelial cells (HuLec) (P.K. Mehta, personal communication). Using this system, bacterial attachment, internalization, intracellular growth of *M. tuberculosis* and passage through the two cell layers into the space below was observed. Both cell lines used in these studies were transformed, though primary cells can be used and are likely more appropriate. Immunological components could also be added to the apical and basal surfaces of this bilayer system. The addition of immune cells might: (1) demonstrate the effect of *M. tuberculosis* on the migration of immune cells from the blood to the surface of the epithelial cell layer; (2) allow us to examine the chronological induction of cytokines and cellular adhesion molecules during the infectious process; and (3) demonstrate the effect of immune cells on the growth and inhibition of *M. tuberculosis*. Additionally, this model might allow the investigator to examine the acute stages of infection that may occur in the human lung, and to follow subsequent stages through granuloma development.

Perhaps the most exciting use of this model system will be for the identification of differentially expressed bacterial virulence-associated genes. Because of the resemblance to the physiology of the human lung, use of this model may be appropriate for the in vivo expression technology (IVET), subtractive hybridization, or differential display technologies discussed below.

Nevertheless, this model remains an in vitro system and, as such, lacks the exquisite complexity of the human lung. All of the subtle interactions present among the tissues of the human lung may not be present, any one of which might be vital to the disease process.

2.2 Molecular and Genetic Techniques

There are a wide variety of methods for the efficient identification and isolation of cellular macromolecules from mycobacteria. Used alone, techniques such as two-dimensional gel electrophoresis, gel filtration, fast-pressure liquid chromatography, and iso-electric focusing are exceedingly powerful. Several vaccine and diagnostic candidates for the mycobacteria have been identified and isolated in sufficient amounts using these methods (AFFRONTI 1988; ANDERSON and BRENNAN 1994). However, unless a particular target macromolecule has been identified previously using other methods (usually genetic or immunologic), only random molecular species can be isolated and examined. This random analysis is a slow and imprecise method in the study of bacterial pathogenesis. Alternatively, the genetic approaches (described below) and immunologic methods (described in other chapters in this volume) are useful primarily for the identification of target molecules and not their isolation and characterization. Therefore, a combination of methods is required for optimal study of potential virulence factors. Though the biochemical approaches are critical for the complete understanding of virulence factors and the corresponding immunological responses, only the molecular and genetic methodologies will be described in this review.

As discussed by JACOBS and BLOOM (1994), three conditions need to be met in fulfilling molecular Koch's postulate: (1) the isolation of virulence mutants, (2) the cloning of the responsible genes, and (3) complementation of the mutant phenotype with introduction of the wild-type gene. Though research has brought us closer to this goal, molecular Koch's postulates have not yet been fulfilled for *M. tuberculosis* and, therefore, no virulence factors or virulence genes have as yet been conclusively identified for this organism. Thus, for the purpose of this review, putative virulence-associated genes will be defined as genes whose inactivation demonstrates decreased virulence phenotype, or which correspond to a known virulence factor from another organism at the DNA or amino acid levels. The putative factors themselves will be discussed in detail later in this chapter, while this section will focus instead on the current methodologies being developed to identify and confirm the virulence-associated status of these genes and gene products of *M. tuberculosis*.

2.2.1 Isolation of Avirulent Mutants

One of the most useful methods for the study of bacterial virulence factors is the identification of avirulent or attenuated mutants that occur either naturally or through laboratory manipulation of the culture conditions. However, successful use of this method with human pathogens presupposes several conditions: (1) that an appropriate virulence model exists that accurately represents human physiology; (2) that biochemical or molecular biological tools of sufficient sensitivity to identify the sometimes minor phenotypic and genotypic modifications are present in the mutant cells; and (3) that the alteration to avirulence is total, irreversible, and present throughout all cells in the population. Though conditions 1 and 3 (and possibly 2) have been met for the vaccine strain *Mycobacterium bovis* bacillus Calmette-Gurin (BCG), no *M. tuberculosis* strain has yet been identified or laboratory-modified which fulfill all three criteria. Even *M. tuberculosis* strain H37Ra has been shown in several animal studies only to be ''less virulent'' (Steenken et al. 1934).

2.2.2 Chemical and Physical Mutagenesis

Analysis of the virulence factors of *M. tuberculosis* will require the generation of strains with specific chromosomal mutants. Chemical agents, irradiation and transposon mutagenesis systems for the induction of mutations have been developed and used with some success with a variety of bacterial organisms (Kathariou et al. 1987; Ngeleka et al. 1992). However, the slow-growing mycobacteria, *M. tuberculosis* and *M. bovis* are not well-suited for traditional mutagenesis protocols and screens. Mutagenesis with DNA-damaging agents has the limitation that each chromosome may receive more than a single mutagenic lesion. This problem is readily overcome with gene transfer systems such as conjugation, transduction, or transformation that enable the transfer of defined chromosomal fragments and the construction of strains with single mutations. Conjugation (Tokunaga et al. 1973) and transduction (Raj and Ramakrishnan 1970) have been described for *M. smegmatis*, but these systems have not been shown to be routinely useful for *M. tuberculosis*. Another essential feature of a mutant screen is the ability to generate a colony derived from a single mutated cell. *M. tuberculosis* presents a problem for such analyses because this organism generally grows in aggregates. Lastly, the slow doubling of *M. tuberculosis* makes viable count experiments difficult to perform (though this can now be overcome by using luminescent vital stains), and the fact that the organism needs to be handled in biosafety level 3 containment areas makes the screening of mutagenized (and thus the large number of virulent unmutagenized) cells disconcerting.

2.2.3 Transposon Mutagenesis and Gene Replacement

Insertional mutagenesis is a technique that uses a gene marker such as antibiotic resistance to select for a mutagenic event. Transposon mutagenesis and allelic exchange are the two techniques that have been developed for use with a number of bacterial genera. Thus, insertional mutagenesis can provide a method to identify and enrich for a pure population of mutated cells.

Transposition was first demonstrated in *M. smegmatis* (GUILHOT et al. 1994) and recently with *M. bovis* BCG to create amino acid biosynthetic auxotrophs (MCADAM et al. 1995). While allelic exchange has been described for *M. smegmatis* (KALPANA et al. 1991), and some evidence suggesting a double crossover event could be obtained in *M. bovis* BCG (NORMAN et al. 1995), no real progress had been made with *M. tuberculosis*. However, recently a very exciting novel method has been developed by BALASUBRAMANIAN et al. (1996) for affecting allelic exchange in *M. tuberculosis* by using long linear recombination substrates. This method has been used insertionally to inactivate a leucine biosynthetic gene in two different strains of *M. tuberculosis*, generating leucine-requiring auxotrophs. This strategy could be readily applied to disrupt any desired gene in any *Mycobacterium* species.

2.2.4 *Escherichia coli* and Other Mycobacterial Species as Surrogate Hosts for Gene Expression

One of the most successful techniques for the study of *M. tuberculosis* genes, including several putative virulence genes, has been the use of *E. coli* and *M. smegmatis* as surrogate expression systems. Initially, success was achieved with the development of recombinant DNA technologies that permitted efficient cloning of mycobacterial genes into vectors in *E. coli*. Numerous genes encoding protein antigens (YOUNG et al. 1985) and enzymes (JACOBS et al. 1986) were cloned from the slow-growing pathogenic mycobacteria. This approach has yielded considerable information about the nature of many important antigens found in mycobacterial infections. Though the G+C content of the genomes of *E. coli* and *M. tuberculosis* are quite disparate (50% and 66%, respectively), several highly conserved *M. tuberculosis* genes have been successfully cloned and stably maintained within various recipient strains of *E. coli*. Putative virulence-associated genes that have been cloned and maintained successfully in *E coli* include: (1) an invasion-associated gene (ARRUDA et al. 1993) that shows 27% homology at the DNA level to the *Listeria monocytogenes* internalin gene; (2) a riboflavin biosynthesis gene (*ribA*) (KIKUTA-OSHIMA et al. 1994 and personal communication); and (3) a *M. tuberculosis* gene homologous to conserved regions of the VirF and VirFy virulence gene regulators of *Shigella flexneri* and *Yersinia* spp., respectively (GUPTA and TYAGI 1993). Other mycobacterial species can also be used as surrogate hosts for *M. tuberculosis* genes. Examples of this include the use of *M. smegmatis* (VERMA et al. 1994; FALCONE et al. 1995; LIM et al. 1995), *M. bovis* BCG (CURCIC et al. 1994), and *M. vaccae* (GARBE et al. 1994).

2.2.5 Polymerase Chain Reaction and Direct Sequencing for the Identification of Virulence Genes

An alternative method to circumvent stability problems when incorrectly modified DNA is cloned into a surrogate host, may be through the use of oligonucleotides, polymerase chain reaction (PCR) amplification of target genes, and direct sequencing of PCR products. However, to use these approaches successfully, the mycobacterial target gene and known virulence gene template must be highly conserved at the DNA level. This is not likely to be the case with the high G+C ratio of mycobacterial DNA. Therefore, several laboratories have developed a degenerate code for the conversion of oligonucleotide sequences to the more G+C-rich *M. tuberculosis* sequences. Using oligonucleotide sequences corresponding to *E. coli* and *L. monocytogenes* sequences, OWENS et al. (personal communication) have successfully identified and sequenced the gene coding for SecA, a critical factor in the transport of most bacterial secreted factors. In addition, KING and SHINNICK (1995) cloned and sequenced a gene resembling the *L. monocytogenes* sulfhydryl-based hemolysins.

2.2.6 Complementation Analyses of Arivulent and Less Virulent Strains

This approach of using complementation analyses of avirulent strains is designed to identify the genes of a pathogen that confer a particular virulence-associated phenotype to a nonpathogenic strain, such as the ability to attach, invade, or grow intracellularly. Vectors and techniques for transferring genes from a virulent strain of *M. tuberculosis* to an avirulent strain (and thus generating a virulence phenotype) have been developed by LEE et al. (1991) and demonstrated by PASCOPELLA et al. 1994, using strains H37Ra and H37Rv, and a mouse virulence model. Briefly, cosmid integrating vector, pYUB178, a 7-kb vector containing the bacteriophage L5 site-specific integration system, has been shown to successfully transfer large chromosomal fragments from a virulent *M. tuberculosis* strain, and integrate into a unique *attB* site in the chromosome of the less virulent strain. The recombinant fragment is stably maintained in the host chromosome without selective pressure. However, the use of this system requires the identification of virulent/avirulent isogenic matched sets of strains, and if there are multiple genetic lesions, these mutations must be already identified, linked, or within a 40-kb proximity. Nevertheless, as isogenic mutants become available, this will prove to be an excellent system for fulfilling molecular Koch's postulates.

2.2.7 In Vivo Gene Expression Assays Using Auxotrophs of *M. tuberculosis*

An alternative approach to the traditional molecular and genetic tools are methodologies based on the direct identification of in vivo differentially expressed genes. There are no toxins, specific phenotypes, or other specified factors that need to be cloned or mutated. These technologies instead focus on previously unknown genes induced in the bacterium during infection in an appropriate virulence model system. The underlying assumption of this technology is that genes expressed only when the organism is growing in vivo are likely to encode products required for survival and multiplication in the host, i.e., virulence factors. Regardless of whether the identified genes code for so-called housekeeping factors, or actual known virulence factors, any identified genes may be necessary to cause disease in a particular virulence model being examined.

MAHAN et al. (1993) recently published the first of these new methodologies. The IVET genetic system does not rely on an observable phenotype for the identification of a virulence factor, but instead directly detects, through the use of a positive selection fusion-reporter system, the induction of any virulence-related bacterial gene during infection in an animal host. Briefly, a promoter fusion library is constructed in the host strain. The desired fusion will only express the auxotrophic marker under proper in vivo conditions. Several genes from *Salmonella typhimurium* that are expressed exclusively during mouse infection have been identified and sequenced. However, for use with *M. tuberculosis*, several components are required before this system can be operated successfully, including the identification of selectable and auxotrophic markers, plasmid systems that can be transformed and stably maintained, and an appropriate virulence model for inducing the bacterial genes under consideration. With *M. tuberculosis*, a selectable marker for kanamycin resistance has been successfully expressed (PASCOPELLA et al. 1994), and a *M. tuberculosis* leucine auxotroph has recently been identified (BALASUBRAMANIAM et al. 1996). However, a confounding problem that remains is the selection of an appropriate virulence model system that will present the bacterium with conditions comparable to those encountered during a human infection. Nevertheless, this technique, or a variation, is currently being performed in several laboratories. It will remain to be seen if the identified *M. tuberculosis* genes are relevant to the study of human tuberculosis.

2.2.8 Identification of Differentially Expressed Genes by Subtractive Hybridization

A technique known as RNA-based subtractive hybridization has been developed specifically for the purpose of identifying eukaryotic and prokaryotic genes expressed differentially under defined conditions (Duguid and Dinauer 1989; Lisitsyn et al. 1993; Mathiopuolos and Sonenshein 1989; Straus and Ausubel 1990; Timblin et al. 1990). For example, an RNA–cDNA based subtractive hybridization procedure identified a single genetic difference between two isogenic strains of *L. monocytogenes* engineered to differ in only one virulence gene (Utt et al. 1995). Recently, two studies with *M. tuberculosis* (Kinger and Tyagi 1993, and Kikuta-Oshima et al. 1994 and unpublished observations) used similar cDNA–cDNA hybridization methods to identify chromosomal fragments containing genes differentially expressed between *M. tuberculosis* H37Rv and H37Ra strains. Another study by Plum and Clark-Curtiss (1994), identified a putative heat shock gene that was expressed by *M. avium* during tissue culture infection.

The RNA subtractive hybridization methodology (RSH), is based upon mRNA homologies rather than DNA–DNA homology and targets differences in gene expression. Briefly, RNA from the positive strain (strain with the phenotype of interest such as virulence) and the negative strain (possessing a negative phenotype such as avirulence) is extracted. The RNA species are subsequently reverse transcribed and hybridized. The remaining cDNA molecules from the positive population contain the differentially expressed target genes of interest. Limitations of this technique such as the requirement for large amounts of starting material (RNA) will be overcome as the bacterial lysis, RNA isolation, and reverse transcription methodologies improve and become more sensitive. For improved sensitivity in the identification of the target DNA, improved methods need to be developed for separating bacterial RNA from contaminating eukaryotic (virulence model) RNA.

Perhaps the most interesting facet of this technique and how this procedure may avoid the debate over the most appropriate virulence model is that differentially expressed bacterial genes now can be identified directly from clinical specimens. Our group has sufficiently increased the sensitivity of this method to allow for the identification of genes induced during human, animal, or tissue culture infection by extracting RNA from the bacteria while present in an infected specimen, sufficiently amplifying the RNA, and subtracting against RNA isolated from the corresponding *M. tuberculosis* strain grown on artificial medium. We are currently processing sputum specimens from patients infected with *M. tuberculosis* and have identified at least one differentially expressed gene using this method (L.C. Kikuta-Oshima et al., personal communication).

2.2.9 Identification of Differentially Expressed Genes Using Arbitrary Primed and Differential Display PCR

Arbitrary primed PCR (AP-PCR) is a modified PCR amplification method that was initially developed to obtain a DNA fingerprint for the analysis of polymorphisms in plant and animal cells (ZHANG and MEDINA 1993). Unlike the typical PCR reaction where conditions are optimized to amplify a single target sequence, in AP-PCR reactions, conditions are modified such that multiple sequences are simultaneously amplified. This fingerprint of multiple amplified bands on a gel is specific (for the primer and template) and reproducible. Thus, in one run up to 90 bands can be seen, each representing a DNA sequence at a specific region of the genome. Another key feature of this technique is that it is sequence-dependent; that is, if either a primer of a different sequence is used or the sequence of the template is altered, the fingerprint will change.

Differential display PCR (ddPCR) developed by LIANG and PARDEE (1992), is also based on the use of random primers and has been used to identify differences in gene expression between two target cell populations. The mRNA is first reverse transcribed into cDNA and the cDNA is subjected to PCR using a 10-nucleotide long arbitrary oligonucleotide primer and a corresponding primer, in the presence of ^{35}S-dATP. The multiple PCR fragments are resolved by polyacrylamide gel electrophoresis. The bands on an autoradiogram represent a random sample of the genome. However, since the fingerprint is different with each primer, using several primers or different reactions with the disease and nondisease associated cDNA will increase the likelihood of detecting a difference in expression.

Both the AP-PCR and dd-PCR procedures are rapid, sensitive, and capable of detecting deletions, amplifications, and translocations. Furthermore, they can be performed with small amounts of sample materials.

3 Virulence Factors and Their Interaction with Host Cells

Mycobacterium tuberculosis is a facultative, intracellular parasite that infects and multiplies, primarily within professional phagocytes, but it can also efficiently infect nonprofessional phagocytic cells in vitro. Little is known about how this bacterium enters eukaryotic cells, what virulence factors exist that enable the bacilli to survive and multiply within host cells, and how this intracellular parasitism relates to the pathogenesis of disease.

Mycobacterial virulence factors can have both direct and indirect effects on host cells, and can ultimately lead to tissue destruction and disease. Direct cell–cell interactions, including attachment, invasion, and intracellular multiplication, and indirect interactions, through secreted bacterial factors such as hemolysin and cytotoxin, can cause lysis of the host cells (McDONOUGH and KRESS 1995;

LEAO et al. 1995; KING et al. 1993). In addition, lipoarabinomannans (LAMs), heat shock proteins (hsp), and mycobacterial products can stimulate host cells to produce inflammatory products or cytokines that can amplify tissue damage in the host.

3.1 Factors Needed for Attachment and Invasion

There are clear differences between virulent and less virulent laboratory strains of *M. tuberculosis* that mediate entry and survival within macrophages, suggesting strain-specific determinants for phagocytosis and ultimately intracellular survival. The mechanisms for this bacterially-mediated uptake are unclear. Although macrophages can phagocytize live and killed *M. tuberculosis*, only live *M. tuberculosis* can enter nonprofessional phagocytic cells such as cultured fibroblast and epithelial cells (FILLEY and ROOK 1991; FILLEY et al. 1992; MEHTA et al. 1996) suggesting a bacterially-mediated signal for entry into host cells. Nonprofessional phagocytes such as fibroblast or epithelial cells require a triggering of a signal transduction pathway for uptake of a foreign particle, and many intracellular bacterial pathogens utilize this pathway as a means to gain entry into these cells (BLISKA et al. 1993). Perhaps *M. tuberculosis* contains cell entry determinants similar to other bacterial species that trigger this signal transduction. This hypothesis has recently been tested, and ARRUDA et al. (1993) isolated a DNA fragment from a less virulent strain of *M. tuberculosis* that, when cloned into *E. coli*, induced entry of the recombinants into HeLa cells and enhanced survival inside human macrophages. One open reading frame within this fragment was found to contain a small degree of homology to several proteins involved in mammalian cell entry from other intracellular bacterial parasites.

More recently, a *Mycobacterium leprae* gene encoding a fibronectin binding protein was cloned, and antibodies against this recombinant protein was found to significantly block attachment and invasion of *M. leprae* in epithelial and schwann cells (SCHOREY et al. 1995). This gene was localized to a single, identically sized *Pst* I fragment in the *M. tuberculosis* genome. *M. tuberculosis*, also shown to contain fibronectin-binding proteins, may adhere to mucosal surfaces and macrophages through fibronectin receptors (ABOU-ZEID et al. 1988), similar to the adhesion of *M. bovis* BCG to fibronectin-coated surfaces (RATLIFF et al. 1987).

Host cell factors must also be important in the entry of *M. tuberculosis* and its intracellular survival. HIRSCH et al. (1994) demonstrated that complement receptor CR4 was the major complement receptor mediating uptake of a less virulent strain of *M. tuberculosis* in human alveolar macrophages, while CR1 and CR3 were important in human blood monocytes. Although this less virulent strain grew intracellularly in both alveolar macrophages and blood monocytes, alveolar macrophages were better at slowing the intracellular growth of this strain; CR4-mediated phagocytosis with increased production of tumor necrosis factor-α [TNF-α] was suggested as the cause of this slowed growth.

M.tuberculosis also has been shown to alter surface monocyte adhesion molecules soon after infection of these cells (LOPEZ et al. 1994).

Once inside the mononuclear phagocyte, the precise mechanism by which *M. tuberculosis* avoids the antibacterial activities of the phagocytic cell may be multifaceted. It has been proposed that the enzymes superoxide dismutase (SOD) and catalse inactivate toxic oxygen molecules produced by macrophages (MIDDLEBROOK and COHN 1953; COHN et al. 1954; ZHANG et al. 1991b). Though the process of phagolysosomal fusion and membrane trafficking of *M. tuberculosis* are not understood, bacterial surfaces and secreted factors are likely to play a role in dictating the ultimate fate of intracellular bacterial location. For example, CLEMENS and HORWITZ (1995) showed *M. tuberculosis* resides in an endosome-like vacuole, and does not enter a lysosomal compartment. These investigators demonstrated that the majority of the phagosomes did not mature into phagolysosomes, but retained late endosomal membrane markers such as transferrin and lysosomal membrane glycoprotein LAMP-1. Alternatively, due to differences in interactions with bacterial factors, *M. avium* phagosomes do contain lysosomal membrane glycoproteins but lack membrane proton-ATPases required for normal acidification of the phagosome (STRUGILL-KOSZYCKI et al. 1994). Therefore, species-specific bacterial factors may dictate the course of bacterial cells through a vacuolar system that will be advantageous to their survival.

These host cell studies indicate that *M. tuberculosis* resides in a novel phagosome different from the normal phagosome that leads to intracellular killing by mononuclear phagocytes (MOULDER 1985). ARMSTRONG and HART (1975) were the first to indicate this in studies that demonstrated that the presence or absence of serum components at the time of infection of macrophages by *M. tuberculosis* affected phagolysosome fusion. Together, these observations suggest a type of bacterially-mediated invasion similar to other intracellular bacterial parasites in which *M. tuberculosis* utilizes uptake pathways that avoid the normal bacteriocidal mechanisms of mononuclear phagocytes (MOULDER 1985; WILSON et al. 1980).

3.2 Toxins and Cytotoxins

M. tuberculosis may be able to escape from the phagosome and reside in the cytoplasm. Two independent studies have shown that the same virulent strain of *M. tuberculosis* was found in the cytoplasm of murine and rabbit macrophages, whereas an avirulent vaccine strain, *M. bovis* BCG, was always observed in membrane-bound vacuoles (MYRVIK et al. 1984; MCDONOUGH et al. 1993). In addition, MCDONOUGH et al. (1993) reported that they observed *M. tuberculosis* but not *M. bovis* BCG budding out of phagosomes into novel tightly apposed vesicles and into the cytoplasm. Although this budding phenomenon was not shown, the resultant *M. tuberculosis* phagosome was proposed to be different from the process observed with one nonvirulent strain of BCG.

Although the final cytoplasmic location of *M. tuberculosis* remains controversial, one could speculate that this mechanism may well be a transient phenomenon in which the bacilli escape from the phagosome by budding into a membrane-bound vesicle similar to the budding phenomenon observed in the cell-to-cell spread of *L. monocytogenes* and *S. flexneri.* These intracellular parasites produce hemolysins and phospholipases that dissolve the phagosomal membrane and permit the bacteria to escape into the host cell cytoplasm (PORTNOY et al. 1988; SANSONETTI 1991). In addition, these cytoplasmic bacteria then spread to adjacent cells by first budding out of the plasma membrane of the infected cell into the plasma membrane of the adjacent cell, forming a double-membrane vesicle. It is interesting to note that *M. tuberculosis*-infected phagosomes apparently also produce small vesicles through a budding phenomenon of the phagosomal membranes (XU et al. 1994), and these small vesicles contain LAM, a major component of the mycobacterial cell (BRENNAN 1989). This suggests that *M. tuberculosis* actively secretes components through the phagosomal membrane. Although the mechanisms for this budding phenomenon through host cell plasma membranes have not been fully explained, *M. tuberculosis* may share common virulence mechanisms with other intracellular pathogens.

M. tuberculosis and other selected mycobacterial species have been shown to possess varying degrees of hemolytic activity (KING et al. 1993; UDOU 1994; FISCHER et al. 1996). Interestingly, KING et al. (1993) found that a virulent strain of *M. tuberculosis*, H37Rv, produced significantly more hemolytic activity than the avirulent vaccine strain *M. bovis* BCG. The differences in hemolytic activity by these two strains positively correlated with the differences in intracellular location of the same strains within macrophages (MCDONOUGH et al. 1993). More recently, LEAO et al. (1995) cloned a phospholipase gene of *M. tuberculosis* that expressed hemolytic activity in *E. coli.* This suggests that the presence of a hemolytic activity may be responsible for the escape of *M. tuberculosis* into the cytoplasm or may be related to the appearance of the novel tightly apposed vesicle surrounding the bacilli within the cytoplasm. Clearly, further studies will be required to determine if hemolytic and phospholipase activities are involved in the intracellular survival of *M. tuberculosis*.

Other roles may exist for these activities. A cytotoxin could mediate intravacuole nutrient acquisition by *M. tuberculosis* through leakage of the *M. tuberculosis* phagosome. Evidence exists that *M. tuberculosis* growth within phagosomes requires intracellular methionine and leucine (JACOBS and BLOOM 1994), and these amino acids could be acquired through phagosomal membrane leakage. Bacterial hemolysins also have been shown to produce profound pleiotropic effects on eucaryotic cells. The *E. coli* hemolysin has a strong effect on energy mebabolism in human monocytes and can alter the response to exogenous signals through interaction with membrane receptors (WELCH et al. 1986). The *L. monocytogenes* hemolysin listeriolysin O has been shown to inhibit antigen processing and presentation in macrophages (CLUFF and ZIEGLER

1987; CLUFF et al. 1990). Listeriolysin O has been found to elicit a protective immune response against *L. monocytogenes* (BERCHE et al. 1987). In a similar manner, one study suggests that protective immunity to *M. tuberculosis* may be mediated in part by presentation of secreted antigens, for example, hemolysin and phospholipase, in conjunction with major histocompatibility-complex (MHC) class I determinants to $CD8^+$ T-cells (FLYNN et al. 1992). Therefore, factors released, by *M. tuberculosis* into the macrophage cytoplasm may play a role in this protective immunity and could be useful in eliciting $CD8^+$ protective responses in future recombinant vaccines against tuberculosis (BLOOM and FINE 1994).

It is possible that the hemolysin phenotype is related to the tissue destruction observed during human disease. *M. tuberculosis* produces liquefaction of granulomas after reactivation of latent infection and also produces lung cavitations during advanced disease (DANNENBERG 1982; DANNENBERG and ROOK 1994). This tissue destruction is believed to be mediated in part through host factors such as TNF-α (ROOK and BLOOM 1994). Recently, evidence has been presented that *M. tuberculosis* produces a cytopathic effect on cultured epithelial cells including cultured human lung alveolar cells (GADEA et al. 1993; MCDONOUGH and KRESS 1995). Additionally, MCDONOUGH and KRESS (1995) observed cytotoxicity in the absence of host factors such as TNF-α, and only observed cytotoxicity with the virulent strain of *M. tuberculosis*, H37Rv, but not the attenuated vaccine strain *M. bovis* BCG. Although this factor has not been identified, this observation correlated with the presence of the hemolytic activity observed by KING et al. (1993) between these two strains, suggesting that hemolysin activity is related to cytotoxicity. Thus, bacterially-mediated cytotoxicity and host factors may both be required for the pathogenesis *M. tuberculosis* in human disease.

3.3 Lipoarabinomannans

Virulence of *M. tuberculosis* has been associated with LAM (BRENNAN and NIKAIDO 1995). LAMs from the more virulent strains of *M. tuberculosis* do not induce as much TNF-α, a cytokine involved in host resistance to mycobacterial infections, when compared with less virulent strains (ROACH et al. 1993; ADAMS et al. 1993; CHATTERJEE et al. 1992). The lack of TNF-α production by virulent mycobacteria correlates with their relative inability to induce the early gene c-*fos*, KC, and JE as well as NK-kB (ROACH et al. 1993, 1994; BROWN and TAFFET 1995). Structural dissimilarities between LAM at their nonreducing termini may be responsible for differences observed in induction of host responses (CHATTERJEE et al. 1992). For example, LAM from the less virulent *M. tuberculosis* H37Ra has extensive arabinan side chains while LAM from the virulent Erdman strain has short mannan segments that mask the arabinan (CHATTERJEE et al. 1991).

Macrophage effector function is downregulated by LAMs. Activation of macrophages by interferon (IFN)-γ and induction of nitric oxide are inhibited by treatment with LAM from virulent Erdman, resulting in decreased microbicidal activity (SIBLEY et al. 1988; CHAN et al. 1991; ROACH et al. 1995). LAM also induces interleukin (IL)-8, a chemotactic factor for neutrophils, that may be involved in the initiation of granuloma formation (ZHANG et al. 1995). Although LAM may stimulate the recruitment of inflammatory cells to the site of infection, it is also a potent oxygen radical scavenger and may inactivate both neutrophil and macrophage-mediated antimicrobial activity (CHAN et al. 1991).

3.4 Heat Shock Proteins

Increased synthesis of HSPs occurs in both bacteria and host cells under conditions of stress, but they are also constitutively expressed under normal conditions, suggesting the HSPs are involved in normal cellular maintenance. HSPs are members of families of proteins subdivided on the basis of their molecular weight into HSP90 HSP70 HSP60, and low molecular weight HSPs (WIKER and HARBOE 1992). Bacterial and eukaryotic HSPs are highly conserved and the 65-kDa HSP of *M. tuberculosis* has a 40%–50% amino acid sequence identity with human mitochondrial HSP60 (JINDAL et al. 1989). Because of the high degree of homology between bacterial and human HSPs, the role of HSPs in autoimmunity became a target for investigation. Recognition of conserved, cross-reacting epitopes of self HSPs by specific T cells primed to nonself HSPs, perhaps mycobacterial in origin, may result in a pathological autoimmune state. Antibodies to *M. bovis* HSP65 are found in patients with multisystem autoimmune disease, rheumatoid arthritis, systemic lupus erythematosus, mycobacterial infections, and even healthy controls (KAROPOULOUS et al. 1995). Despite these findings, several laboratories suggest the use of a recombinant mycobacterial HSP65 as a vaccine formulation for use against mycobacterial infections because it was found efficacious in mice against a H37Rv infection (SINGH et al. 1995; SILVA 1995). No significant differences in mononuclear cell proliferation responses to the *M. bovis* HSP65 were observed between healthy donors that were positive for the PPD skin test, and tuberculosis patients, suggesting that the HSP65 does not play a major role in resistance to *M. tuberculosis* infection (MENDEZ-SAMPERIO et al. 1995).

Mycobacterial HSPs may play a role in virulence by inducing inflammatory cytokines that contribute to host tissue damage. The mycobacterial HSPs *M. tuberculosis* HSP70, *M. leprae* HSP65, and *M. bovis* BCG HSP65, when added to cultures of murine macrophages, induce the expression of mRNA for IL-1α, IL-1β, IL-6, tumor necrosis factor (TNF)-α and granulocyte/macrophage-colony stimulating factor (GM-CSF), but addition of *M. tuberculosis* hsp10, does not affect cytokine induction (RETZLAFF et al. 1994). In human macrophages, *M. bovis* BCG hsp65 also stimulates production of TNF-α and IL-1β and increases expression of the complement receptor CR3 (PEETERMANS et al. 1994).

3.5 Other Mycobacterial Factors

One of the first virulence factors associated with *M. tuberculosis* was cord factor, trehalose 6,6′-dimycolate (TDM) (BLOCH 1950; MIDDLEBROOK et al. 1947). Removal of TDM from BCG decreases its ability to survive in the lungs of mice (SILVA et al. 1985), while the addition of TDM to BCG infected mice enhances the infection, causing the mice to die more rapidly (BLOCH and NOLL 1953). Additionally, TDM concentrations are higher on the surface of virulent *M. tuberculosis* H37Rv than with the less virulent H37Ra (MIDDLEBROOK et al. 1947). There are, however, data that indicate cord factor may not be a virulence factor. TDM is present on all mycobacteria including nonpathogenic *M. smegmatis*, and the amount of cord factor has not been shown to correlate with pathogenicity of the organisms (GOREN and BRENNAN 1979). Toxicity of TDM is dependent on its physical form when presented to host cells. One inoculation of 10 μg of TDM in mineral oil in mice causes granuloma formation and hemorrhagic pneumonitis, while three injections cause death (RETZINGER et al. 1982; PEREZ et al. 1994; BEHLING et al. 1993). This same TDM preparation without mineral oil has no toxic effects and even inoculation of 10,000 μg in the absence of oil does not demonstrate toxicity (MIDDLEBROOK et al. 1947). The toxicity of TDM is dependent on its structure in the presence of 30% trehalose and 70% exposed mycolic acids and forms cords similar to those seen with *M. tuberculosis* (RETZINGER et al. 1981). Trehalose-6 monomycolate (TMM), which differs from TDM by the absence of one mycolic acid, does not form cords. In addition, TMM does not induce as many granulomas nor as much TNF-α or macrophage procoagulant activity as TDM (BEHLING et al. 1993). These findings suggest the surface cording structures of TDM may be responsible for virulence observed with *M. tuberculosis* infections.

Sulfolipid 1 is an acidic lipid isolated from virulent *M. tuberculosis* that primes neutrophils to produce super oxide (ZHANG et al. 1991a). Production of sulfolipids is associated with virulence of mycobacteria (GOREN et al. 1974), and although sulfolipid 1 stimulates antimicrobial activity in both neutrophils and monocytes, it may also act as a virulence factor by inducing inflammatory proteins that can cause damage to host cells (ZHANG et al. 1988). It has been demonstrated that mycobacteria possess a SOD gene and that SOD is secreted by *M. tuberculosis* as it grows (KOBAYASHI et al. 1988; ANDERSEN et al. 1991; DESHPANDE et al. 1993), suggesting that these bacteria have numerous mechanisms for evading or neutralizing antimicrobial activity of host cells.

The urease gene of *M. tuberculosis* recently has been cloned, characterized, and found to be similar to ureases from other species of bacteria (REYRAT et al. 1995; CLEMENS et al. 1995). It has been hypothesized that intracellular mycobacterial urease may generate ammonia that inhibits phagosome–lysosome fusion and acidification of lysosomes within macrophages (CLEMENS et al. 1995). Increased urease activity, however, is not associated with relative virulence of mycobacterial species as demonstrated by urease from

M. smegmatis that has 11 times more activity than the urease of *M. tuberculosis* (CLEMENS et al. 1995).

Free iron is very limited in the host, particularly in extracellular sites. To obtain iron at sites where it is limited, many pathogens have developed high affinity iron-binding molecules of their own called siderophores, which can remove iron from host iron-binding molecules. *M. tuberculosis* has been shown by Barclay and Ratledge (1988) and Gobin et al. (1995) to produce siderophores called exochylins. It has been proposed that exochelins bind iron in the extracellular agueous environment and transport the metal to mycobactin, another molecule in the mycobacterial cell wall, which facilitate transport of iron into the cell cytoplasm.

Other factors contributing to virulence of *M. tuberculosis* are not necessarily isolated proteins or gene products, but intrinsic properties of the organism such as the ability to multiply quickly. Studies by North and Izzo show that the growth rate of pathogenic *M. tuberculosis* Erdman and H37Rv is significantly greater in the lungs of hydrocortisone-treated severe combined immunodeficiency (SCID) mice than the growth rate of BCG and H37Ra (NORTH and IZZO 1993). Another mechanism of virulence may be the ability of the different strains of *M. tuberculosis* to resist the effects of reactive nitrogen intermediates. *M. tuberculosis* strains that survive exposure to sodium nitrate for 24 h correlate to strains that are virulent in guinea pigs (O'BREIN et al. 1994). Virulence of *M. tuberculosis* strains does not, however, appear to correlate with drug resistance (ORDWAY et al. 1995). Drug-resistant *M. tuberculosis* strains, including some that are multi-drug resistant, were used to infect immunocompetent, healthy mice; however, no pattern of virulence emerged from the experiments.

4 Conclusions

In this chapter, we described some of the ways in which *M. tuberculosis* interacts with the host to cause disease and various means in which to assess these interactions. Many questions remain, however, as to the pathogenesis of tuberculosis and how these organisms are able to undermine host inflammatory and immune responses to cause disease.

References

Abou-Zeid C, Ratliff TL, Wiker HG, Harboe M, Bennedsen J, Rook GAW (1988) Characterization of fibronectin-binding antigens released by *Mycobacterium tuberculosis* and *Mycobacterium bovis* BCG. Infect Immun 56: 3046–3051

Adams LB, Fukutomi Y, Krahenbuhl JL (1993) Regulation of murine macrophage effector functions by lipoarabinomannan from mycobacterial strains with different degrees of virulence. Infect Immun 61: 4173–4181

Affronti LF (1988) Mycobacterial antigens: reagents for tuberculin skin testing and serodiagnosis of tuberculosis. In: Bendinelli M, Friedman H (eds) *Mycobacterium tuberculosis*. Interactions with the immune system. Plenum, New York, pp 1–37

Andersen P, Askgaard D, Ljungqvist L, Bennedsen J, Heron I (1991) Proteins released from *Mycobacterium tuberculosis* during growth. Infect Immun 59: 1905–1910

Andersen AB, Brennan P (1994) Proteins and antigene of *Mycobacterium tuberculosis*. In: Bloom BR (ed) Tuberculosis pathogenesis, protection, and control. ASM Press, Washington DC, pp 307–332

Armstrong JA, Hart PD (1971) Response of cultured macrophages to *Mycobacterium tuberculosis* with observations on fusion of lysosomes with phagosomes. J Exp Med 134: 713–740

Armstrong JA, Hart PD (1975) Phagosome–lysosome interactions in cultured macrophages infected with virulent tubercle bacilli. Reversal of the usual nonfusion pattern and observations on bacterial survival. J Exp Med 142: 1–16

Arruda S, Bomfim G, Knights R, Huima-Byron T, Riley LW (1993) Cloning of an *M. tuberculosis* DNA fragment associated with entry and survival inside cells. Science 261: 1454–1457

Balasubramanian V, Wiegeshaus EH, Smith DW (1994) Mycobacterial infection in guinea pigs. Immunobiology 191 (4–5): 395–401

Balasubramanian V, Pavelka MS Jr, Bardarov SS, Martin J, Weisbrod TR, McAdam RA, Bloom BR, Jacobs WR Jr (1996) Allelic exchange *Mycobacterium tuberculosis* with long linear recombination substrates. J Bacteriol 178: 273–279

Barclay R, Ratledge C (1988) Mycobactins and exochelins of *Mycobacterium tuberculosis, M. bovis, M. africanum*, and other related species. J Gen Microbiol 134: 771–776

Behling CA, Perez RL, Kidd MR, Staton GW Jr, Hunter RL (1993) Induction of pulmonary granulomas, macrophage procoagulant activity and tumor necrosis factor-alpha by trehalose glycolipids. Ann Clin Lab Sci 23: 256–266

Berche P, Gaillard JL, Sansonneti RL (1987) Intracellular growth of *Listeria monocytogenes* as a prerequisite for in vivo induction of T cell-mediated immunity. J Immunol 138: 2266–71

Birkness KA, Swisher BL, White EH, Long EG, Ewing EP Jr, Quinn FD (1995) A tissue culture bilayer model to study the passage of *Nisseria meningtidis*. Infect Immun 63: 402–409

Bliska J B, Galan JE, Falkow S (1993) Signal transduction in the mammalian cell during bacterial attachment and entry. Cell 73: 903–920

Bloch H (1950) Studies on the virulence of tubercle bacilli: isolation and biological properties of a constituent of virulent organisms. J Exp Med 91: 197–210

Bloch H, Noll H (1953) Studies on the virulence of tubercle bacilli. Variations in the virulence effect elicited by Tween 80 and thiosemicarbazone. Br J Exp Pathol 97: 1–16

Bloom BR, Fine PE (1994) The BCG experience: implications for future vaccines against tuberculosis. In: Bloom BR (ed) Tuberculosis: Pathogenesis, protection and control. ASM Press, Washington DC, pp 531–557

Brennan PJ (1989) Structure of mycobacteria: recent developments in defining cell wall carbohydrates and proteins. J Infect Dis 11: S420–S430

Brennan PJ, Nikaido H (1995) The envelope of mycobacteria. Annu Rev Biochem 64: 29–63

Brown MC, Taffet SM (1995) Lipoarabinomannans derived from different strains of *Mycobacterium tuberculosis* differentially stimulate the activation of NF-kappaB and KBF1 in murine macrophages. Infect Immun 63: 1960–1968

Chan J, Fan X, Hunter SW, Brennan PJ, Bloom BR (1991) Lipoarabinomannan, a possible virulence factor involved in persistence of *Mycobacterium tuberculosis* within macrophages. Infect Immun 59: 1755–1761

Chatterjee D, Bozic CM, McNeil M. Brennan PJ (1991) Structural features of the arabinan component of the lipoarabinomannan of *Mycobacterium tuberculosis*. J Biol Chem 266: 9652–9660

Chatterjee D, Roberts AD, Lowell K, Brennan PJ, Orme IM (1992) Structural basis of capacity of lipoarabinomannan to induce secretion of tumor necrosis factor. Infect Immun 60: 1249–1253

Clemens DL, Horwitz MA (1995) Characterization of the *Mycobacterium tuberculosis* phagosome and evidence that phagosomal maturation is inhibited. J Exp Med 181: 257–270
Clemens DL, Lee BY, Horwitzz MA (1995) Purification, characterization, and genetic analysis of *Mycobacterium tuberculosis* urease, a potentially critical determinant of host-pathogen interaction. J Bacteriol 177: 5644–5652
Cluff CW, Ziegler HK (1987) Inhibition of macrophage-mediated antigen presentation by hemolysin-producing *Listeria monocytogenes* [published erratum appears in J Immunol 1988 Apr 1; 140(7): 2477]. J Immunol 139: 3808–3812
Cluff CW, Garcia G, Ziegler HK (1990) Intracellular hemolysin-producing *Listeria monocytogenes* strains inhibit macrophage-mediated antigen processing. Infect Immun 58: 3601–12
Cohn ML, Kovitz C, Oda U, Middlebrook G (1954) studies on isoniazia and tubercle bacilli: II. The growth requirements, catalase activities, and pathogenic properties of isoniazid-resistant mutants. AM Rev Tuberc 70: 641–664
Cooper AM, Roberts AD, Rhoades ER, Callahan JE, Getzy DM, Orme IM (1995) The role of interleukin-12 in acquired immunity to *Mycobacterium tuberculosis* infection. Immunol 84: 423–432
Curcic R, Dhandayuthapani S, Deretic V (1994) Gene expression in mycobacteria: transcriptional fusions based on *xy1E* and analysis of the promoter region of the response regulator *mtrA* from *Mycobacterium tuberculosis*. Mol Microbiol 13: 1057–1064
Dannenberg AM Jr (1982) Pathogenesis of pulmonary tuberculosis. Am Rev Infect Dis 125: 25–29
Dannenberg AM, Jr, Rook GAW (1994) Pathogenesis of pulmonary tuberculosis: an interplay of tissue-damaging and macrophage-activating immune responses-dual mechanisms that control bacillary multiplication. In: Bloom BR (ed) Tuberculosis: pathogensis, protection, and control. ASM Press, Washington DC, pp 459–483
Deshpande RG, Khan MB, Savariar LS, Windham YZ, Navalkar RG (1993) Superoxide dismutase activity of *Mycobacterium avium* complex (MAC) strains isolated from AIDS patients. Tubercle Lung Dis 74: 305–309
Dhillon J, Mitchison DA (1994) Effect of vaccines in a murine model of dormant tuberculosis. Tubercle Lung Dis 75(1): 61–64
Duguid JR, Dinauer JC (1989) Library subtraction in vitro cDNA libraries to identify differentially expressed genes in scrapie infection. Nucl Acid Res 18: 2789–2792
Falcone V, Bassey E, Jacobs W Jr, Collins F (1995) The ummunogenicity of recombinant *Mycobacterium smegmatis bearing* BCG genes. Microbiol 141: 1239–1245
Filley EA, Rook GAW (1991) Effect of mycobacteria on sensitivity to the cytotoxic effects of tumor necrosis factor. Infect Immun 59: 2567–2572
Filley EA, Bull HA, Dowd PM, Rook GAW (1992) The effect of *Mycobacterium tuberculosis* on the susceptibility of human cells to the stimulatory and toxic effects of tumor necrosis factor. Immunology 77: 505–509
Fishcer LJ, Quinn FD, White EH, King CH (1996) Intracellular growth and cytoxicity of *Mycobacterium haemophilum* in a human epithelial cell line (Hec-1-B). Infect Immun 64: 269–276
Flynn JL, Goldstein MM, Triebold KJ, Koller B, Bloom BR (1992) Major histocompatibility complex class I-restricted T cells are required for resistance to *Mycobacterium tuberculosis* infection. Proc Natl Acad Sci USA 89: 12013–12017
Gadea I, Zapardiel J, Ruiz P, Gegundez MI, Estaban J, Soriano F (1993) Cytopathic effect mimicking virus culture due to *Mycobacterium tuberculosis*. J Clin Microbiol 31: 2517–2518
Garbe TR, Barathi J, Barnini S, Zhang Y, Abou-Zeid C, Tang D, Mukherjee R, Young DB (1994) Transformation of mycobacterial species using hygromycin resistance as selectable marker. Microbiology 140: 133–138
Gilbert S, Steinbrech DS, Landas SK, Hunninghake GW (1993) Amounts of angiotensinconverting enzyme mRNA reflect the burden of granulomas in granulomatous lung disease AM Rev Respir Dis 148: 483–486
Gobin J, Moore CH, Reeve JR Jr, Wong DK, Gibson BW, Horwitz MA (1995) Iron acquisition by *Mycobacterium tuberculosis*: Isolation and characterization of a family of iron-binding exochelins. Proc Natl Acad Sci USA 92: 5189–5193
Goren MB, Brennan PJ (1979) Mycobacterial lipids: chemistry and biologic activities. In: Youmans GP (ed) Tuberculosis. Saunders, Philadelphia, pp 63–193
Goren MB, Broki O, Schaefer WB (1974) Lipids of putative relevance to virulence in *Mycobacterium tuberculosis*: correlation of virulence with elaboration of sulfatides and strongly acidic lipids. Infect Immun 9: 142–149

Greenberg SS, Xie J, Kolls J, Mason C, Didier P (1995) Rapid induction of mRNA for nitric oxide synthase II in rat alveolar macrophages by intratracheal administration of *Mycobacterium tuberculosis* and *Mycobacterium avium*. Proc Soc Exp Biol Med 209: 46–53

Guilhot C, Otal I, VanRompaey I, Martin C, Gicquel B (1994) Efficient transposition in mycobacteria: construction of *Mycobacterium smegmatis* insertional mutant libraries. J Bacteriol 176: 535–539

Gupta S, Tyagi AK (1993) Sequence of a newly indentified *Mycobacterium tuberculosis* gene encoding a protein with a sequence homology to virulence-regulating proteins. Gene 126: 157–158

Hirsch CS, Ellmer JJ, Russell DG, Rich EA (1994) Complement receptor-mediated uptake and tumor necrosis factor-alpha-mediated growth inhibition of *Mycobacterium tuberculosis* by human alveolar macrophages. J Immunol 152: 743–753

Jacobs WR Jr, Bloom BR (1994) Molecular genetic strategies for identifying virulence determinants of *Mycobacterium tuberculosis*. In: Bloom BR (ed) Tuberculosis: pathogenesis, protection, and control. ASM Press, Washington DC, pp 253–268

Jacobs WR, Doherty MA, Curtiss R III, Clark-Curtiss JE (1986) Expression of *Mycobacterium leprae* genes from a *Streptococcus mutans* promoter in *Escherichia coli* K–12. Proc Natl Acad Sci USA 83: 1926–1930

Jindal S, Dudani AK, Singh B, Harley CB, Gupta RS (1989) Primary structure of a human mitochondrial protein homologous to the bacterial and plant chaperonins and to the 65-kilodalton mycobacterial antigen. Mol Cell Biol 9: 2279–2283

Kalpana GV, Broom BR, Jacobs WR Jr (1991) Insertional mutagenesis and illegitimate recombination in mycobacteria. Proc Natl Acad Sci USA 88: 5433–5437

Karopoulous C, Rowley MJ, Handley CJ, Strugnell RA (1995) Antibody reactivity to mycobacterial 65kDa heatshock protein – relevance to autoimmunity. J Autoimmunity 8: 235–248

Kathariou S, Metz P, Hof H, Goebel W (1987) Tn916-induced mutations in the hemolysin determinant affecting virulence of *Listria monocytogenes*. J Bacteriol 169: 1291–1297

Kikuta-Oshima LC, King CH, Shinnick TM, Quinn FD (1994) Methods for the identification of virulence genes expressed in *Mycobacterium tuberculosis* strain H37Rv. Ann NY Acad Sci 730: 263–265

King CH, Shinnick TM (1995) Isolation of putative hemolysin gene from *Mycobacterium tuberculosis*. J Cell Biochem [Suppl 19B]: 85

King CH, Mundayoor S, Crawford JT, Shinnick TM (1993) Expression of contactdependent cytolytic activity by *Mycobacterium tuberculosis* and isolation of the genomic locus that encodes the activity. Infect Immun 61: 2708–2712

Kinger AK, Tyagi JS (1993) Identification and cloning of genes differentially expressed in the virulent strain of *Mycobacterium tuberculosis*. Gene 131: 113–117

Kobayashi Y, Apella E, Yamada M. Copeland TD, Oppenheim JJ, Matsushima K (1988) Phosphorylation of intracellular precursors of human IL-1. J Immunol 140: 2279–2287

Leao SC, Rocha CL, Murillo LA, Parra CA, Patarroyo ME (1995) A species specific nucleotide sequence of *Mycobacterium tuberculosis* encodes a protein that exhibits hemolytic activity when expressed in *Escherichia coli*. Infect Immun 63: 4301–4306

Lee MH, Pascopella L, Jacobs WR Jr, Hatfull GF (1991) Site-specific integration of mycobacteriophage L5: integration-proficient vectors for *Mycobacterium smegmatis*, BCG, and *M. tuberculosis*. Proc Natl Acad Sci USA 88: 3111–3115

Liang P, Pardee AB (1992) Differential display of eukaryotic messenger RNA by measn of the polymerase chain reaction. Science 257: 967–971

Lim EM, Rauzier J, Timm J, Torrea G, Murray A, Gicquel B, Portnoi D (1995) Identification of *Mycobacterium tuberculosis* DNA sequences encoding exported proteins by using *phoA* gene fusions. J Bacteriol 177: 59–65

Lisitsyn N, Lisityn N, Wigler M (1993) Cloning the difference between two complex genomes. Science 259: 946–51

Lopez GM, Rom WN, Ciotoli C, Talbot A, Martiniuk F, Cronstein B, Reibman J (1994) *Mycobacterium tuberculosis* alters expression of adhesion molecules on monocyte cells. Infect Immun 62: 2515–2520

Lukacs NW, Kunkel SL, Allen R, Evanoff HL, Shaklee CL, Sherman JS, Burdick MD, Strieter RM (1995) Stimulus and cell specific expression of C-X-C and C-C chemokines by pulmonary stromal cell populations. Am J Physiol 268: L856–L861

Mahan MJ, Slauch JM, Mekalanos JJ (1993) Selection of bacterial virulence genes that are specifically induced in host tissues. Science 259: 686–688

Malpani BL, Kadival GV, Samuel AM (1992) Radio immunoscintigraphic approach for the in vivo detection of tuberculomas–a preliminary study in a rabbit model. Int J Rad Applic Instru–Part B., Nucl Med Bio 19(1): 45–53

Mathiopoulos C, Sonenshein AL (1989) Identification of *Bacillus subtilis* genes expressed early during sporulation. Mol Microbiol 3: 1071–1081

McAdam RA, Weisbrod TR, Martin J, Scuderi JD, Brown AM, Cirillo JD, Bloom BR, Jacobs WR Jr (1995) In vivo growth characteristics of leucine and methionine auxotrophic mutants of *Mycobacterium bovis* BCG generated by transposon mutagenesis. Infect Immun 63: 1004–1012

McDonough KA, Kress Y (1995) cytotoxicity for lung epithelial cells is virulence is a virulence-associated phenotype of *Mycobacterium tuberculosis*. Infect Immun 63: 4802–4811

McDonough KA, Kress Y, Bloom BR (1993) Pathogenesis of tuberculosis: interaction of *Mycobacterium tuberculosis* with macrophages [published erratum appears in Infect Immun 1993 Sep; 61(9): 4021–4]. Infect Immun 61: 2763–73

Mehta PK, King CH, White EH, Murtaugh JJ Jr, Quinn FD (1996) Comparing in vitro models for the study of *Mycobacterium tuberculosis* invasion and intracellular replication. Infect Immun (in press)

Mendez-Samperio P, Gonzalez-Garcia L, Pineda-Fragoso PR, Ramos-Sanchez E (1995) Specificity of T cells in human resistance to *Mycobacterium tuberculosis* infection. Cell Immunol 162: 194–201

Middlebrook G, Cohn ML (1953) some observations on the pathogenicity of isoniazid-resistant variants of tubercle bacilli Science 118: 297–299

Middlebrook G, Dubos RJ, Pierce CH (1947) Virulence and morphological characteristics of mamalian tubercle bacilli. J Exp Med 86: 175–184

Moulder JW (1985) Comparative biology of intracellular parasitism. Microbiol Rev 49: 298–337

Myrvik QN, Leake ES Wright MJ (1984) Disruption of phagosomal membranes of normal alveolar macrophages by the H37Rv strain of *Mycobacterium tuberculosis*. A correlate of virulence. Am Rev Respir Dis 129: 322–328

Ngeleka M, Harel J, Jacques M, Fairbrother JM (1992) Characterization of a polysaccharide capsular antigen of septicemic *Escherichia coli* 0115: K "V165": F165 and evaluation of its role in pathogenicity. Infect Immun 60: 5048–5056

Norman E Dellagostin OA, McFadden J, Dale JW (1995) Gene replacement by homologous recombination in *Mycobacterium bovis* BCG. Mol Microbiol 16:755–760

North RJ, Izzo AA (1993) Mycobacterial virulence. Virulent strains of *Mycobacteria tuberculosis* have faster in vivo doubling times and are better equipped to resist growth inhibiting functions of macrophages in the presence and absence of specific immunity. J Exp Med 77: 1723–1733

O'Brien L, Carmichael J, Lowrie DB, Andrew PW (1994) Strains of *Mycobacterium tuberculosis* differ in susceptibility to reactive nitrogen intermediates in vitro. Infect Immun 62: 5187–5190

Ordway DJ, Sonnenberg MG, Donahuye SA, Belisle JT, Orme IM (1995) Drug resistant strains of *Mycobacterium tuberculosis* exhibit a range of virulence for mice. Infect Immun 63: 741–743

Pascopella L, Collins FM, Martin JM, Lee MH, Hatfull GF, Stover CK, Bloom BR, Jacobs WR Jr (1994) Use of in vivo complementation in *Mycobacterium tuberculosis* to identify a genomic fragment associated with virulence. Infect Immun 62:1313–1319

Peetermans WE, Raats CJI, Langermans JAM, van Furth R (1994) Mycobacterial heat shock protein 65 induces proinflammatory cytokines but does not activate human mononuclear phagocytes. Scand J Immunol 39: 613–617

Perez RL, Roman J, Staton GW Jr, Hunter RL (1994) Extravascular coagulation and fibrinolysis in murine lung inflammation induced by the mycobacterial cord factor trehalose-6,6'-dimycolate. Am J Respir Crit Care Med 149: 510–518

Perosio PM, Frank TS (1993) Detection and species identification of mycobacteria in paraffin sections of lung biopsy specimens by the polymerase chain reaction. Am J Clin Pathol 100 (6): 643–647

Plum G, Clark-Curtiss JE (1994) Induction of *Mycobacterium avium* gene expression following phagocytosis by human macrophages. Infect Immun 62:476–483

Portnoy DA, Jacks PS, Hinrichs DJ (1988) Role of hemolysin for the intracellular growth of *Listeria monocytogenes*. J Exp Med 167: 1459–1471

Raj CV, Ramakrishnan T (1970) Transduction in *Mycobacterium smegmatis*. Nature 228: 280–281

Ratliff TL, Palmer JO, McGarr J, Brown EJ (1987) Intravesical bacillus Calmette- therapy for murine bladder tumors: initiation of the response by fibronectin-mediated attachment of bacillus Calmette-Guerin. Cancer Res 47: 1762–1766

Retzinger GS, Meredith SC, Takayama K, Hunter RL (1981) The role of surface in the biological activities of trehalose 6,6' dimycolate: surface properties and development of a model system. J Biol Chem 256: 8208–8216

Retzinger GS, Meredith SC, Hunter RL, Takayama K, Kezdy FJ (1982) Identification of the physiologically active state of the mycobacterial glycolipid trehalose 6,6' dimycolate monolayers. J Immunol 129: 735–744

Retzlaff C, Yamamoto Y, Hoffman PS, Friedman H, Klein TW (1994) Bacterial heat shock proteins directly induce cytokine mRNA and interleukin-1 secretion in macrophage cultures. Infect Immun 62: 5689–5693

Reyrat JM, Berthet FX, Gicquel B (1995) The urease locus of *Mycobacterium tuberculosis* and its utilization for the demonstration of allelic exchange in *Mycobacterium bovis* bacillus Calmette-Guerin. Proc Natl Acad Sci USA 92: 8768–8772

Roach TIA, Barton CH, Chatterjee D, Blackwell JM (1993) Macrophage activation: Lipoarabinomannan from avirulent and virulent strains of *Mycobacterium tuberculosis* differentially induces the early genes c-fos, KC, JE, and tumor necrosis factor-α. J Immunol 150: 1886–1896

Roach TIA, Barton CH, Chatterjee D, Liew FY, Blackwell JM (1995) Opposing effects of interferon-gamma on iNOS and interleukin-10 expression in lipopolysaccharide-and mycobacterial lipoarabinomannan-stimulated macrophages. Immunol 85: 106–113

Roach TIA, Chatterjee D, Blackwell JM (1994) Induction of early response genes KC and JE by mycobacterial lipoarabinomannans: regulation of KC expression in murine macrophages by *LshItyBcg* (candidate *Nramp*). Infect Immun 62: 1176–1184

Rock FM, Landi MS, Meunier LD, Morrsi TH, Rolf LL, Warnick CL, McCreedy BJ, Hughes HC (1995) Diagnosis of a case of *Mycobacterium tuberculosis* in a cynomolgus (Macaca fascicularis) monkey colony by polymerase chain reaction and enzyme-linked immunosorbent assay. Lab Ani Sci 45: 315–319

Rook GAW, Bloom BR (1994) Mechanisms of pathogenesis in tuberculosis, In: Bloom BR (ed) Tuberculosis: pathogenesis, protection, and control. ASM press, Washington DC, pp 485–501

Sansonetti PJ (1991) Genetic and molecular basis of epithelial cell invasion by Shigella species. Rev Infect Dis 13: S85–92

Schlesinger LS (1993) Macrophage phagocytosis of virulent but not attentuated strains of *Mycobacterium tuberculosis* is mediated by mannose receptors in addition to complement receptors. J Immunol 150:2920–2930

Schlesinger LS, Bellinger-Kawahare CG, Payne NR, Horowitz MA (1990) Phagocytosis of *Mycobacterium tuberculosis* is mediated by human monocyte complement receptors and complement component C3. J Immunol 144: 2771–2780

Schorey JS, Li Q, McCourt DW, Bong-Mastek M, Clark-Curtiss JE, Ratliff TL, Brown EJ (1995) A *Mycobacterium leprae* gene encoding a fibronectin binding protein is used for efficient invasion of epithelial cells and Schwann cells. Infect Immun 63: 2652–2657

Sibley LD, Hunter SW, Brennan PJ, Krahenbuhl JL (1988) Mycobacterial lipoarabinomannan inhibits gamma interferon mediated activation of macrophages. Infect Immun 56: 1232–1236

Silva CL (1995) New vaccines against tuberculosis. Braz J Med Biol Res 28: 843–851

Silva CL, Ekizlerian SM, Fazioli RA (1985) Role of cord factor in the modulation of infection caused by mycobacteria. Am J Pathol 118: 238–247

Singh NB, Srivastava K, Malaviya B, Srivastava A, Gupta HP (1995) The 65 kDa protein of *Mycobacterium habana* and its putative role in immunity against experimental tuberculosis. Immunol Cell Biol 73: 372–376

Smith H (1968) Biochemical challange of microbial pathogenicity. Bacteriol Rev 32: 164–184

Stadnyk AW (1994) Cytokine production by epithelial cells. FASEB J 8: 1041–1047

Steenken W Jr, Oatway WH, Petroff SA (1934) Biological studies of the tubercle bacillus. III. Dissociation and pathogenecity of the R and S variants of the human tubercle bacillus (H37). J Exp Med 60: 515–525

Stephens DS (1989) Gonococcal and meningococcal pathogenesis as defined by human cell, cell culture, and organ culture assays. Clin Microbiol Rev 2 (suppl.): S104–S111

Straus D, Ausubel FM (1990) Genomic subtractions for cloning DNA corresponding to deletion mutation. Proc Natl Acad Sci 87: 1889–1893

Sturgil-Koszycki S, Schlesinger PH, Chakraborty P, Haddix PL, Collins HL, Fok AK, Allen RD, Gluck SL, Heuser J, Russell DG (1994) Lack of acidification in Mycobacterium phagosomes produced by exclusion of the vesicular proton-ATPase. Science 263: 678–681

Timblin C, Battey J, Kuehl WM (1990) Application for PCR technology to subtractive cDNA cloning: identification of genes expressed specifically in murine plasmacytoma cells. Nucl Acids Res 18: 1587–1593

Tokunaga T, Mizuguchi Y, Suga K (1973) Genetic recombination in mycobacteria. J Bacteriol 113: 1104–1111

Udou T (1994) Extracellular hemolytic activity in rapidly growing mycobacteria. Can J Microbiol 40: 318–321

Utt EA, Brousal JP, Kikuta-Oshima LC, Quinn FD (1995) The identification of bacterial gene expression differences using mRNA-based isothermal subtractive hybridization. Can J Microbiol 41: 152–156

Verma A, Kinger AK, Tyagi JS (1994) Functional analysis of transcription of the *Mycobacterium tuberculosis* 16s rDNA-encoding gene. Gene 148: 113–118

Welch RA, Felmlee T, Pellett F, Chenoweth DE (1986) The *Escherichia coli* haemolysin: Its gene organization and interaction with neutrophil receptors. In: lark DL et al. (eds) Protein-carbohydrate interactions in biological systems: the molecular biology of microbial pathogenicity. New York, pp 431–438

Wiker HG, Harboe M (1992) The antigen 85 complex: a major secretion product of *Mycobacterium tuberculosis*. Microbiol Rev 56: 648–661

Weinberg ED (1993) The iron-withholding defense system. ASM News 59: 559–562

Wilson CB, Tsai V, Remington JS (1980) Failure to trigger the oxidative burst of normal macrophages. Possible mechanisms for survival of intracellular pathogens. J Exp Med 151: 328–346

Xu S, Cooper A, Sturgill-Koszycki S, van Heyningen T, Chatterjee D, Orme I, Allen P, Russell DG (1994) Intracellular trafficking in *Mycobacterium tuberculosis* and *Mycobacterium avium*-infected macrophages. J Immunol 153: 2568–2578

Young RA, Bloom BR, Grosskinsky CM, Ivanyi J, Thomas D, Davis RW (1985) Dissection of *Mycobacterium tuberculosis* antigens using recombinant DNA. Proc Natl Acad Sci USA 82: 2583–2587

Zhang L, Medina D (1993) Gene expression screening for specific genes associated with mouse mammary tumor development. Mol Carcino 8: 123–136

Zhang L, Goren MB, Holzer TJ, Andersen BR (1988) Effect of *Mycobacterium tuberculosis* derived sulfolipid I on human phagocytic cells. Infect Immun 56: 2876–2883

Zhang L, English D, Andersen BR (1991a) Activation of human neutrophils by *Mycobacterium tuberculosis* derived sullfolipid-I. Immunol 146: 2730–2736

Zhang Y, Lathigra R, Garbe T, Catty D, Young D (1991b) Genetic analysis of superoxide dismutase, the 23 kilodalton antigen of *Mycobacterium tuberculosis*. Mol Microbiol 5: 381–391

Zhang Y, Broser M, Cohen H, Bodkin M, Law K, Reibman J, Rom WN (1995) Enhanced interleukin-8 release and gene expression in macrophages after exposure to *Mycobacterium tuberculosis* and its components. J Clin Invest 95: 586–592

Pathogenesis of Experimental Tuberculosis in Animal Models

D.N. McMurray[1], F.M. Collins[2], A.M. Dannenberg, Jr.[3], and D.W. Smith[4]

1 Introduction

Pathogenesis, in the context of an infectious disease, can be defined as the temporal interplay between the microbe and its host which ultimately results in clinical illness. An understanding of pathogenesis requires knowledge of the precise aggressive and/or evasive strategies employed by the microbe and the elucidation of the beneficial and detrimental responses which the host develops upon infection. The kinetics of the host-microbe interaction are at least as important as the isolated activities of either protagonist, since the effectiveness of the strategies and counterstrategies is likely to depend upon the context and environment in which the confrontation occurs. For this reason, the pathogenesis of a disease like tuberculosis will only be revealed by studies of the course of infection in a suitable host.

Although much has been learned about the pathogenesis of tuberculosis from studies of human patients, many fundamental observations have been made in experimental animal models. The current global tuberculosis epidemic has stimulated intense activity at the clinical and basic research levels to generate the data upon which more successful control of this disease will eventually

[1]Department of Medical Microbiology and Immunology, Texas A&M University Health Science Center, College Station, TX 77843, USA
[2]Mycobacteriology Laboratory, Food and Drug Administration, Bethesda, MD 20892, USA
[3]Department of Environmental Health Sciences, School of Hygiene and Public Health, The Johns Hopkins University, Baltimore, MD 21205, USA
[4]Department of Medical Microbiology and Immunology, University of Wisconsin, Madison, WI 53706, USA

be based. The world-wide resurgence of this ancient scourge has revealed significant gaps in our knowledge about the host-parasite relationship, as well as exposing the weaknesses in our current strategies for prevention, diagnosis and treatment (BLOOM and MURRAY 1992; COLLINS 1993). Thus, improved vaccines or vaccination regimens, more rapid and accurate diagnostic tests, and new chemotherapeutic and immunotherapeutic modalities will likely all contribute to better tuberculosis control (ELLNER et al. 1993). It is clear that many of these new tools will require development and testing in animal models of tuberculosis before there is sufficient justification to deploy them in humans.

Since the first published studies of experimental infection with *Mycobacterium tuberculosis* by Koch more than a century ago (KOCH 1882), many animal models have been employed (MCMURRAY 1994; ORME and COLLINS 1994; SMITH and WIEGESHAUS 1989). In fact, one of the difficulties encountered when one attempts to synthesize a common understanding of the host-parasite relationship in experimental tuberculosis is the tremendous heterogeneity in the animal models used around the world in the past 40 years (SMITH and HARDING 1977b). The choice of an animal model may have a profound effect on the conclusion drawn from the experiments performed, and it is quite clear that different animal models may yield dramatically different results when identical experimental questions are asked. Although one might argue that a rationale exists upon which the selection of the most rational or relevant test system for human tuberculosis could be made (SMITH and HARDING 1977a), only a parallel comparison of various models with humans tested under identical conditions would reveal which model was truly the most relevant. The inherent advantages and disadvantages of existing animal models of tuberculosis suggest that different models might better serve different test purposes. Thus, the greatest utility of animal models in the current struggle against tuberculosis comes from their selective application to the different experimental questions we wish to answer.

Since the 1940s, several animal species have been used in experimental tuberculosis research, including the rat (GRAY 1961; WESSELS 1941), mouse (ORME and COLLINS 1994), rabbit (ALLISON et al. 1962; DANNENBERG et al. 1968; LURIE 1964), guinea pig (SMITH and WIEGESHAUS 1989; SMITH and HARDING 1977a), and nonhuman primate (BARCLAY et al. 1970). Currently, most active laboratories employ mice, guinea pigs, or rabbits. A significant body of literature exists describing the host-parasite relationship in these latter species following infection with virulent human and bovine tubercle bacilli. Descriptions of the models themselves, and the nature of the host immune response to *Mycobacterium tuberculosis* in each species, have been published recently (DANNENBERG 1994a; DANNENBERG and ROOK 1994; MCMURRAY 1994; ORME and COLLINS 1994). In this chapter, we will focus on the pathogenesis of infection with *M. tuberculosis* in the guinea pig, mouse and rabbit. We will attempt to draw comparisons between these animal models and humans, with special emphasis on the principal features of tuberculosis and the determinants of disease outcome in susceptible and resistant hosts.

2 Experimental Tuberculosis in the Guinea Pig

Early work with guinea pigs revealed that they were exquisitely susceptible to infection with virulent strains of *M. tuberculosis* following either injection or inhalation (LONG et al. 1931; RATCLIFFE and PALLADINO 1953). The ability to study the pathogenesis of tuberculosis in the guinea pig in a consistent and systematic fashion, however, required the development of an aerosol infection device in which animals could be exposed uniformly and reproducibly to aerosols containing low levels of virulent mycobacteria as single-cell droplet nuclei (MIDDLEBROOK 1952; WIEGESHAUS et al. 1970). Because the design of the infection chamber and the construction of the nebulizer for aerosol infection is essential to the performance of studies employing this route of challenge, and because recent experience has indicated that problems have existed with chamber design and nebulizer performance, it is appropriate to include a brief update on equipment for aerosol infection of animals. MIDDLEBROOK (1952), developed an aerosol infection chamber, the TRI-R chamber, which is the most widely employed infection chamber. WIEGESHAUS et al. (1970) extensively modified the TRI-R chamber to improve its performance (the modifications also apply to the Glas-Col chamber, developed to replace the TRI-R chamber). For example, modifications in the Madison chamber (illustrated in diagrammatic form in WIEGESHAUS et al. 1970) include: shift to horizontal configuration; an increase in the airflow rate in order to reduce the temperature and humidity in the chamber; introduction of a mixing chamber before and after the animal exposure basket; and four fans mounted in the chamber to increase mixing. The most recent modification of the Madison chamber was to replace the all glass Nebulizer-Venturi unit of the TRI-R chamber with a modified MRE type, Collison-Venturi Unit (GUSSMAN 1984). The low variance between animals and between runs of the Madison chamber (HO et al. 1978) was due to the design changes incorporated in the chamber and the use of a frozen suspension to prepare the challenge inoculum (GROVER et al. 1967).

Using this technology, studies of the fundamental aspects of the host-parasite relationship in pulmonary tuberculosis have been carried out over the past 25 years. It was demonstrated that guinea pigs could be infected reproducibly by the respiratory route with only a few (one to three) viable, virulent *M. tuberculosis* H37Rv or Erdman (SMITH et al. 1970). It was observed that the number of viable mycobacteria cultured from the whole lung immediately after aerosol exposure to a virulent strain was not significantly different from the number of primary tubercles which developed over the subsequent 4–6 weeks (unpublished observations). Following implantation of droplet nuclei containing virulent *M. tuberculosis* in the alveolar spaces of the immunologically naive guinea pig lung, the following series of events were shown to occur. Tubercle bacilli grew exponentially in the infected lung lobes between 3 and 14 days postinfection. At about 14 days, the first viable organisms appeared in the lymph nodes draining the lung fields, and somewhat later (18–21 days postinfection) the mycobacteria were first detected in the spleen, which had been seeded by organisms in the blood which entered from the bronchotracheal and hilar lymph nodes via the thoracic ducts (SMITH and WIEGESHAUS 1989; SMITH et al. 1970). Virtually concomitant with detectable bacillemia, the guinea pigs converted their tuberculin skin tests to positive. Infected guinea pigs developed intense, indurated delayed hypersensitivity reactions 24 h following intradermal injection of low doses (1–5 TU) of purified protein derivative (PPD) which were remarkably similar to human tuberculin reactions histologically. Meanwhile, bacterial accumulation in the lung continued unabated until 21–24 days postinfection, at which point the bacillary load in the lung exceeded 1 million viable *M. tuberculosis*. The primary tubercle in the guinea pig was observed to consist of a typical granulomatous lesion composed of lymphocytes, blood monocytes, epithelioid cells and multinucleated giant (Langhans) cells (SMITH and WIEGESHAUS 1989). At this point, the onset of an intense cell-mediated immune response to antigens from the large number of bacilli resulted in massive local tissue destruction, the onset of caseation necrosis and progressively compromised respiratory function which ultimately led to death (WIEGESHAUS et al. 1970; SMITH and HARDING 1977a). Primary lung lesion often calcified, producing a very characteristic pattern on X-ray examination (SMITH et al. 1991).

Cellular and bacterial interactions in the lung during the early phase of pulmonary tuberculosis in this model, however, are not as simple as that. One of the advantages of the low dose of infection is that not all lung lobes receive virulent mycobacteria initially via the aerosol. Smith and colleagues compared the bacterial loads of separate lung lobes, those containing primary lesions and those with no primary lesion (FOK et al. 1976; HO et al. 1978). Lobes initially not infected remained free of bacilli for more than 2 weeks when, simultaneously with the extrapulmonary spread of mycobacteria via the lymph to the hilar and bronchotracheal lymph nodes and then via the blood stream to the liver and spleen, viable mycobacteria first began to appear in previously sterile lung lobes (FOK et al. 1976; HO et al. 1978). Subsequently, levels of viable *M. tuberculosis* increased in those previously lesion-free lobes until reaching approximately the same levels as those observed in primary lobes. These experiments proved that a "silent bacillemia," which is thought to be an important aspect of pathogenesis in human tuberculosis, also occurs in guinea pigs (BALASUBRAMANIAN et al. 1992a, 1994; PRABHAKAR et al. 1987; STEAD 1989). The results also suggest that events in the lung are dynamic and complicated, with a few mature granulomas in which the expression of the developing immune response will result in some control of bacillary accumulation and tissue necrosis, and many hematogenous foci in which the mycobacteria are attempting to thrive in the face of preexisting cell-mediated immunity (BALASUBRAMANIAN et al. 1994; SMITH and WIEGESHAUS 1989). These metastatic lung foci are thought to be the ones which reactivate to produce endogenous tuberculosis in the elderly and in HIV-infected individuals (SMITH and WIEGESHAUS 1989; STEAD 1989).

The significance of this bacillemic phase of pulmonary tuberculosis, and our ability to study it in the guinea pig, cannot be overstated. The propensity of a human isolate of *M. tuberculosis* to result in hematogeneous seeding of the spleen and lung may be a measure of the intrinsic virulence of that strain. MITCHISON (1964) reported that isolates of tubercle bacilli from south India differed in virulence for guinea pigs after intramuscular infection. PRABHAKAR et al. (1987) reported assays based on intramuscular infection or airborne infection. For both groups of investigators, virulence was assessed by the extent of gross disease in the organs and by the number of bacilli recovered from the spleen. BALASUBRAMANIAN et al. (1992a) described a quantitative modification of the "root index of virulence" assay employed by the Mitchison group. BALASUBRAMANIAN et al. (1992b) reported that south India isolates differed in efficiency to initiate an infection via the respiratory route, a difference which was correlated with the rate of multiplication at sites of hematogenous seeding. Moreover, metastatic foci which developed in the lungs were fewer in number for the low virulence isolates. Thus, the low virulence isolates seem to have a reduced probability of establishing an apical lesion in the lungs of humans via the bloodstream (BALASUBRAMANIAN et al. 1994). They probably grow less well both in the primary focus and in the metastatic apical focus. Therefore, fewer bacilli would reach the apex, and fewer would accumulate there.

Whether a human strain of *M. tuberculosis* is highly virulent (i.e., can produce metastatic lung foci on first infection) or a low virulence (i.e., requires multiple exposures to airborne bacilli to result in a viable focus in the apex of the lung) may have profound effect on subsequent events and disease expression in the host (BALASUBRAMANIAN et al. 1994; SMITH and WIEGESHAUS 1989; WIEGESHAUS et al. 1989). Exogenous reinfection tuberculosis results from infection by inhalation of a different external inoculum of bacilli from that which caused the primary infection. Different vaccination strategies may be required to effectively protect people from endogenous reactivation than from exogenous reinfection. World tuberculosis conferences (FOGARTY 1978; PITTSFIELD 1986; WHO 1976, 1982) have encouraged research on the development of tuberculosis vaccines designed to inhibit some aspect of exogenous re-infection tuberculosis, (reviewed by BALASUBRAMANIAN et al. 1994). There is a need for experimental test systems in which to evaluate vaccines which prevent seeding or inhibit development of aerosol implants. Such animal models have been referred to as antireactivation models (WIEGESHAUS and SMITH 1989). Research toward an antireactivation model has been reported by LAGRANDERIE et al. (1993), who evaluated the protective potency of BCG administered intradermally and aerogenically to guinea pigs in low and high doses and showed evidence that BCG given in a high dose via the respiratory route affected the growth of the virulent challenge strain at the sites of implantation in the lung. This is perhaps the first evidence of a protective effect of a vaccine which could influence either the endogenous reactivation or the exogenous reinfection pathways.

The pathogenesis of pulmonary tuberculosis as just described in naive guinea pigs can be modified by altering the immunological status of the animal prior to infection . Previous vaccination with *Mycobacterium bovis* BCG vaccine results in a significant alteration in the course of events. BCG vaccination has been shown to be very effective at very low doses (10^1–10^2 cfu) when given parenterally (JESPERSEN 1956) or at higher doses by the intranasal (unpublished observation) or aerosol routes (LAGRANDERIE et al. 1993). BCG-vaccinated guinea pigs have a significantly higher long-term survival than nonvaccinated animals following low dose virulent pulmonary challenge, both in terms of percent survival 18 month postchallenge and mean days to death for those dying before the termination of the experiment (WIEGESHAUS et al. 1970). The experiment was terminated at a point when the guinea pigs reached their normal longevity and deaths due to natural aging were expected to occur. At this time, more than 60% of the BCG-vaccinated animals were still alive and in apparent good health (WIEGESHAUS et al. 1970). Vaccinated animals necropsied at earlier intervals (3–5 weeks postchallenge) revealed that BCG did not alter the number of primary pulmonary tubercles, but that the lesions were smaller and less necrotic (SMITH et al. 1970; SMITH and WIEGESHAUS 1989). Aerosol vaccination with BCG has been demonstrated to cause a reduction in the number of primary tubercles in guinea pigs (LAGRANIDIERE et al. 1993). Moreover , the lungs of BCG-vaccinated guinea pigs contained nearly 100-fold fewer viable *M. tuberculosis* than the lungs of control animals, demonstrating that one important function of vaccination is to control the early growth of tubercle bacilli at the primary sites of implantation (SMITH et al. 1970).

A comparison of the kinetics of infection in BCG-vaccinated and nonvaccinated guinea pigs during the first 5 weeks revealed that no effect of preexisting immunity on the accumulation of virulent mycobacteria in the lung could be detected until 14–18 days, when bacterial loads had risen to more than 10^3 cfu (SMITH et al. 1970). This observation was somewhat surprising, since the vaccinated animals exhibited strong tuberculin hypersensitivity at the time of infection. However, the results of human BCG vaccine trials have revealed little relationship between postvaccination tuberculin hypersensitivity and protection (COMSTOCK 1988). The results suggest that a critical combination of bacillary load and antimicrobial immunity is required for the expression of resistance. The higher the level of preexisting immunity, the lower the bacillary load required to trigger its recall. A corollary of this theorem is that the higher the bacillary load in a naive animal at the time that antimycobacterial immunity first appears, the more tissue damage will result. Thus, it is likely that nonvaccinated guinea pigs eventually develop the same antimycobacterial immunity, qualitatively, in response to a virulent infection as do the vaccinated animals to the BCG plus virulent infection. The difference between death (for the former) and life (for the latter) may be the difference between 10^6 and 10^3 viable mycobacteria, respectively, in the lung at the time when effective antimycobacterial resistance begins to be expressed within the lung (DANNENBERG 1993, 1994b).

A second major effect of vaccination on the pathogenesis of tuberculosis in the guinea pig is the reduction of the bacillemic phase of this disease. Thus, in kinetic studies such as the one just described, the effect of BCG vaccination on levels of virulent mycobacteria within the spleens following hematogeneous seeding was a 10000-fold reduction in vaccinated compared to nonvaccinated guinea pigs (SMITH et al. 1970). Single lung lobe studies (described earlier) revealed that BCG vaccination reduces the extent of extrapulmonary dessemination and prevents or retards hematogenous establishment of metastatic foci in the lung (FOK et al. 1976; HARDING and SMITH 1977; HO et al. 1978). SMITH et al. (1979) compared the potency of ten BCG vaccines for their influence on the bacillemic phase of experimental airborne tuberculosis in guinea pigs. The result revealed that all vaccines except one, an experimental vaccine no longer used in humans, reduced the bacillemia which accompanied the virulent infection. WIEGESHAUS and SMITH (1989) reviewed the recent literature on evaluation of the protective potency of tuberculosis vaccines and reported data on the protective potency of two vaccines developed from new technologies. They called attention to the disagreements in the published literature about the protective potency of living and nonliving vaccines and suggested that animal models should be developed from an understanding of the key events in the pathogenesis of tuberculosis. In their review, these authors concluded that in most assays of tuberculosis vaccines, potency is based on the extent to which vaccines inhibit the bacillemia which occurs during the first infection with virulent tubercle bacilli: i.e., protection against the endogenous reactivation pathway. Because animal models for evaluation of vaccines which inhibit the bacillemic phase of endogenous tuberculosis are the only animal models currently available, we have not evaluated the new vaccines for their ability to protect against exogenous reinfection tuberculosis. Two studies in the literature demonstrate that, under some circumstances, BCG vaccination can prevent the development of grossly visible primary tubercles in the lungs following inhalation of virulent tubercle bacilli (LURIE et al. 1952b; LAGRANDERIE et al. 1993). These observations provide a basis upon which models for the evaluation of vaccines against exogenous reinfection tuberculosis could be developed.

In addition to vaccination, other factors have been shown to modify the pathogenesis of pulmonary tuberculosis in the guinea pig. In the mouse, studies with several inbred strains have revealed both MHC-linked and non-MHC-linked genetic effects on resistance to infection with mycobacteria (BUSCHMAN et al. 1988; NICKONENKO et al. 1985). A direct comparison of the two commercially available inbred strains of guinea pigs (strains 2 and 13) revealed no significant differences in susceptibility to low dose aerosol exposure to virulent *M. tuberculosis*. Vaccination with BCG protected both strains equally well. The inbred strains responded immunologically to mycobacteria in a manner indistinguishable from outbred, Hartley strain guinea pigs (COHEN et al. 1987).

Another determinant of disease outcome in the guinea pig is nutritional status. Chronic, moderate dietary protein deprivation has been shown to result in significant alterations in the pathogenesis of pulmonary tuberculosis in BCG-vaccinated animals. Malnourished, vaccinated guinea pigs are much less successful at controlling the growth of virulent mycobacteria in the lungs and spleens (MCMURRAY et al. 1985, 1986a,b). Exhibit many more metastatic foci in the lungs, and fail to develop mature, well-circumscribed tubercles (Mainali and McMurray, submitted). Several measures of T cell-mediated immunity are impaired in protein deprived, tuberculous guinea pigs including tuberculin responses in the skin and PPD-induced lymphoproliferation and cytokine production in virto (McMurray and Bartow 1992; McMurray et al. 1989). Thus, malnourished guinea pigs appear to mimic anergic human patients at the nonresponsive pole of the clinical spectrum of tuberculosis (McMurray and Echeverry 1978).

The contributions of specific cell populations and cytokines to the pathogenesis of tuberculosis in the guinea pig have been difficult to define for lack of reagents specific for this species. Successful antimycobacterial resistance appears to require populations of $CD2^{+}$, $CD4^{+}$ antigen-specific T lymphocytes which are capable of producing high levels of interleukin- 2 (IL2) and which can recirculate and traffic into inflammatory foci (Mainali and McMurray, submitted; McMurray and Bartow 1992). These studies have been facilitated by the development of adoptive lymphocyte transfer protocols between syngeneic strain 2 guinea pigs, and by studying tuberculous pleuritis in the guinea pig (Phalen and McMurray 1993a). Other studies have revealed an immunomodulatory role for circulating immune complexes containing mycobacterial antigens and T cell subsets bearing functional Fc receptors (Bartow and McMurray 1989). Preliminary experiments suggest a protective role for tumor necrosis factor-α (TNFα) in this model, inasmuch as high levels of this cytokine are found in pleural fluid (Phalen and McMurray 1993b). Treatment of infected guinea pigs with drugs, such as thalidomide, which are thought to suppress TNFα production, have a moderately beneficial effect of reducing local tissue damage in hematogenous foci in the liver, but do not significantly modify bacterial loads in the lung or spleen, or PPD-induced delayed hypersensitivity and lymphoproliferation in vitro (Phalen and McMurray, manuscript in preparation).

3 Experimental Tuberculosis in the Mouse

Rodents are not naturally subject to tuberculous disease (with the possible exception of the vole which may be infected with *M. microti* (Wells 1937). However, mice can be readily infected in the laboratory with a variety of mycobacterial species (Collins 1984) and, in recent years, inbred mice have been used increasingly in tuberculosis research, partly for economic reasons but also because of the availability of a variety of strains exhibiting different degrees of innate resistance to mycobacterial infection (Buschman et al. 1988). Most of the early studies used a lethal intravenous or intraperitoneal challenge, in which protection was assessed as the increase in mean survival time in vaccinated mice compared to unvaccinated controls. This type of test protocol can demonstrate the presence of an immune response, yet tells the investigator very little about the mechanism(s) involved, since all of the challenged animals eventually die of tuberculosis (Collins 1984; Siebenmann and Barbara 1974). This experimental protocol was greatly improved by the use of a small, sublethal

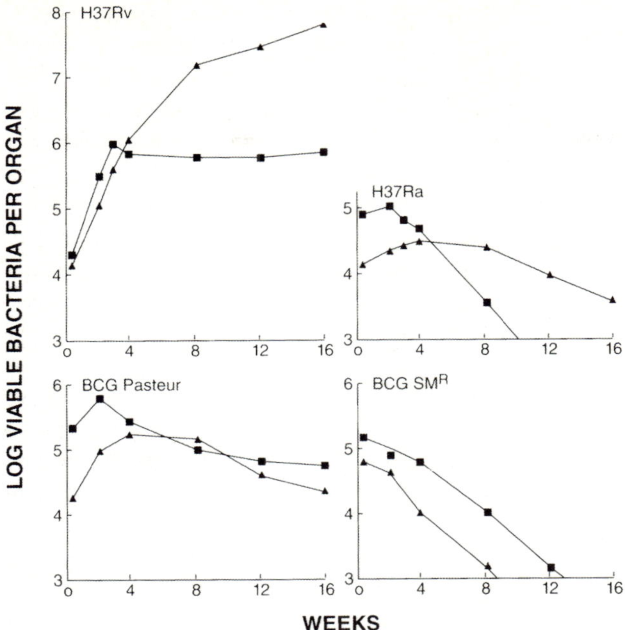

Fig. 1. Growth of Mycobacterium tuberculosis H37Rv (*top left*), H37Rv (*top right*), M. bovis BCG Pasteur (*bottom left*) or BCG SM^R (*bottom right*) in intravenously infected C57 BL/6xDBA/2 F1 hybird mice spleen (■) lungs (▲). The counting error for five determinations was usually less than 10% of the mean

inoculum which could be quantitated as the number of viable tubercle bacilli present in the lungs and spleens of the vaccinated host after an arbitrarily selected time interval (Dubos and Schaefer 1953). This time-consuming and tedious experimental procedure has the advantage that it assesses the immune response in terms of the growth or inactivation of the virulent challenge within the lungs of the vaccinated mouse rather than merely recording its survival or death. This approach also provided a measure of the relative virulence of different strains of *M. tuberculosis* (e.g., H37Rv vs H37Ra) or the immunogenicity of different strains of *M. bovis* BCG (e.g., Pasteur vs Glaxo) by quantitating the amount of bacterial growth within the lung and spleen (Dubos and Pierce 1956). When mycobacteria are injected intravenously into normal mice, the inoculum partitions between the liver, spleen, and lungs in a 90:9:1 ratio. There the organisms multiply for 2–4 weeks before the emerging immune response limits further growth within the liver and spleen. However, virulent organisms (H37Rv or Erdman) continue to multiply within the lung, the amount of growth varying according to its virulence for the mouse (Fig. 1). By contrast, both the H37Ra and BCG Pasteur lung counts declined during this period while an avirulent BCG SM^R mutant failed to grow or even survive within the lung at all. Earlier studies have shown that the latter strains fails to induce an antituberculous immune response, despite the presence of the necessary antigens (Collins 1991).

This challenge procedure was further improved as a model of human tuberculosis by the use of a small aerogenic dose delivered directly into the alveolar tissues by means of an infectious aerosol generated in a Middlebrook chamber (MIDDLEBROOK 1952). This final modification was by no means a trivial one, since the virulence of *M. tuberculosis* varies widely depending upon the experimental conditions (COLLINS and SMITH 1969; COLLINS et al. 1974). In particular, the manner of preparation of the challenge inoculum (the growth medium, the growth phase, the relative viability and the extent of clumping), together with its size, route of inoculation and the test organs selected for culture can all affect the growth of the organisms in vivo (COLLINS and MONTALBINE 1975). Thus, mice infected by the intravenous, subcutaneous, aerogenic and intragastric routes may receive inocula differing by as much as one million-fold, yet develop very similar splenic infections (Fig. 2). Although the mean time to death for the four groups will differ significantly, with time they all eventually succumb to the infection (COLLINS 1984). Only the BCG-vaccinated host has the level of acquired resistance within the lung to prevent a fatal infection, although this too may

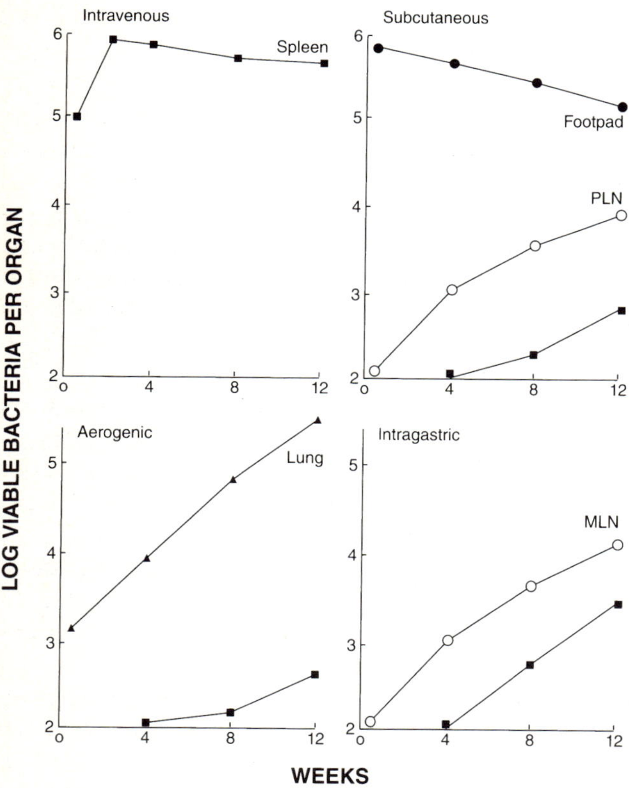

Fig. 2. Growth of M. tuberculosis H37Rv following infection of C57BL/6 mice by the intravenous (10^6CFU), subcutaneous (10^6CFU), aerogenic (10^3CFU) or intragastric ($5X10^8$CFU) routes; spleen (■), footpad (●), lungs (▲), mesenteric lymph node (○)

depend upon the amount of growth and persistence by the BCG vaccine in the spleen (COLLINS 1984; COLLINS and MONTALBINE 1975). Thus, it seems hardly surprising that different investigators have assessed the same vaccine with wildly differing levels of effectiveness when each tested it by the assay method of their choice (WIEGESHAUS et al. 1971).

One of the most important variables in this type of immunogenicity assay will be the experimental host itself. Different strains of mice have long been known to vary in their innate resistance to tuberculous challenge (LYNCH et al. 1965). In the case of BCG, this characteristic seems to be related to the presence of a *Bcg* gene on chromosome 1 which affects the level of macrophage activation in the BCG infected host (BUSCHMAN et al. 1988). This effect is best seen in mice subjected to a small intravenous challenge with BCG Montreal, although a number of nontuberculous mycobacteria may also be controlled by this gene. However, the Bcg gene is not involved in the expression of innate resistance to *M. tuberculosis*, which rather depends upon the presence of the *Tbc-1* gene (NICKONENKO et al. 1985).

A number of experimental animal models have been used to study the role played by different T cell subsets in the expression of acquired antituberculous resistance. Infections carried out in neonatally or adult thymectomized mice indicate a pivotal role for the CD4+ T lymphocyte population, both in the expression of tuberculin hypersensitivity and antituberculous resistance (NORTH 1973). The T cell-depleted mice eventually die from an increasing lung consolidation induced by the unresolved mycobacterial granuloma. This effect can be reversed by an infusion of CD4+ T cells harvested from a suitably vaccinated syngeneic donor (ORME and COLLINS 1983). Similar studies have also been carried out using congenitally athymic nude (nu+/nu+) mice which develop life-threatening systemic infections following inoculation with *M. tuberculosis* (COLLINS and STOKES 1987). Recently, a number of experiments have been carried out in BALB/c mice with severe combined immunodeficiency ($SCID^+/SCID^+$) and which lack both humoral and cellular defenses against systemic infection (SCHULTZ and SIDMAN 1987). These mice are exquisitely susceptible to mycobacterial infections and will develop a rapidly lethal disease following inoculation with a small dose of live BCG (NORTH and IZZO 1993). T cell separation techniques make it possible to replace specific subpopulations in the immunosuppressed host in order to determine their role in limiting the growth and persistence of the mycobacteria in the different host organs (ORME et al. 1992)

Although CD4+ T cells are generally considered to be the mediators of acquired antituberculous resistance, depletion studies indicate that cytotoxic CD8+ T cells may also be involved in this process, particularly during the later stages of infection (MULLER et al. 1990). The involvement of cytotoxic T cells in the expression of antituberculous resistance has recently been confirmed using gene knock-out (GKO) mice lacking the β_2 microglobulin gene which was disrupted by homologous recombination (KOLLER et al. 1990). These GKO mice lack the ability to produce the β_2 microglobulin component of the class I major histocompatibility complex (MHC) and fail to activate their $CD8^+$ T cell defenses.

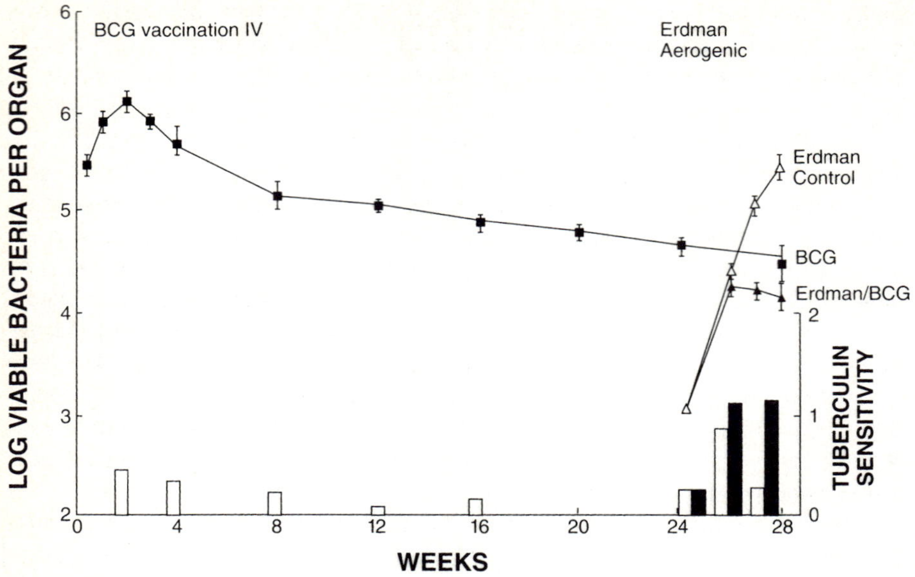

Fig. 3. Growth of M. bovis BCG Pasteur in the (■) of C57BL/6 mice following intravenous inoculation with 5×10^6 CFU. The vaccinated mice were challenged after 6 months with 10^3 CFU *M. tuberculosis* Erdman by the aerogenic route; levels of virulent bacilli in the lungs of control (△) or BCG-vaccinated (▲) mice. The *bars* represent the footpad swelling 24 h after injection with 2.5 μg PPD

This genetic defect renders the GKO mouse exquisitely susceptible to a tuberculous challenge (Flynn et al. 1992). Other GKO mouse strains have now been developed with a disrupted gene for interferon-γ (IFN-γ) production (Dalton et al. 1993). These mice are also highly susceptible to *M. tuberculosis* infection, developing a uniformly fatal systemic disease (Flynn et al. 1993) which resembles the miliary tuberculosis seen in young children (Schaaf et al. 1993) and in AIDS patients (Bouza et al. 1993). These knock-out mouse studies emphasize the crucial role played by specific lymphokines in the expression of acquired resistance to this pathogen (Orme and Collins 1994). Other GKO mouse strains are being developed and this type of experimental approach holds great promise as the method of choice in studies into the role played by molecular mediators in the modulation of the immune response to a variety of intracellular parasites (Rook 1990).

Although countless protection tests have been carried out in BCG vaccinated mice (and guinea pigs), few attempts were made to quantitate the in vivo growth behavior of the vaccine in the lymphoreticular organs of the host which could be related to the level of antituberculous resistance expressed against a subsequent virulent challenge. In many cases, the vaccine was standardized in terms of the wet weight of cells rather than the number of viable bacteria. Furthermore, the host response was assessed in terms of tuberculin hypersensitivity induction determined immediately before the animals were challenged with a lethal inoculum of *M. tuberculosis*. An examination of BCG growth data obtained both in vitro and in vivo (COLLINS 1984; LAGRANGE et al. 1976; PIERCE et al. 1956) clearly demonstrates that not all BCG subspecies are equivalent in terms of growth, persistence, or immunogenicity, presumably as a result of varying degrees of attenuation developed while they were maintained under slightly differing conditions for extended periods of time (GRANGE et al. 1983). The effectiveness of a given vaccine preparation depends on its growth and persistence in the host tissues (particularly the spleen and lymph nodes) so that a detailed knowledge of the in vivo growth characteristics of each vaccine is an essential prerequisite to a rational assessment of its protective value. The combined in vivo growth data for a typical vaccination/challenge protection study carried out in SPF C57BL/6 (Bcg^S) mice shown in Fig. 3, together with the relevant delayed hypersensitivity determinations, is typical of this type of study. Mice were vaccinated with 5×10^6 cfu BCG Pasteur i.v. (or s.c., not shown). After an initial increase in viability, the BCG counts slowly declined, presumably as a result of the developing cell-mediated immune response. Spleen cells harvested during this period contained a population of CD4+ T cells which adoptively protected naive syngeneic recipients against a lethal aerogenic challenge with *M. tuberculosis*, Erdman (ORME and COLLINS 1983). The vaccinated animals developed a borderline significant tuberculin hypersensitivity to PPD (Fig. 3) which was recalled by the small aerogenic challenge with *M. tuberculosis*, Erdman (ca 10^3 cfu per lung). Unlike the unvaccinated controls, this enhanced tuberculin hypersensitivity was not followed by an anergic state. Viable counts carried out on the lung homogenates from the vaccinated mice included residual BCG unless they were plated on Middlebrook 7H11 agar containing 2 µg/ml of 2-thiophenecarboxylic acid hydrazide which inhibited the growth of the BCG (COLLINS and MACKANESS 1970). Differential counts of both vaccinating and challenge populations can also be carried out with isoniazid or acriflavine-resistant organisms (COLLINS et al. 1977). The virulent challenge inoculum in the lungs of vaccinated and control mice increased equally over the first 2 weeks of the protection experiment (Fig. 3). Subsequent Erdman counts in the BCG vaccinated mice indicated a prolonged bacteriostasis, so that by day 28 the difference between the counts in the vaccinated vs control mice was highly significant ($p < 0.01$). Furthermore, the vaccinated mice survived the challenge for at least 3 months, whereas the controls all died. This bacteriostatic phase is characteristic of the BCG induced response in both actively and adoptively immunized mice (ORME 1987; ORME and COLLINS 1983) and is presumably similar to the mechanisms responsible for protection against the miliary and meningeal forms of the disease in BCG-vaccinated children living in highly endemic areas where they will be heavily exposed to tuberculous infection (FINE 1989). These enhanced responses (both skin hypersensitivity and protection) are also important considerations when considering the option of BCG vaccination for tuberculin-negative health care workers who are in close and prolonged contact with patients suffering from multidrug-resistant tuberculosis, as well as family members who are likely to be exposed to such infection while living under conditions of overcrowding , poor nutrition, and inadequate medical care.

4 Experimental Tuberculosis in the Rabbit

When infected by virulent human-type tubercle bacilli (*M. tuberculosis*), the rabbit shows more resistance than the mouse and much more resistance than the guinea pig (DANNENBERG 1984). However, rabbits are fully susceptible to infection with virulent bovine-type tubercle bacilli *(M. bovis).* In fact, the rabbit is the classic animal used to distinguish between virulent human-type bacilli and virulent bovine-type bacilli. These animals always die when infected with the bovine-strain and always live when infected with the human strain. Mice are generally considered to be more resistant to tuberculosis than the other two species, because they show less tissue destruction during the disease, probably as a result of their low sensitivity to the tuberculin-like components of the bacillus. Among the three animal models, a careful comparison of the rates of growth and inhibition of both human and bovine strains of bacilli has yet to be made.

LURIE (1964) studied the pathogenesis of tuberculosis in rabbits with both strains of tubercle bacilli (summarized by LURIE and DANNENBERG 1965). He chose this species because tuberculosis in rabbits more closely resembled human tuberculosis than that produced in any other animal model (DANNENBERG 1984, 1990, 1994b). LURIE (1964) selectively bred susceptible and resistant rabbits. The susceptible rabbits developed a hematogenously spread tuberculosis resembling that found in infants and immunosuppressed adults (including those with AIDS). The resistant rabbits developed a cavitary-type of tuberculosis with bronchogenic spread, resembling that found in immunocompetent human beings. Guinea pigs exposed by inhalation to low doses of virulent tubercle bacilli may also develop cavities, but bronchogenic spread of the disease in these animals is rare (D.W. Smith, personal communication). With higher doses of bacilli, guinea pigs usually die before such cavities develop. Mice never develop typical cavities followed by bronchogenic spread of the disease.

Lurie's susceptible rabbits always lived when infected with the virulent human strain, H37Rv, but these susceptible rabbits developed primary and secondary foci (of hematogenous origin) after inhaling H37Rv, and this disease was not eliminated even after 12 months (LURIE 1964; DANNENBERG 1994a). The resistant rabbits even formed cavities after such H37Rv infection, but these did not progress, and secondary tubercles of bronchial origin were rare. New Zealand white rabbits available commercially more closely resemble Lurie's resistant rabbits than they do his susceptible strains (Dannenberg, unpublished observations). One would expect, however, that such market rabbits would show more interrabbit variation than did Lurie's selectively bred strains.

These studies clearly showed that rabbits develop pulmonary tuberculosis when they inhale virulent human-type strains of tubercle bacilli even though these strains are not fully virulent for rabbits. This fact enabled LURIE et al. (1952a,b, 1955) to develop one of the most precise methods available to measure the efficacy of tuberculosis vaccines. He simply exposed BCG-vaccinated and control rabbits to an aerosol of H37Rv and counted the number of primary tubercles in the lung 5 weeks later. Vaccination with BCG stopped the progression of many primary pulmonary tubercles while they were still microscopic in size. Therefore, fewer gross primary tubercles were visible at 5 weeks. In other words, BCG vaccination can prevent X-ray-visible pulmonary lesions as well as the clinical disease resulting from such lesions.

All of Lurie's selectively bred rabbit strains are extinct today. However, one race III strain of rabbits has been further inbred by L.F.M. van Zutphen in the Netherlands (Dannenberg 1994a). It is distantly related to Lurie's resistant race III rabbits and in one of our experiments (with virulent bovine-type bacilli) has been shown to produce the resistant form of this disease (Dannenberg, unpublished experiments). To our knowledge, no living susceptible strain of rabbits has been discovered to date as a substitute for Lurie's susceptible strains.

The most complete experimental study of the pathogenesis of pulmonary tuberculosis in rabbits was made by Lurie (1964). The onset of pulmonary tuberculosis involves the deposition in the alveoli of droplet nuclei containing one to three viable tubercle bacilli (Lurie 1964; Dannenberg and Tomashefski 1988). In the alveoli, the bacilli are ingested by alveolar macrophages (AMS). In the rabbit, most AMS are highly activated cells that can readily destroy the majority of virulent human-type bacilli. Rabbit AM cannot destroy virulent bovine-type bacilli, but humans are rather resistant to both types of bacilli.

We do not know how many virulent bacilli of either type must be inhaled on the average to generate one primary pulmonary tubercle in human beings (Riley et al. 1962). Some experts believe that a single, fully virulent tubercle bacillus deposited into an alveolus in a susceptible host is capable of causing infection. Other experts believe that infection occurs only when highly virulent bacilli are ingested by a weak alveolar macrophage. Although such a combination can occur after inhaling one bacillary unit, in most cases multiple units must be inhaled before any such unit will be ingested by an alveolar macrophage weak enough to allow it to multiply intracellularly.

If such multiplication occurs, the alveolar macrophage will eventually die and release the bacilli it contains. The free bacilli are chemotactic for other AMS and also for blood-borne monocyte/macrophages. The latter are nonactivated cells that allow the bacillus to multiply readily within phagocytic vacuoles, and the bacillus does so in a logarithmic fashion at the same rate in both resistant and susceptible rabbits. However, before this logarithmic intracellular bacillary growth occurs, a 20- to 30- fold difference in the number of human-type bacilli in the lungs of resistant and susceptible rabbits is present (Lurie et al. 1955). About a five fold difference in the number of bovine-type bacilli is observed (Allison et al. 1962). Susceptible rabbits evidently contain many weak AMS that allow more primary tubercles to develop and also allow intracellular bacillary multiplication to occur earlier.

In the developing primary tubercle, the logarithmic growth of the bacilli in the nonactivated macrophages is called the symbiotic stage of tuberculosis (Lurie 1964; Dannenberg 1993). Tubercle bacilli are facultative intracellular parasites ideally adapted to grow intracellularly in nonactivated macrophages. These phagocytes divide locally and continually enter the lesions from the blood-stream (reviewed in TSUDA et al. 1976). Thus, not only do the bacilli proliferate in the developing tubercle, but the number of macrophages (and to a lesser degree lymphocytes) also increases there. Thus, there is symbiosis with respect to the accumulation of both macrophages and bacilli.

During the symbiotic stage, which lasts about 21 days (LURIE et al. 1955; ALLISON et al. 1962; DANNENBERG 1993), the proliferating bacilli secrete tuberculin-like antigens that produce a T lymphocyte immune response. We call this initial response "tissue-damaging" or "cytotoxic" delayed-type hypersensitivity (DTH) (DANNENBERG and ROOK 1994; DANNENBERG 1968, 1989, 1994b) because it kills the bacilli-laden macrophages (and nearby tissues). In other words, the tissue-damaging DTH causes the caseous necrosis found in the center of primary tubercles. In this caseous center, some of the bacilli live in a dormant state, frequently for many years, but as long as the caseous material remains solid, the bacilli apparently do not multiply therein.

From time to time, some of the bacilli escape from the edge of the caseous focus and again are ingested by local macrophages. If these bacilli are ingested by highly activated macrophages, the bacilli cannot multiply and further development of the tubercle is arrested (ANDO et al. 1977). Highly activated perifocal macrophages are readily produced in resistant rabbits due to the strong cell-mediated immunity (CMI) that they develop. To produce good CMI, T lymphocytes (especially Th1 lymphocytes) are required. These cells produce IFN-γ and other macrophage-activating cytokines. The disease in such rabbits can be arrested at this point, especially with human-type bacilli.

In most immunocompetent human beings, the disease is usually arrested at this stage (DANNENBERG and TOMASHEFSKI 1988). The person becomes tuberculin-positive, but there is usually no X-ray evidence of the disease. As a rule, such people remain free from active tuberculosis for the rest of their lives, unless some immunosuppressive event allows endogenous reactivation tuberculosis to develop.

If, however, the bacilli escaping from the edge of a solid caseous focus are ingested by poorly activated macrophages (as in Lurie's susceptible rabbits), the bacilli again multiply intracellularly and again tissue-damaging DTH is used to destroy the macrophages, including some of the surrounding tissues. Unless highly activated macrophages surround the caseous center, the progression of the disease will continue. Much of the lung is destroyed, and hematogenous spread produces enlarging secondary lesions. With bovine-type of tubercle bacilli, such susceptible rabbits die of the disease in a relatively few months. This type of disease is also found in infants and in immunocompromised individuals, because they cannot develop good CMI.

In resistant (but not in susceptible) rabbits, tubercles may progress to liquefaction and cavitation (LURIE 1964; LURIE and DANNENBERG 1965). The liquefied caseum serves as an excellent growth medium in which the tubercle bacillus grows extracellularly, in tremendous numbers, for the first time during the course of the disease. Antimicrobial resistant strains may develop in this large population of bacilli. In fact, if liquefaction were prevented, the occurrence of drug resistant bacilli might be rare. Macrophages apparently do not survive in liquefied caseum and, therefore, are ineffective in controlling the extracellular replication of mycobacteria within the cavity (DANNENBERG and ROOK 1994).

If the liquefied tubercle is located near an airway and it erodes the bronchial wall, an open cavity forms. The bacilli can then spread to other parts of the lung and into the surrounding air. Cavitary tuberculosis is, therefore, particularly contagious. In human beings (and in rabbits infected with human-type tubercle bacilli), cavities may heal spontaneously by collapse and firbrosis (DANNENBERG and TOMASHEVSKI 1988). However, viable tubercle bacilli may persist for years in arrested caseous foci or in a "healed" cavity (DANNENBERG and ROOK 1994).

Vaccines cannot prevent the establishment of a primary tubercle in the lung. They can only prevent the progression of microscopic pulmonary lesions into clinical (or X-ray visible) disease (see above), AMs cannot recognize specific mycobacterial antigens. Only lymphocytes have "memory," i.e., expansion of an antigen-specific population following vaccination.

Following BCG vaccination, rabbits develop populations of sensitized T lymphocytes capable of recognizing mycobacterial antigens when the virulent challenge occurs. Thus, during the initial phase of bacillary replication, sufficient antigenic stimulus is produced by relatively few virulent mycobacteria to activate these preexisting sensitized T lymphocytes. The vaccinated rabbits then show "accelerated tubercle formation," i.e., a rapid local accumulation of lymphocytes and activated macrophages. This early and vigorous antimicrobial response suppresses the development of the lesion by restricting the intracellular growth of mycobacteria or killing the mycobacteria outright. Such a vaccine-induced beneficial outcome is thought to be mediated by an array of macrophage-activating cytokines, produced principally by CD4 T lymphocytes, and a cytotoxic T cell response (mediated by CD8 T lymphocytes) which allows the host to drive mycobacteria from the shelter of nonactivated phagocytic cells (KAUFMANN 1988). The result is a smaller lesion histopathologically, with much less tissue damage and necrosis, which either stabilizes or heals completely. The infection is truncated before it can progress to disease. In a sense, then, prior vaccination lowers the "threshold" antigenic load which is required to trigger antimicrobial resistance and circumvents the intense tissue destruction that would result if the same immunological mechanisms were to be induced in the presence of a much larger bacillary mass (DANNENBERG and ROOK 1994). These events are strikingly similar to those in BCG-vaccinated guinea pigs described earlier in this chapter.

5 Summary and Conclusions

Studies of experimental tuberculosis in guinea pigs, mice and rabbits have revealed some fundamental insights into the host-parasite relationship between *M. tuberculosis* and its host. Resistance to the initial implantation of fully virulent mycobacteria into the lung, whether innate or vaccine-induced, is not expressed until the organism has replicated sufficiently to achieve some critical minimal bacillary (or antigenic) mass. The phagocytic cell responsible for this initial response (in the first several days to 2 weeks of infection) is the blood-derived monocyte. The minimal antigenic load necessary to trigger the anitmicrobial response depends upon the preexisting state of immunity, i.e., the greater the level of CMI at the time of infection, the lower the level of viable mycobacteria required. Conversely, in an immunologically naive host, much more extensive replication of the infecting bacilli (corresponding to 3–4 weeks of infection) is necessary to induce and then elicit successful control of mycobacterial growth in the developing tubercle.

The higher the bacillary load in the tissues at the time when the first effective cell-mediated immune response occurs, the greater is the resulting tissue necrosis. One important beneficial effect of BCG vaccination in these models, therefore, is to effectively lower the threshold number of virulent bacilli required to trigger resistance, which is then achieved at minimal cost to the host in terms of lung damage. However, even under optimal circumstances resistance is not absolute. In none of the animal species, including humans, does successful vaccination induce a sterilizing immunity.

Another common theme revealed by a comparative analysis of the pathogenesis of tuberculosis in these animal models is the fundamental importance of extrapulmonary dissemination. Spread of virulent tubercle bacilli beyond the sites of initial implantation in the lungs occurs relatively early in the course of infection by lymphatic drainage to thoracic lymph nodes, and then to the blood stream via the thoracic duct. Metastatic foci are established in several organs, including the lung.

Control of the hematogeneous phase of infection may be a crucial determinant of the ultimate fate of the infected host. Another important beneficial effect of successful BCG vaccination is to retard and limit the dissemination of tubercle bacilli and the seeding of the lung with metastic tubercles. A large part of the reduction in secondary (hematogenously derived) lesions is due to a decrease in ability of bacilli to grow at the new site, due to an established immune response (Lurie et al. 1955). Therefore, BCG-vaccinated hosts exhibit a reduction in the incidence of miliary and meningial tuberculosis.

In cavitary disease, however, the bacilli seed the lung directly via the airways to establish secondary lesions. When the bacilli-laden bronchial secretions are swallowed, the lymphoid tissues associated with the gastrointestinal tract may also be infected.

Studies in the mouse and the rabbit have revealed the importance of genetically determined innate resistance as a determinant of disease outcome. Resistance is expressed nonspecifically relatively early in infection in naturally activated AMs (in the rabbit) or splenic macrophages (in the mouse). The precise mechanism(s) of genetic resistance in these species remain to be elucidated, however, at least two genes *(Bcg and Tbc)* appear to be involved in the mouse, while the corresponding genetic factor(s) in rabbits and humans remains elusive.

At later times, blood-derived macrophages become activated immunologically to inhibit or destroy the bacilli, and the degree of such activation may also be under genetic control.

A number of crucial questions about the pathogenesis of tuberculosis remains to be answered in these model systems: (1) What is the relationship between delayed hypersensitivity and resistance? (2) What are the precise mechanisms of innate resistance? (3) What are the precise mechanisms of vaccine-induced resistance? (4) How do immunological events in primary and metastatic tubercles differ? (5) How do mycobacteria persist for long periods in the tissues of a "resistant" host? (6) What are the factors which control the evolution of the granuloma, leading to caseation and cavitation? (7) How can exogenous reinfection tuberculosis be modeled in experimental animals? These questions and others must be addressed experimentally before we can truly comprehend the nature of tuberculosis in human beings. Such studies will yield a body of knowledge upon which novel therapeutic and preventative strategies can be based.

References

Allison MJ, Zappasodi P, Lurie MB (1962) Host-parasite relationships in natively resistant and susceptible rabbits on quantitative inhalation of tubercle bacilli: their significance for the nature of genetic resistance. Am Rev Respir Dis 85: 553–569

Ando M, Dannenberg AM Jr, Sugimoto M, Tepper BS (1977) Histochemical studies relating the activation of macrophages to the intracellular destruction of tubercle bacilli. Am J Pathol 86: 623–634

Balasubramanian V, Guo-zhi, W, Wiegeshaus E, Smith D (1992a) Virulence of *Mycobacterium tuberculosis* for guinea pigs: a quantitative modification of the assay developed by Mitchison. Tubercle Lung Dis 73: 268–272

Balasubramanian V, Wiegeshaus E, Smith D (1992b) Growth characteristics of recent sputum isolates in guinea pigs infected by the respiratory route. Infect Immun 60: 4762–4767

Balasubramanian V, Wiegeshaus E, Smith D (1994) Pathogenesis of tuberculosis: pathway to apical localization. Tubercle Lung Dis 75: 168–178

Barclay WR, Anacker RL, Brehmer W, Lief W, Ribi E (1970) Aerosol-induced tuberculosis in subhuman primates and the course of disease after intravenous BCG-vaccination. Infect Immun 2: 574–582

Bartow RA, McMurray DN (1989) Vaccination with *Mycobacterium bovis* BCG affects the distribution of Fc receptor-bearing T lymphocytes in experimental pulmonary tuberculosis. Infect Immun 57: 1374–1379

Bloom BR, Murray CJ (1992) Tuberculosis: commentary on a reemergent killer. Science 257: 1055–1064

Bouza E, Diaz-Lopez MD, Moreno S, Bernaldo de Quiros JC, Vincente T, Berenguer J (1993) *Mycobacterium tuberculosis* bacteremia in patients with and without HIV infection. Arch Intern Med 153: 496–500

Buschman E, Apt AS, Nickonenko BV, Moroz AM, Averbakh MH, Skamene E (1988) Genetic aspects of innate resistance and acquired immunity in inbred mice. Springer Semin Immunopathol 10: 319–336

Cohen MK, Bartow RA, Mintzer CL, McMurray DN (1987) Effects of diet and genetics on *Mycobacterium bovis* BCG vaccine efficacy in inbred guinea pigs. Infect Immun 55: 314–319

Collins FM (1984) Protection against mycobacterial disease by means of live vaccines tested in experimental animals. In: Kubica GP, Wayne LG (eds). The mycobacteria: a sourcebook Dekker, New York, pp 787–839

Collins FM (1991) Antituberculous immunity: new solutions to an old problem. Rev Infect Dis 13: 940–950

Collins FM (1993) Tuberculosis: the return of an old enemy. Crit Rev Microbiol 19: 1–16

Collins FM, Mackaness GB (1970) The relationship of delayed hypersensitivity to acquired antituberculous immunity. I. Tuberculin sensitivity and resistance to reinfection In BCG-vaccinated mice. Cell Immunol 1: 253–265

Collins FM, Montalbine V (1975) Relative immunogenicity of streptomycin-resistant and sensitive strains of BCG. II. Effect of route of inoculation on growth and immunogenicity. Am Rev Respir Dis 111: 43–51

Collins FM, Smith MM (1969) A comparative study of the virulence of *Mycobacterium tuberculosis* measured in mice and guinea pigs. Am Rev Respir Dis 100: 631–639

Collins FM, Stokes RW (1987) *Mycobacterium aium*-complex infections in normal and immunodeficient mice. Tubercle 68: 127–136

Collins FM, Wayne LG, Montalbile V (1974) The effect of cultural conditions on the distributiion of *Mycobacterium tuberculosis* in the spleens and lungs of specific pathogen-free mice. Am Rev Respir Dis 110: 147–156

Collins FM, Auclair L, Mackaness GB (1977) *Mycobacterium bovis* (BCG) infections of the lymph nodes of normal, immune and cortisone-treated guinea pigs. J Natl Cancer Inst 59: 1527–1535

Comstock GW (1988) Identification of an effective vaccine against tuberculosis. Am Rev Respir Dis 138: 479–480

Dalton DK, Pitts-Meek S, Keshav S, Figari IS, Bradley A, Stewart TA (1993) Multiple defects of immune cell function in mice with disrupted IFN-γ genes. Science 259: 1739–1742

Dannenberg AM Jr (1968) Cellular hypersensitivity and cellular immunity in the pathogenesis of tuberculosis; specificity, systemic and local nature, and associated macrophage enzymes. Bacteriol Rev 32: 85–102

Dannenberg AM Jr (1984) Pathogenesis of tuberculosis: native and acquired resistance in animals and humans. In: Leive L, Schlessinger D (eds) *Micribiology* - 1984. American Society for Microbiology, Washington D C, PP 344–354

Dannenberg AM Jr (1989) Immune mechanisms in the pathogenesis of pulmonary tuberculosis. Rev Infect Dis 11 [Suppl 2]: S369–S378

Dannenberg AM Jr (1990) Controlling tuberculosis: the pathologist's point of view, Res Microbiol 141: 192–196

Dannenberg AM Jr (1993) Immunopathogenesis of pulmonary tuberculosis. Hosp Pract 28: 51–58

Dannenberg AM Jr (1994a) Rabbit model of tuberculosis. In: Bloom BR (ed) Tuberculosis: pathogenesis, protection and control. ASM, Washington D C, pp 149–156

Dannenberg AM Jr (1994b) Roles of cytotoxic delayed-typed hypersensitivity and macrophage-activating cell-mediated immunity in the pathogenesis of tuberculosis. Immunobiology 191: 461–473

Dannenberg AM Jr, Rook GAW (1994) Pathogenesis of pulmonary tuberculosis: an interplay of tissue-damaging and macrophage-activating immune responses-dual mechanisms that control bacillary multiplication. In: Bloom BR (ed) Tuberculosis: pathogenesis, protection and control. ASM, Washington D C, pp 459–483

Dannenberg AM Jr, Tomashefski JF Jr (1988) Pathogenesis of pulmonary tuberculosis. In: Fishman AP (ed) Pulmonary diseases and disorders, 2nd edn: vol 3. McGraw-Hill, New York, pp 1821–1842

Dannenberg AM Jr, Meyer OT, Esterly JR, Kambara T (1968) The local nature of immunity in tuberculosis, illustrated histochemically in dermal BCG lesions. J Immunol 100: 931–941

Dubos RJ, Pierce CH (1956) Differential characteristics *in vitro* and *in vivo* of several substrains of BCG. IV. Immunizing effectiveness. Am Rev Tuberc 74: 699–717

Dubos RJ, Schaefer WB (1953) Antituberculous immunity induced in mice by vaccination with living cultures of attenuated tubercle bacilli. J Exp Med 97: 207–220

Ellner JJ, Hinman AR, Booley SW, Fischt MA, Sepkowitz KA, Goldberger MJ, Schinnick TM, Iseman MD, Jacobs WR (1993) Tuberculosis symposium: emerging problems and promise. J Infect Dis 168: 537–551

Fine PEM (1989) The BCG story: lessons from the past and implications for the future. Rev Infect Dis 57 [Suppl 2]: S353–359

Flynn JL, Goldstein MM, Tribold KJ, Koller B, Bloom BR (1992) Major histocompatibility complex class I-restricted T cells are required for resistance *Mycobacterium tuberculosis* infection. Proc Natl Acad Sci USA 89: 12013–12017

Flynn JL, Chan J, Triebold KJ, Dalton DK, Stewart TA, Bloom BR (1993) An essential role for IFN-γ in resistance to *Mycobacterium tuberculosis*. J Exp Med 178: 2248–2253

Fogarty Center Workshop (1978) Summary, conclusions and recommendations from the international workshop on "research toward global control and prevention of tuberculosis: with an emphasis on vaccine development". J Infect Dis 158: 248–253

Fok JS, Ho RS, Arora PK, Harding GE, Smith DW (1976) Host-parasite relationships in experimental airborne tuberculosis. V. Lack of hematogeneous dissemination of *Mycobacterium tuberculosis* to the lungs of animals vaccinated with bacille-Calmette-Guerin. J Infect Dis 133: 137–144

Grange JM, Gibson J, Osborn TW, Collins CH, Yates MD (1983) What is BCG? Tubercle 64: 129–139

Gray DF (1961) The relative natural resistance of rats and mice to experimental pulmonary tuberculosis. J Hyg 59: 471–477

Grover AA, Kim HK, Wiegeshaus EH, Smith DW (1967) Host-parasite relationships in experimental airborne tuberculosis II. Reproducible infection by means of an inoculum preserved at - 70C. J Bacteriol 94: 832–835

Gussman RA (1984) Notes on the particle size ouput of Collison Nebulizers. Am Indust Hyg Assoc J 45: B8–B12

Harding GE, Smith DW (1977) Host-parasite relationships in experimental airbone tuberculosis. VI. Influence of vaccination with bacille-Calmette-Guerin on the onset and/or extent of hematogeneous dissemination of virulent *Mycobacterium tuberculosis* to the lungs. J Infect Dis 136: 439–443

Ho RS, Fok JS, Harding GE, Smith DW (1978) Host-parasite relationships in experimental airborne tuberculosis. VII. Fate of *Mycobacterium tuberculosis* in primary lung lesions and in primary lesion-free lung tissue infected as a result of bacillemia. J Infect Dis 138: 237–241

Jespersen A (1956) Studies on tuberculin sensitivity and immunity in guinea pigs induced by vaccination with varying doses of BCG vaccine. Acta Pathol Microbiol Scand 38: 203–209

Kaufmann SHE (1988) $CD8^+$T lymphocytes in intracellular microbial infections. Immunol Today 9: 168–174

Koch R (1882) Aetiologie der Tuberculose. Berl Klin Wochenschr 19: 221–230

Koller B, Marrack P, Kappler JW, Smithies O (1990) Normal development of mice deficient in β2 M, MHC Class I proteins and $CD8^+$T-cells. Science 248: 1227–1230

Lagranderie M, Ravisse P, Marchal G, Gheorghiu M, Balasubramanian V, Wiegeshaus E, Smith DW (1993) BCG-induced protection in guinea pigs vaccinated and challenged via the respiratory route. Tubercle Lung Dis 74: 38–46

Largrange PH, Miller TE, Mackaness GB (1976) Parameters conditioning the potentiating effect of BCG on the immune response. IN: Lamoureaux G, Turcotte R, Portelance V (eds) BCG in cancer immunotherapy. Grune and Stratton, New York, pp 23–26

Long ER, Vorwald AJ, Donaldson L (1931) Early cellular reaction to tubercle bacilli: a comparison of this reaction in normal and tuberculous guinea pigs and in guinea pigs immunized with dead bacilli. Arch Pathol 12: 956–969

Lurie MB (1964) *Resistance to tuberculosis: experimental stuides in native and acquired defensive mechanisms*. Harvard University Press, Cambridge MA

Lurie MB, Dannenberg AM Jr (1965) Macrophage function in infectious disease with inbred rabbits. Bacteriol Rev 29: 466–476

Lurie MB, Abramson BS, Heppleston AG (1952a) On the response of genetically resistant and susceptible rabbits to the quantitative inhalation of human type tubercle bacilli and the nature of resistance to tueberculosis. J Exp Med 95: 119–134

Lurie MB, Zappasodi P, Cardona-Lynch E, Dannenberg AM Jr (1952b) The response to the intracutaneous inoculation of BCG as an index of native resistance to tuberculosis. J Immunol 68: 369–387

Lurie MB, Zappasodi P, Tickner C (1955) On the nature of genetic resistance of tuberculosis in the light of the host-parasite relationships in natively resistant and susceptible rabbits. Am Rev Tuberc Pulm Dis 72: 297–329

Lynch CL, Pierce-Chase CH, Dubos RJ (1965) A genetic study of susceptibility to experimental tuberculosis in mice infected with mammalian tubercle bacilli. J Exp Med 121: 1051–1070

Mainali ES, McMurray DN (submitted for publication) Adoptive transfer of resistance to pulmonary tuberculosis in guinea pigs is altered by protein deficiency

McMurrary DN (1994) Guinea pig model of tuberculosis. In: Bloom BR (ed) *Tuberculosis: pathogenesis, protection, and control.* ASM, Washington DC, pp 135–147

McMurray Dn, Bartow RA (1992) Immunosuppression and alternation of resistance to pulmonary tuberculosis in guinea pigs by protein undernutrition. J Nutr 122: 738–743
McMurray DN, Echeverry A (1978) Cell-mediated immunity in anergic patients with pulmonary tuberculosis. Am Rev Respir Dis 118: 827–834
McMurray DN, Carlomagno MA, Mintzer CL, Tetzlaff CL (1985) *Mycobacterium bovis* BCG vaccine fails to protect protein-deficient guinea pigs against respiratory challenge with virulent *Mycobacterium tuberculosis.*, Infect Immun 50: 555–559
McMurray DN, Kimball MS, Tetzlaff CL, Mintzer CL (1986a) Effects of protein deprivation and BCG vaccination on alveolar macrophage function in pulmonary tuberculosis. Am Rev Respir Dis 133: 1081–1085
McMurray DN, Mintzer CL, Tetzlaff CL, Carlomagno MA (1986b) Influence of dietary protein on the protective effect of BCG in guinea pigs. Tubercle 67: 31–39
McMurray DN, Mintzer CL, Bartow RA, Parr RL (1989) Dietary protein deficiency and *Mycobacterium bovis* BCG affect interleukin 2 activity in experimental pulmonary tuberculosis. Infect Immun 57: 2606–2611
Middlebrook GM (1952) An apparatus for airborne infection of mice. Proc Soc Exp Biol Med 80: 105–110
Mitchison DA (1964) The virulence of tubercle bacilli from patients with pulmonary tuberculosis in India and other countries. Bull Int Union Against TB 35: 287–306
Muller I, Cobbold S, Waldmann H, Kaufmann SHE (1990) Impaired resistance to *Mycobacterium tuberculosis* infection after selective *in vivo* depletion of L3T4+ and Lyt-2+ T cells. Infec Immun 55: 2037–2041
Nickonenko BV, Apt AS, Moroz AM, Averbakh MM, Skamene E (1985) Genetic analysis of susceptibility of mice to H37Rv tuberculosis infection: sensitivity versus relative resistance. Prog Leukoc Biol 3: 291–296
North RJ (1973) Importance of thymus-derived lymphocytes in cell-mediated immunity to infection. Cell Immunol 7: 166–176
North RJ, Izzo AA (1993) Granuloma formation in severe combined immunodeficient (SCID) mice in response to progressive BCG infection. Am J Pathol 142: 1959–1966
Orme IM (1987) The kinetics of emergence and loss of mediator T lymphocytes in response to infection with *M. tuberculosis*. J Immunol 138: 293–298
Orme IM, Collins FM (1983) Protection against *Mycobacterium tuberculosis* infection by adoptive immunotherapy. J Exp Med 158: 74–83
Orme IM, Collins FM (1994) Mouse model of tuberculosis. In: Bloom BR (ed) *Tuberculosis: pathogenesis, protection and control.* American Society for Microbiology, Washington DC, pp 113–134
Orme IM, Miller ES, Roberts AD, Furney SK, Griffin JP, Dobos KM, Chi D, Rivoire B, Brennan PJ (1992) T lymphocytes mediating protection and cellular cytolysis during the course of *Mycobacterium tuberculosis* infection. J Immunol 148: 189–196
Phalen SW, McMurray DN (1993a) T lymphocyte response in a guinea pig model of tuberculous pleuritis. Infect Immun 61: 142–145
Phalen SW, McMurray DN (1993b) Production of tumor necrosis factor (TNFα) in experimental tuberculous pleuritis. J Immunol 150: 66A
Pierce CH, Dubos RJ and Schaefer WB (1956) Differential characteristics in *vitro* and *in vivo* of several substrains of BCG. III. Multiplication and survival *in vivo.* Am Rev Tuberc 74: 683–698
Pittsfield Conference (1986) Supplement of future research in tuberculosis. Prospects and priorities for elimination (Pittsfield, MA). Am Rev Respir Dis 134: 401–423
Prabhakar R, Venkataraman P, Vallishayee RS, Resser P, Musa S, Hashim R, Kim Y, Dimmer C, Wiegeshaus E, Edwards M, Smith DW (1987) Virulence for guinea pigs of tubercle bacilli isolated from the sputum of persons included in the BCG trial, Chingleput district, south India. Tubercle 68: 3–17
Ratcliffe HL, Palladino VS (1953) Tuberculosis induced by droplet nuclei infection: initial homogeneous response of small mammals (rats, mice, guinea pigs and hamsters) to human and to bovine bacilli and the rate and pattern of tubercle development. J Exp Med 97: 61–68
Riley RL, Mills CC, O'Grady F, Sultan LU, Wittstadt F, Shivpuri DN (1962) Infectiousness of air from a tuberculosis ward. Ultraviolet irradiation of infected air: comparative infectiousness of different patients. Am Rev Respir Dis 85: 511–525
Rook GAW (1990) Mycobacteria, cytokines and antibiotics. Pathol Biol 38: 276–280

Schaaf HS, Gie RP, Beyers N, Smuts N, Donald PR (1993) Tuberculosis in infants less than 3 months of age. Arch Dis Child 69: 371–374
Schultz LD, Sidman CL (1987) Genetically determined murine models of immunodeficiency. Annu Rev Immunol 5: 367–403
Siebenmann CO, Barbara C (1974) Quantitative evaluation of the effectiveness of Connaught freeze-dried BCG vaccine in mice and guinea pigs. Bull World Health Organ 51: 283–290
Smith DW, Harding GE (1977a) Animal model: experimental airborne tuberculosis in the guinea pig. Am J Pathol 89: 273–276
Smith DW, Harding GE (1977b) Approaches to the validation of animal test systems for assay of the protective potency of BCG vaccines. J Biol Stand 5: 131–138
Smith DW, Wiegeshaus EH. (1989) What animal models can teach us about the pathogenesis of tuberculosis in humans. Rev Infect Dis 11: S385–S393
Smith DW, McMurray DN, Wiegeshaus EH, Grover AA, Harding GE (1970) Host-parasite relationships in experimental airbrone tuberculosis. IV. Early events in the course of infection in vaccinated and nonvaccinated guinea pigs. Am Rev Respir Dis 102: 937–949
Smith DW, Harding G, Chan J, Edwards M, Hank J, Muller D, Sobhi F (1979) Potency of 10 BCG vaccines as evaluated by their influence on the bacillemic phase of experimental airborne tuberculosis in guinea-pigs. J Biol Stand 7: 179–197
Smith DW, Balasubramanian V, Wiegeshaus E (1991) A guinea pig model of experimental airborne tuberculosis for evaluation of the response to chemotherapy: the effect on bacilli in the initial phase of treatment. Tubercle 72: 223–231
Stead WW (1989) Pathogenesis of tuberculosis: clinical and epidemiological perspective. Rev Infect Dis 11: 366–368
Tsuda T, Dannenberg AM, Jr Ando M, Abbey H, Corrin AR (1976) Mononuclear cell turn-over in chronic inflammation. Studies on tritiated thymidine-labeled cells in the blood, tuberculin traps, and dermal BCG lesions of rabbits. Am J Pathol 83: 255–268
Wells AQ (1937) Tuberculosis in wild voles. Lancet 232: 1221
Wessels CC (1941) Tuberculosis in the rat. I. Gross organ changes and tuberculin sensitivity in rats infected with tubercle bacilli. II. The fate of tubercle bacilli in the various organs of the rat. III. The correlation between the histological changes and the fate of living tubercle bacilli in the organs of the albino rat. Am Rev Tuberc Pulm Dis 43: 449–458; 459–474; 637–644
Wiegeshaus EH, Smith DW (1989) Evaluation of the protective potency of new tuberculosis vaccines. Rev Infect Dis 11 [Suppl]: S484–S490
Wiegeshaus EH, McMurray DN, Grover AA, Harding GE, Smith DW (1970) Host-parasite relationships in experimental airborne tuberculosis III: relevance of microbial enumeration to acquired cellular resistance in guinea pigs. Am Rev Respir Dis 102: 422–429
Wiegeshaus E, Harding G, McMurray D, Grover AA, Smith DW (1971) A cooperative evaluation of test systems used to assay tuberculosis vaccines. Bull World Health Organ 45: 543–550
Wiegeshaus EH, Balasubramanian V, Smith DW (1989) Immunity to tuberculosis from the perspective of pathogenesis. Infect Immun 57: 3671–3676
World Health Organization (1976) The role of the individual and the community in the research, development, and used of biologicals with critieria for guidelines: a memorandum. Bull WHO 54: 645–655
World Health Organization (1982) Immunological research in tuberculosis: a memorandum Bull. WHO 60: 723–727

Immune Responses in Animal Models

I.M. Orme

1 Introduction

Animal models have made major contributions to our understanding of the nature of the immune response to tuberculosis. Although each of the most popular models tend to give us a different perspective of the disease process, the collective information gained has given us considerable insight into the basis of susceptibility and resistance to disease as well as the nature of both protective immunity and immunopathology.

It is generally believed that human beings tend to be "resistant" to tuberculosis, in that most people [perhaps 75%–80%] exposed to aerogenic exposure to a tubercle bacillus destroy the organism by means of innate immunologic mechanisms, without drawing upon the resources of the acquired response. Under these conditions no T cells are sensitized and the individual remains tuberculin negative. The remaining 25% or so do become tuberculin positive, but most people do not usually do not show any evidence of active disease, indicating that the infection has been efficiently controlled and contained by the T cell response. Only a few of these individuals go on to develop active tuberculosis.

Mycobacteria Research Laboratories, Department of Microbiology, Colorado State University, Fort Collins, CO 80523, USA

Thus, animal models are chosen depending upon which of these facets are to be investigated. The mouse and the rat are resistant models, which allow investigation into the nature of the protective immune response and mechanisms of containment and killing of the mycobacterial infection. Guinea pigs and rabbits are "susceptible," in that even small inocula of *Mycobacterium tuberculosis* will eventually kill them. Thus, these latter two models are appropriate for the study of the progressive pulmonary disease and the pathology that subsequently ensues.

Both the mouse and rat develop strong immunity to tuberculosis infection, and for a while at least the rat was utilized as a method of studying the lymphocyte response, given the ease by which these cells could be collected by thoracic duct drainage. Characterization of these cells was in terms of size (small, resting vs blast), replication, and proliferation to antigen, but unfortunately the model had fallen out of use by the time that monoclonal antibodies against major T cell markers in the rat had finally been developed to any extent.

In terms of susceptibility to disease, the guinea pig is an important model of tuberculosis. It shares a number of physiological characteristics with humans (McMurray 1994); in fact, it may not even be a true rodent (Graur et al. 1991). It has been widely used throughout the history of the field and has provided important insights into the nature of the disease in the lungs. It retains the disadvantage that few immunologic reagents are available (the actual basis of susceptibility defined at the T cell level is unknown, for example), although this situation may be gradually improving (Schafer and Burger 1992).

The rabbit is also an important model of susceptibility, thanks to the pioneering work in this model by Lurie et al. (1952), and more recently by Dannenberg (1991). Unfortunately, to date there are no well-defined immunologic markers in rabbits, and hence the definition of the disease process in this animal has to be phrased in purely pathological terms.

For a detailed study of the cell-mediated immune response to tuberculosis infection, the mouse is the model of choice. As the reader is of course aware, there is a massive repertoire of immunologic reagents now available for use in defining murine lymphocytes, cytokines, etc, which are mostly lacking in the other animal models. For this reason, this brief review will concentrate mainly on the murine model of tuberculosis.

2 The Kinetics of the T Cell Response

The mouse model has allowed us to define in considerable detail the emergence of the T cell response to experimental tuberculosis. As a result we have been able to ascribe functional or operational terms to apparent stages of the response, namely, protection, delayed-type hypersensitivity (DTH), and memory immunity. Despite this, however, we still lack definitive proof that these functions are indeed separate or whether they representing overlapping characteristics of a broad pool of mycobacteria antigen-reactive T cells. In addition, it is the purpose of animal modeling to try to extrapolate these ideas to the human disease, but it is probably true to say that the identification of parallel T cell antigen-specific subsets in the human model is still in its infancy.

Where there is obvious agreement, however, is in the pivotal role of the CD4 T cell subset. In mice infected with *M. tuberculosis*, the emergence of a protective T cell subset can be detected by classical cell transfer techniques about a week into the infection. As the course of the infection continues, peak T cell activity occurs at about 3 weeks, a time associated with a progressive downturn in bacterial numbers in target organs (ORME 1987). The cell population mediating protection at this time is in a state of cellular proliferation, as evidenced by its susceptibility to radiation or the drug cyclophosphamide (ORME 1988).

With the availability of lytic monoclonal antibodies, it was then shown that cells mediating the early protective effect in the mouse bore the CD4 phenotype. More recent work has expanded this definition by following changes in expression of the CD44 and CD45RB markers during the course of the infection (GRIFFIN and ORME 1994). These markers confirm the concept of the rapid emergence of a dividing cell population following infection by demonstrating the emergence of a CD4+, physically large, CD44hi, CD45RBhi subset of cells. Over the next 2–3 weeks of the infection there is gradual reduced expression of the CD45RB marker, resulting in increasing numbers of CD44hiCD45lo/neg cells.

Because of the considerable difficulty of sorting these cell populations, we cannot as yet ascribe any functions to each population, or for that matter definitively prove if each population is even actually involved in host resistance. If we may be permitted to speculate, however, we favour the following model.

Soon after infection, bacilli within macrophages begin to secrete a number of proteins which we believe are mirrored to a great extent by those that can be harvested from culture filtrates in vitro (see below). These are removed somehow from the phagosome and enter acidic endosomes where they are processed into peptides incorporated into class II MHC molecules. These are then presented to T cells, amongst which there will be clones capable of specific recognition.

Upon sensitization, the T cell will enter an activated state in which it will increase membrane expression of CD44 and CD45RB and become physically larger. Cell division will begin, rapidly expanding the pool of cytokine-secreting "protective" T cells. During the second week of infection CD45RB expression begins to drop on many cells; whether this reflective of bacterial control and a drop in antigen concentration or is a completely independent T cell event is completely unknown. By the fourth week of infection there is now a significant population of CD44hiCD45RBlo/neg cells. It is interesting to note that this particular phenotype is often associated with memory immunity, and we ourselves have observed that T cells that appear to be associated with memory in the mouse tuberculosis model begin to emerge at around this time (ORME 1988).

The precise role of CD8 T cells in immunity to tuberculosis is less clear, although the accumulating evidence certainly points to an important requirement for CD8 activity in protection against pulmonary infections. The first observation in this regard was made in this laboratory (ORME and COLLINS 1984), when we showed that Lyt-2 enriched cells were marginally protective against lethal dose aerosol infections in irradiated recipient mice. This was then later confirmed using CD8-enriched cells (ORME 1987).

In this latter study we found that both CD4 and CD8 cells were protective if the aerosol dose was sublethal (Table 1), with CD4 cells in fact producing better protection than the CD8 cells. This indicates that CD4 T cells are fully capable of being protective in the lungs, but that they mediate this protection at a certain rate that cannot protect if the infectious challenge is unrealistically high. This supports the idea, therefore, that CD4 mediated immunity in the lungs is not quickly expressed.

One logical explanation for this is that either CD4 cells home poorly in the lungs (probably unlikely) or that antigens are not well expressed in the lungs by class II molecules, since initially at least cells capable of expressing these molecules will not be present at the sites of infection. We favor this latter idea, in that accumulation of antigen-presenting cells in infected lesions may delay the ability of CD4 cells to mediate their protective function, thus explaining their apparent inability to protect against an acute aerosol challenge. Under these conditions, therefore, CD8 cells may be crucial in protecting the host in view of their ability to lyse infected target cells. The latter include alveolar endothelial cells, into which some bacilli may have eroded, thus releasing bacilli for subsequent uptake by professional phagocytic cells that are entering the lesion, resident in blood sinuses, or draining afferent lymphatics elsewhere (ORME 1993).

If this hypothesis is correct, then CD8 cells may be specifically required as an essential adjunct to CD4 cells in the protection of pulmonary tissues. This would also predict that, in animals lacking CD8 cells, local absesses could potentially form in the lungs causing widespread damage. This hypothesis is thus particularly strengthened by the observations of FLYNN and her colleagues (1992), who have observed extensive necrosis following infection of β2-microglobulin

Table 1. Both CD4 and CD8 T cells protect against sublethal aerosol infections

Challenge dose (log 10)	Cells infused	Assay day (postchallenge)	Bacteria in lungs (log 10)	Resistance
10^4	Normal T	30	8.5	
	Immune CD4		8.1	0.4
	Immune CD8		7	1.5[a]
10^1	Normal T	40	5.1	
	Immune CD4		3.2	1.9
	Immune CD8		4.5	0.6

[a]All mice died within 90 days.

gene "knock-out" mice, which lack sensitized CD8 cells. It is very important to note, moreover, that in that study the *M. tuberculosis* infection was given intravenously and yet the destructive pathology was seen in the lungs which possess low numbers of resident class II MHC presenting cells, and not in the liver or spleen, which are rich in such cells.This further suggests that CD8 T cells have specialized protective abilities within the lung tissues.

3 Analysis of T Cell Functions

To date, T cell functions in immunity to tuberculosis have to be expressed in operational terms. Thus, we can ascribe functions such as protection, DTH effector, cytolytic, and memory to differently emerging populations of cells.

At present, we tend to regard these populations in this laboratory as representing two distinct phases of immunity, to wit "protection" and "immunosurveillance." In the primary phase, the animal generates populations of protective T cells that are relatively short-lived, and which secrete large amounts of interferon-γ (IFN-γ). These populations are primarily CD4 T cells, but the protective role of other cells with the potential to also secrete this cytokine, namely CD8, $\gamma\delta$, and (natural killer (NK) cells, cannot be ruled out.

The second phase, one of immunosurveillance, is probably designed to hunt down bacilli that may have escaped the initial focus of infection and disseminated elsewhere and to guard against the possibility of the infection recrudescing. Thus, DTH effector T cells may represent a more longer-lived cousin of the CD4 protective T cell, which is able to recirculate and which can move quickly into extravascular sites of inflammation. Here this cell population recognizes mycobacterial antigens, including the secreted/export proteins of the bacillus (Orme et al. 1992), and releases cytokines promoting the influx of bloodborne monocytes.

Cytolytic CD4 T cells have also been observed in several laboratories (Mustafa and Godal 1987; Hancock et al. 1989; Boom et al. 1991; Orme et al. 1992); in the mouse model they appear 3–6 weeks into the intravenous infection model (Orme et al. 1992). It remains unclear as to their precise role, but a reasonable hypothesis is that they are scavenger cells that lyse heavily infected macrophages, thus releasing resident bacilli into the tissue where they can be destroyed by incoming activated monocytes (Kaufmann 1988).

Perhaps the most important cell population in this second phase is the true memory T cell, which appears to be capable of a very extensive lifespan. This cell is also presumably recirculating, and gives rises to a very rapid recall of resistance to a secondary challenge infection (Orme 1988). It preferentially recognizes members of the secreted/export proteins of the bacillus (see below), which may suggest that it arises from the initial protective T cell population. This is far from clear, however, nor is it yet known what are the specific antigenic targets of this population. As this information becomes available, the objective of improving current vaccination against tuberculosis should become closer to reality.

4 Role of Cytokines in Expression of Protective Immunity

Macrophages infected with *M. tuberculosis* in vitro release a large number of inflammatory, acute-phase, and leukopoietic cytokines. These promote a variety of events, including changes in local permeability; recruitment of mediator cells to the lesion site; and activation of the infected cells themselves. Certain cytokines, such as interleukin (IL)-1, IL-6, IL-10, and IL-12, almost certainly influence the emerging T cell response, presumably by helping promote the needed short-lived protective Th1-like response.

At the macrophage level, there is emerging evidence that tumor necrosis factor (TNF) and IL-12 may be key regulatory molecules. TNF is known to be important in contributing to the integrity of the granuloma (Kindler et al. 1989), in addition to well-documented antimicrobial properties (Wallis and Ellner 1994). In fact, TNF may synergize with IFN-γ to promote the destruction of intracellular bacteria (Appelberg et al. 1994).

A interesting twist to this story that has emerged recently is the observation that the more virulent strains of mycobacteria do not induce a particularly strong TNF response from infected macrophages and that this event may be directly related to the structure of the mycobacterial lipoarabinomannan (LAM), a lipoglycan that copiously covers the bacterial cell wall (Chatterjee et al. 1991, 1992a,b, 1993; Prinzis et al. 1993). Thus, it has been shown (Chatterjee et al. 1992c) that the capped variety of LAM (several units of mannose capping the ends of the arabinose side chains of the molecule), which is possessed by strains that can grow in macrophages (*M. tuberculosis*, BCG, etc.), are poor inducers of LAM, whereas LAM molecules from avirulent (usually "fast-growing") mycobacterial strains such as several of the 'non-tuberculosis" varieties are potent inducers of the TNF response, thus promoting the intracellular destruction of the bacillus (Fig. 1). These differences have been shown to be directly related to the capacity of these molecules to trigger a number of early response genes in stimulated macrophages (Roach et al. 1993); moreover, very recent evidence that similar effects occur in human monocytes has now been obtained (Z. Toossi; personal communication).

Macrophages infected with live *M. tuberculosis* also secrete large amounts of IL-12 (Cooper and Orme; unpublished data). We have recently observed that in vivo administration of this cytokine to infected mice reduces the bacterial load over the first 20–30 days of the infection (Cooper et al. 1995). Whilst much of this effect can be attributed to enhanced IFN-γ secretion by protective antigen-specific CD4 T cells, an effect on granuloma integrity seems also to be involved.

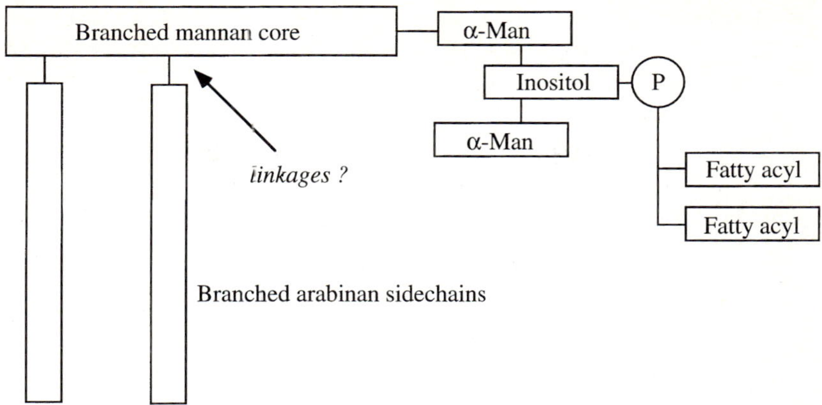

Inositol-P-mannan "capping"?

Fig. 1. Basic structure of lipoarabinomannan. The major features are a mannose-rich phosphoinositol core region, and extensive branched arabinan sidechains. Virulent strains also appear to possess substantial capping of these sidechains with additional mannose units (which may themselves be phosphorylated). Regions of the molecule at both the fatty acyl and nonreducing terminals can potentially interact with macrophage surface and intracellular receptors

This is suggested by the observations that: (a) T cell deficient mice such as SCID mice can be partially protected by IL-12 and that lesions in the treated mice, while containing very large numbers of bacteria, also contain substantially increased numbers of infiltrating macrophages compared to controls; (b) there is a substantial decrease in apparent granuloma integrity in normal infected mice given anti-IL-12 monoclonal antibody in vivo. Thus, these data suggest to us that IL-12 plays a role at the nonspecific level, by aiding the recruitment of cells into the granuloma and modulating the integrity of this structure. In fact, TNF and IL-12 may act in tandem in this role.

At the CD4 T cell level, the accumulating evidence all support the long-held contention that IFN-γ plays the pivotal role in the expression of specific resistance to tuberculosis infection. The newest wave of this evidence is based upon the use of mice in which the gene encoding IFN-γ has been specifically disrupted (Dalton et al. 1993). These mice were found to poorly express class II MHC molecules and were unable to significantly produce reactive oxygen or nitrogen radicals; moreover, such mice were highly susceptible to BCG infections. When infected with *M. tuberculosis*, either aerogenically (Cooper et al. 1993) or intravenously (Flynn et al. 1993), a similar pattern of events was observed, in which the infection disseminated widely throughout the animal whilst reaching lethal loads in primary target organs. Histologic examination revealed widespread caseous necrosis with infiltrations of granulocytes, particularly eosinophils, but with a complete absence of any mononuclear granulomatous response. Such studies spectacularly reveal the pivotal role of IFN-γ in resistance to tuberculosis.

5 Preferential Recognition of Secreted/Export Protein Antigens

A central paradox in the acquired response to tuberculosis infection in the mouse model was the observation that T cell mediated immunity could be detected well before the time at which the infection had been prevented from growing progressively in the animal. As such, this tended to argue against the idea that the mycobacterial antigens initially being presented were structural/constitutive antigens, given the fact that the organism would have to be destroyed and substantially processed for this to occur. Despite this, however, substantial effort was directed towards characterization of the heat shock proteins of *M. tuberculosis*, with a series of reports (almost invariably involving immunization of mice with dead bacilli in a strong adjuvant) pointing to the hsp60 molecule as the immunodominant antigen of the organism (Kaufmann 1990).

This designation, at least in the context of protective immunity, has not stood the test of time. Instead, interest has now centered on the role of secreted or export proteins of the bacillus, given increasing evidence that they are strongly recognized by T cells harvested from mice given live infections (Orme et al. 1993a).

In this regard, it can be shown (Orme et al. 1993b) that immune CD4 T cells secret IFN-γ in considerable amounts if overlaid in vitro on macrophages pulsed with purified culture filtrate proteins (CFPs). Cells harvested as early as day 5 of the infection can do this, and production of this cytokine peaks between days 10 and 20, at a time when the infection is contained and is being destroyed (Orme 1987). Similarly, the CFP antigen pool is also strongly recognized following rechallenge of memory immune animals.

The CFP is a complex mixture of probably at least 50–100 proteins. Some progress in characterizing immunoreactive proteins within this pool is now underway, and pioneering work by Andersen and his colleagues (1991a,b, 1992) has begun to define those proteins strongly recognized by immune T cells. These investigators have, further shown that vaccinated mice can be protected from a virulent challenge infection (Andersen 1994).

As a result of such studies, Andersen has identified two regions using short-term CFP pools, one around 5–12 kDa and the other at 28–32 kDa, that are very strongly recognized. We have also observed this same profile in our laboratory, although we note with caution here that when we use a more long-term CFP pool the spectrum of T cell recognition appears to be much broader. Thus whether the "key protective antigens" of *M. tuberculosis* can be narrowed to one or a few proteins in the CFP or whether they are represented by a larger number of proteins (which may also differ from isolate to isolate) remains to be determined.Certainly it remains to be shown that isolated proteins or pools of proteins from our "laboratory strains" can protect animals from aerosol challenge infections with a variety of clinical isolates.

6 Role of Cells Other Than αβ T Cells

Both γδ T cells and cells can be observed in mycobacterial lesions under certain experimental conditions, but hard evidence remains lacking as to whether these cell populations are mediating a specific protective function or whether they are responding to local inflammatory conditions (perhaps in a way analogous to the way in which neutrophils accumulate at such sites).

This laboratory would tend to lean towards the latter viewpoint. In early reports large accumulations of γδ cells were observed in mice exposed to aerosols of tuberculin (a very inflammatory material when delivered in this manner); AUGUSTIN et al. 1989), and similar results were obtained when mice were inoculated in footpads with dead bacilli in Freund's adjuvant (JANIS et al. 1989) or with adjuvant alone (GRIFFIN et al. 1991). In clinical situations, γδ cell accumulations have been seen in patients with the lepromatous form of leprosy (FALANI et al. 1989) or with tuberculous lymphadenitis (MODLIN et al. 1989), but not in patients in which bacterial numbers are contained by adequate granuloma formation (TAZI et al. 1991).

In this regard, a popular viewpoint has emerged that γδ cells provide a "first line of defense" by accumulating rapidly at sites of infection. But is this hypothesis actually true? A rapid influx of γδ cells were observed in the peritoneal cavity of mice inoculated with a large dose of BCG (INOUE et al. 1991), but this was not observed in a study in which *M. tuberculosis* was injected at this site (GRIFFIN et al. 1991) and in which peak accumulation of γδ cells was observed after about 3 weeks of infection. In this latter study, moreover, γδ cell accumulation was observed in mice infected intravenously, but this accumulation occurred at the same rate as αβ cells. In another infection model, using *Listeria*, γδ cells appeared to be a late appearing compensatory device in the absence of CD8 T cells (ROBERTS et al. 1993), whereas in a model of viral infection γδ cells appear to be involved in the late, inflammatory stage of protection (CARDING et al. 1990). Similarly, a role for γδ cells has been proposed in nude mice, in which both the bacterial load and the degree of inflammation are severe (IZZO and NORTH 1992).

None of this is helped by the equally confusing picture regarding the antigens that γδ cells supposedly recognize. These cells certainly can secrete a range of cytokines when stimulated with bacterial sonicates (FOLLOWS et al. 1992), and recent work has concentrated on the possibility that a major target of γδ is a mycobacterial compound which is not a protein (PFEFFER et al. 1990). A very recent report from Fournie's laboratory now appears to strengthen this possibility further (CONSTANT et al. 1994).

In addition, however, there is substantial data to support the hypothesis that certain clones of γδ can recognize the hsp60 stress protein of *M. tuberculosis* (O'BRIEN et al. 1992), leading to the speculation that such cells may play a "scavenger" role by lysing heavily laden macrophages (inside which stressed bacilli would be expected to be producing hsp60). Given the limited variable gene element repertoire of $\gamma\delta$ cells, a focus on stress-induced molecules has a certain attraction.

If one is to take a holistic approach to the question, perhaps it should be noted that not all interactions with T cells occur through the T cell receptor. Is it not possible, for instance, that small molecular weight, perhaps nonproteinaceous, inflammatory molecules can preferentially trigger $\gamma\delta$ cells to migrate to inflammatory sites and secrete cytokines without the T cell receptor being involved at all? As an example, one can point to older literature that has shown that some T cells express the H2-type of the histamine receptor (Rocklin et al. 1980). Since this molecule plays a pivotal role in recruiting other leukocytes, perhaps it also influences "primeval" $\gamma\delta$ cells?

Although NK cells have been clearly implicated in very slow growing infections caused by members of the *M. avium* complex, there is little direct evidence at this time to suggest that they play any important role in tuberculosis infections. In work on this topic in our own laboratory we have observed that NK cells only represent a very minor population in mouse spleens (about 2%) and that this does not increase when the animal is infected intravenously. However, treatment of infected mice with IL-12, which is known to induce IFN-γ secretion by NK cells, can increase their resistance to the infection to some extent (Cooper et al. 1995). This increase also occurs in beige mice, but this still does not completely rule out NK activity, since although NK cells in beige mice have no lytic activity, they possibly may still be able to produce IFN-γ. Thus, while our current feeling is that NK cells do not contribute significantly to protection in immunocompetent mice, this remains far from proven.

7 Mechanisms of Mycobacterial Killing

Despite its substantial acquired response the mouse cannot completely destroy experimental tuberculosis infections. The question is therefore how does it kill these bacilli, and why do some of them survive?

There is much more information concerning the first part of this question. In fact, there may be a number of ways in which bacterial destruction is achieved. A classical explanation has been that the activated macrophage is more metabolically active, which may reduce the phagosomal pO2 (mycobacteria are aerobes), or that phagolysosomal fusion is increased, flooding the phagosome with acid hydrolases. While there is no evidence either way for the first idea (although it remains attractive to this author, particularly in the context of bacilli in the centers of active granulomas), the second concept appears to be more complicated.

The reason for this is the increasingly popular notion that a key destructive mechanism revolves around reduction of the phagosomal pH. It is now known that the phagosomal pH is not acidic, as once believed, but close to neutral

(CROWLE et al. 1991); presumably the pH drops when the macrophage is activated, and in fact work in this laboratory has shown that reducing the pH of the surrounding medium enhances the antimicrobial activity of IFN-γ (APPELBERG and ORME 1993).

The actual mechanism involved is now much clearly, due to recent work by Russell and his colleagues (STURGILL-KOSZYCKI et al. 1994). They found that, in contrast to other pathogens in which invasion of the phagosome resulted in a significant reduction in local pH, only a minor reduction was seen when macrophages were infected with mycobacteria. Isolation of the phagosomes from these cells and examination by electron microscopy and immunogold staining showed clearly that those containing mycobacteria lacked the presence on the membrane of the vesicular ATP-ase associated proton pumps normally seen in response to other infections (Fig. 2).

These data help explain why phagolysosomal fusion, although seen, may not be very effective, given the neutral pH of the phagosome. They also allow us to speculate that the mycobacteria have adapted to live in the neutral phagosome and would like it to stay that way; the fact that mycobacteria produce ammonia to absorb protons also supports this idea (GORDON et al. 1980). In this regard it will be fascinating to eventually see if the failure to construct the proton pumps on the phagosomal membrane is an active mechanism (the bacillus produces something that prevents this) or passive (the bacillus sneaks in without triggering this host cell mechanism).

A second area of much interest is the destruction of the bacillus by reactive nitrogen radicals and nitrous oxide produced by activated cells. These molecules have been implicated in the destruction of a number of bacterial and protozoal pathogens in the mouse (NATHAN and HIBBS 1991), including *M. tuberculosis* (CHAN et al. 1992).

Here this author must inject some healthy skeptiscism. It is my prejudice that I do not find the "NO story" all that convincing. The reasons for this are: (1) Much of the data concerning growth of bacteria in vitro have been obtained using a radioactive uracil assay. When the culture well contains NO the uracil count drops; this is interpreted as killing, but often as not CFU counts do not appear to be that significantly reduced, suggesting the effect is static rather than cidal. (2) The uracil assay involves lysis of the culture, so that the NO and other materials (hydrolases, etc.) are all freely mixed with the bacilli; in the intact cell, the NO is measured in the supernatant, but how much is in the phagosome? Moreover, one might expect the NO to give rise to nitrous acid in the lysate, thus dropping the pH (which mycobacteria do not appear to like). (3) In assays based upon CFU counts, NO does not appear to be active against other strains of mycobacteria (APPELBERG and ORME 1993).

Another puzzling aspect is whether NO is part of an innate or acquired response. The knowledge that infected macrophages only make NO when exposed to large doses of IFN-γ clearly suggests the latter. Nonetheless, some workers have recently associated NO production to the phenotype of the mouse *Bcg* gene, believed to play a role in the control of innate resistance to myco-

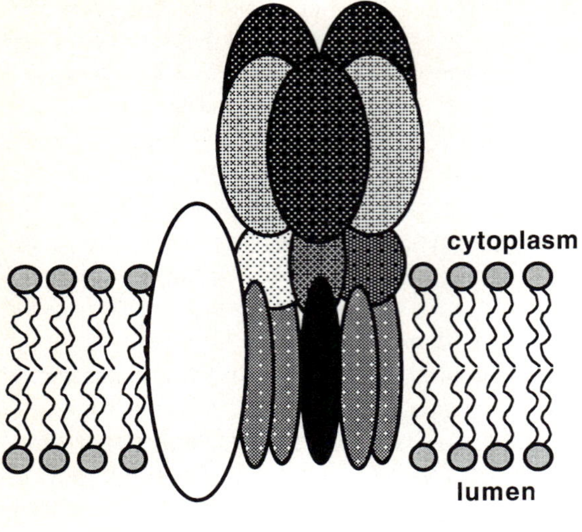

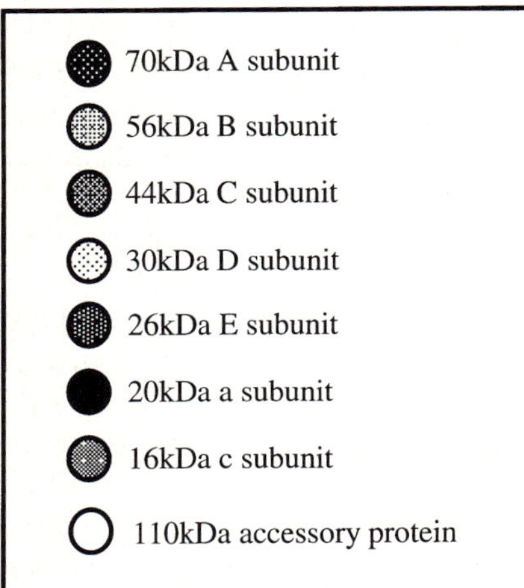

Fig. 2. Basic structure of vesicular ATP-ase dependent proton pump. Phagosomes containing mycobacteria do not appear to incorporate these pumps into phagosomal membrane; as a result the phagosome luminal pH remains close to neutral allowing the bacillus to survive. (Model courtesy of Dr. David Russell)

bacterial infections. The sequence of the *Nramp* gene, which maps to the *Bcg* locus appears to resemble sequences for membrane transporters, and in view of this finding VIDAL and her colleagues (1993) have suggested that this protein directly transports NO into the phagosome. However, why a transporter would be needed for a molecule that should readily diffuse across membranes anyway is puzzling.

Again, to be holistic (as well as reactionary), it is quite possible that the potential antimicrobial activities of NO are only a minor part of the story, and that maybe the primary reason this molecule is secreted out of the macrophage in such quantities is to promote local arteriolar relaxation, making it easier for other cells to egress into the local site of infection. It should also be noted that the NO pathway does not appear to be operative in human macrophages.

8 Closing Comments: Are We Reinventing the Tubercle Bacillus?

Very recent progress in the field has centered on events within the macrophage phagosome containing the invading bacillus. These studies seem to further confirm what many of us already suspected; namely, that the mycobacterium is beautifully adapted to become phagocytosed by macrophages and to subsequently reside and survive within the phagosome. As discussed above, a central facet to this survival appears to be retention of the phagosomal pH to close to neutral.

In view of this, therefore, it is fascinating that recent reports seem to suggest that mycobacteria wish to escape this friendly environment. It has also been suggested that virulent strains of *M. tuberculosis* can escape into the macrophage cytoplasm (McDonough et al. 1993), and a molecule derived from the avirulent strain H37Ra that acts as an invasin (Arruda et al. 1993) has also been cloned into *E. coli*.

Our own findings, however, (Xu et al. 1994), suggest that the phagosomal membrane is very tightly opposed to the bacillus, which may not always be evident, depending upon the type of fixation method employed. While the "escape into the cytoplasm" hypothesis is attractive in the context of explaining the CD8 T cell sensitization that occurs in infected mice, it could equally well be explained by leakage of secreted/export mycobacterial proteins into the cytoplasm as the bacilli divide (each occupies its own discrete phagosome) Alternatively there may be leakage of these materials as a result of phagosomal damage in heavily laden cells an idea consistent with the observation that CD8 cells first appear relatively late during the primary immune response to the infection when such an event might be expected to occur (Orme 1987).

This author, for one, would like to see these findings pursued further, since it seems rather paradoxical that, having adapted so well to survival within the phagosome, the mycobacterium would then want to destroy it.

Acknowledgements. The author would like to acknowledge the contributions of many of his colleagues in the Mycobacteria Research Laboratories, CSU, to the ideas expressed in this article.

References

Anderson P (1994) Effective vaccination of mice against *Mycobacterium tuberculosis* infection with a soluble mixture of secreted mycobacterial proteins. Infect Immun 62: 2536–2544

Andersen P, Askgaard D, Ljungqvist L, Bennedsen J, Heron I (1991a) Proteins released from *Mycobacterium tuberculosis* during growth. Infect Immun 59: 1905–1910

Andersen P, Askgaard D, Ljunqvist L, Bentzon MW, Heron I (1991b) T-cell proliferative response to antigens secreted from *Mycobacterium tuberculosis*. Infect Immun 59: 1558–1563

Andersen P, Askgaard D, Gottschau A, Bennedsen J, Nagai S, Heron I (1992) Identification of immunodominant antigens during infection with *Mycobacterium tuberculosis*. Scand J Immunol 36: 823–831

Appelberg R, Orme IM (1993) Effector mechanisms involved in cytokine-mediated bacteriostasis of *Mycobacterium avium* infections in murine macrophages. Immunology 80: 352–359

Appelberg R, Castro AG, Pedrosa J, Silva RA, Orme IM, Minoprio P (1994) The role of gamma interferon and tumor necrosis factor-alpha during the T cell independent and dependent phases of *Mycobacterium avium* infection. Infect Immun 62: 3962–3977

Arruda S, Bomfim G, Knights R, Huima-Byron T, Riley LW (1993) Cloning of an *M. tuberculosis* DNA fragment associated with entry and survival inside inside cells. Science 261: 1454–1457

Augustin A, Kubo RT, Sim G (1989) Resident pulmonary lymphocytes expressing the γδ receptor. Nature 340: 239–241

Boom WH, Wallis RS, Chervenak KA (1991) Human *Mycobacterium tuberculosis* reactive CD4+ T cell clones: heterogeneity in antigen-recognition, cytokine production and cytoxicity for mononuclear phagocytes. Infect Immun 59: 2737–2743

Carding SR, Allan W, Kyes S, Hayday A, Bottomly K, Doherty PC (1990) Late dominance of the inflammatory process in murine influenza by γδ+ T cells. J Exp Med 172: 1225–1231

Chan J, Xing Y, Magliozzi RS, Bloom BR (1992) Killing of virulent *Mycobacterium tuberculosis* by reactive nitrogen intermediates produced by activated murine macrophages. J Exp Med 175: 1111–1122

Chatterjee D, Bozic CM, McNeil M, Brennan PJ (1991) Structural features of the arabinan component of the lipoarabinomannan of *Mycobacterium tuberculosis*. J Biol Chem 266: 9652–9660

Chatterjee D, Hunter SW, McNeil M, Brennan PJ (1992a) Lipoarabinomannan. Multiglycosylated form of the mycobacterial mannosylphophatidylinositols. J Biol Chem 267: 6228–6233

Chatterjee D, Lowell K, Rivoire B, McNeil M, Brennan PJ (1992b) Lipoarabinomannan of *Mycobacterium tuberculosis*. Capping with mannosyl residues in some strains. J Biol Chem 267: 6234–6239

Chatterjee D, Roberts AD, Lowell K, Brennan PJ, Orme IM (1992c) Structural basis of capacity of lipoarabinomannan to induce secretion of tumor necrosis factor. Infect Immun 60: 1249–1253

Chatterjee D, Khoo K-H, McNeil MR, Dell A, Morris HR, Brennan PJ (1993) Structural definition of the non-reducing termini of mannose-capped LAM from *Mycobacterium tuberculosis* through selective enzymatic degradation and fast atom bombardment-mass spectrometry. Glycobiology 3: 497–506

Constant P, Davodeau F, Peyrat MA, Poquet Y, Puzo G, Bonneville M, Fournie JJ (1994) Stimulation of human γδ T cells by nonpeptidic mycobacterial ligands. Science 264: 267–270

Cooper AM, Dalton DK, Stewart TA, Griffin JP, Russell DG, Orme IM (1993) Disseminated tuberculosis in interferon-γ gene-disrupted mice. J Exp Med 178: 2243–2247

Cooper AM, Roberts AD, Rhoades ER, Callaham JE, Getzy DM, Orme IM (1995) The role of IL-12 in immunity to *Mycobacterium tuberculosis*. Immunology 84: 423–432

Crowle AJ, Dahl R, Ross E, May MH (1991) Evidence that vesicles containing living, virulent *Mycobacterium tuberculosis* or *Mycobacterium avium* in cultured human macrophages are not acidic. Infect Immun 59: 1823–1831

Dalton D, Pitts-Meek S, Keshav S, Figari IS, Bradley A, Stewart TA (1993) Multiple defects of immune cell function in mice with disrupted interferon γ genes Science 259: 1739–1742

Dannenberg AM (1991) Delayed-type hypersensitivity and cell-mediated immunity in the pathogenesis of tuberculosis. Immunol Today 12: 228–233

Falini B, Flenghi L, Pileri S, Pelicci P, Fagioli M, Martelli MF, Moretta L, Ciccone E (1989) Distribution of T cells bearing different forms of the T cell receptor type γ/δ in normal and pathological human tissues. J Immunol 143: 2480–2488

Flynn JL, Goldstein MM, Triebold KJ, Koller B, Bloom BR (1992) Major histocompatability complex class I-restricted T cells are required for resistance to Mycobacterium tuberculosis infection. Proc Natl Acad Sci USA 89: 12013–12017

Flynn JL, Chan J, Triebold KJ, Dalton DK, Stewart TA, Bloom BR (1993) An essential role for IFN-γ in resistance to *M. tuberculosis* infection. J Exp Med 178: 2249–2254

Follows GA, Munk ME, Gatrill AJ, Conradt P, Kaufmann SHE (1992) Gamma-interferon and interleukin-2, but not interleukin-4, are detectable in gamma-delta T-cell cultures after activation with bacteria. Infect Immun 60: 1229–1231

Gordon AH, Hart PD, Young MR (1980) Ammonia inhibits phagosome-lysosome fusion in macrophages. Nature 286: 79–81

Graur D, Hide WA, Wen-Hsiung (1991) Is the guinea pig a rodent? Nature 351: 649–652

Griffin JP, Orme IM (1994) Evolution of CD4 T-cell subsets following infection of naive and memory mice with *Mycobacterium tuberculosis*. Infect Immun 62: 1683–1690

Griffin JP, Harshan KV, Born WK, Orme IM (1991) Kinetics of accumulation of γδ receptor-bearing T lymphocytes in mice infected with live mycobacteria. Infect Immun 59: 4263–4265

Hancock GE, Cohn ZA, Kaplan G (1989) The generation of antigen-specific, major histocompatability complex-restricted cytotoxic T lymphocytes of the CD4+ phenotype. J Exp Med 169: 909–919

Inoue T, Yoshikai Y, Matsuzaki G, Nomoto K (1991) Early appearing γδ-bearing T cells during infection with Calmette Guerin bacillus. J Immunol 146: 2754–2762

Izzo AA, North RJ (1992) Evidence for an αβ T-cell independent mechanism of resistance to Mycobacteria. Bacillus calmette-guerin causes progressive infection in severe combined immunodeficient mice, but not in nude mice or in mice depleted of CD4+ and CD8+ T-cells. J Exp Med 176: 581–586

Janis EM, Kaufmann SHE, Schwartz RH, Pardoll DM (1989) Activation of γδ T cells in the primary immune response to *Mycobacterium tuberculosis*. Science 244: 713–717

Kaufmann SHE (1988) CD8+ T lymphocytes in intracellular microbial infections. Immunol Today 9: 168–174

Kaufmann SHE (1990) Heat shock proteins and the immune response. Immunol Today 11: 129–136

Kindler V, Sappino AP, Grau GE, Piguet PF, Vassalli P (1989) The inducing role of tumor necrosis factor in the development of bactericidal granulomas during BCG infection. Cell 56: 731–740

Lurie MB, Abramson S, Heppleston AG (1952) On the response of genetically resistant and susceptible rabbits to the quantitative inhalation of human-type tubercle bacilli and the nature of resistance to tuberculosis. J Exp Med 95: 119–134

McDonough KA, Kress Y, Bloom BR (1993) Pathogenesis of tuberculosis: interaction of *Mycobacterium tuberculosis* with macrophages. Infect Immun 61: 2763–2773

McMurray DN (1994) Guinea pig model of tuberculosis. In: Bloom BR (ed) Tuberculosis: pathogenesis, protection, and control. ASM, Washington DC, pp 135–147

Mustafa AS, Godal T (1987) BCG-induced CD4+ cytotoxic T cells from BCG vaccinated healthy subjects: relation between cytotoxicity and suppression in vitro. Clin Exp Immunol 69: 255–242

Modlin RL, Pirmez C, Hofmann FM, Torigian V, Uyemura K, Rea TH, Bloom BR, Brenner MB (1989) Lymphocytes bearing antigen-specific γ/δ T-cell receptors accumulate in human infectious disease lesions. Nature 339: 544–548

Nathan CF, Hibbs JB (1991) Role of nitric oxide synthesis in macrophage antimicrobial activity. Curr Opin Immunol 3: 65–70

O'Brien RL, Fu Y, Cranfill R, Dallas D. Ellis C, Reardon C, Lang J, Carding SR, Kubo R, Born W (1992) Heat shock protein Hsp60-reactive γδ cells: a large, diversified T-lymphocyte subset with highly focused specificity. Proc Natl Acad Sci USA 89: 4348–4352

Orme IM (1987) The kinetics of emergence and loss of mediator T lymphocytes acquired in response to infection with *Mycobacterium tuberculosis*. J Immunol 138: 293–298

Orme IM (1988) Characteristics and specificity of acquired immunologic memory to *Mycobacterium tuberculosis* infection. J Immunol 140: 3589–3593

Orme IM (1993) The role of CD8 T cells in immunity to tuberculosis infection. Trends Microbiol 1: 77–78

Orme IM, Collins FM (1984) Adoptive protection of the *Mycobacterium tuberculosis*-infected lung. Cell Immunol 84: 113–120

Orme IM, Miller ES, Roberts AD, Furney SK, Griffin JP, Dobos KM, Chi D, Rivoire B, Brennan PJ (1992) T lymphocytes mediating protection and cellular cytolysis during the course of *Mycobacterium tuberculosis* infection. J Immunol 148: 189–196

Orme IM, Andersen P, Boom WH (1993a) T cell response to *Mycobacterium tuberculosis*. J Infect Dis 167: 1481–1497
Orme IM, Roberts AD, Griffin JP, Abrams JS (1993b) Cytokine secretion by CD4 T lymphocytes acquired in response to *Mycobacterium tuberculosis* infection. J Immunol 151: 518–525
Pfeffer K, Schoel B, Gulle H, Kaufmann SHE, Wagner H (1990) Primary responses of human T cells to mycobacteria: a frequent set of γδ T cells are stimulated by protease-resistant ligands. Eur J Immunol 20: 1175–1179
Prinzis S. Chatterjee D, Brennan PJ (1993) Structure and antigenicity of lipoarabinomannan from *Mycobacterium bovis* BCG. J Gen Microbiol 139: 649–2658
Roach TIA, Barton CH, Chatterjee D, Blackwell JM (1993) Macrophage activation: lipoarabinomannan from avirulent and virulent strains of *Mycobacterium tuberculosis* differentially induces the early genes *c-fos*, KC, JE, and tumor necrosis factor-α. J Immunol 150: 1886–1894
Roberts AD, Ordway DJ, Orme IM (1993) *Listeria monocytogenes* infection in β2 microglobulin-deficient mice. Infect Immun 61: 1113–1116
Rockilin RE, Sheffer AL, Greineder DK, Melmon KL (1980) Generation of antigen-specificsuppressor cells during allergy desensitization. N Engl J Med 302: 1213–1215
Schafer H, Burger R (1992) Analysis of mature guinea pig T cells with a monoclonal antibody directed against a framework determinant of the T-cell receptor for antigen. Scand J Immunol 36: 587–595
Sturgill-Koszycki S, Schlesinger PH, Chakraborty P, Haddix PL, Collins HL, Fok AK, Allen RD, Gluck SL, Heuser J, Russell DG (1994) Lack of acidification in *Mycobacterium* phagosomes produced by exclusion of the vesicular proton-ATPase. Science 263: 678–681
Tazi A, Fajac I, Soler P, Valeyre D, Battesti JP, Hance AJ (1991) Gamma/delta T lymphocytes are not increased in number in granulomatous lesions of patients with tuberculosis or sarcoidosis. Am Rev Respir Dis 144: 1373–1375
Vidal SM, Malo D, Vogan K, Skamene E, Gros P (1993) Natural resistance to infection with intracellular parasites: isolation of a candidate for *Bcg*. Cell 73: 469–485
Wallis RS, Ellner JJ (1994) Cytokines and tuberculosis. J Leukoc Biol 55: 676–681
Xu S, Cooper AM, Sturgill-Koszycki S, van Heyningen T, Chatterjee D, Orme IM, Allen P, Russell DG (1994) Intracellular trafficking in *Mycobacterium tuberculosis* and *Mycobacterium avium*-infected macrophages. J Immunol 153: 2568–2578

Human Cellular Immune Responses to *Mycobacterium tuberculosis*

P.F. Barnes[1] and R.L. Modlin[2]

1 Introduction

Human infection with *Mycobacterium tuberculosis* results in a broad spectrum of outcomes ranging from asymptomatic infection to widespread and rapidly fatal disease. It is generally believed that this spectrum of clinical manifestations

[1]University of Southern California School of Medicine, HMR 904, 2025 Zonal Avenue, Los Angeles, CA 90033, USA
[2]UCLA School of Medicine, 52–121 CHS, 10833 LeConte Avenue, Los Angeles, CA 90024–1750, USA

reflects a complex interaction between *M. tuberculosis* and human immune defenses. The immune response to tuberculosis is a double-edged sword that can contribute both to clearance of infection and tissue damage. This chapter summarizes our current understanding of human cellular immune responses to *M. tuberculosis*.

2 Spectrum of Manifestations of Tuberculous Infection

Most persons who become infected with *M. tuberculosis* mount a protective immune response and remain clinically well, the only evidence of infection being development of a positive tuberculin skin test. A minority develop tuberculosis disease within the first 2 years after infection (primary tuberculosis) or thereafter (reactivation tuberculosis). Miliary tuberculosis is the most serious form, characterized by hematogenous dissemination of large numbers of organisms throughout the body and severe disease that is almost invariably fatal if untreated. These manifestations reflect an ineffective immune response, as evidenced by a high frequency of negative tuberculin skin tests and failure of T lymphocytes to proliferate in response to *M. tuberculosis* antigens. Between the extremes of healthy tuberculin reactors and patients with miliary tuberculosis, several other common manifestations of tuberculosis can also be considered to reflect the efficacy of the immune response. Tuberculous pleuritis results when a small focus of organisms ruptures into the pleural space, triggering an exudative pleural effusion from a vigorous delayed-type hypersensitivity response. Patients with pleuritis mount a resistant immune response to infection, reflected by resolution of pleuritis without therapy in most cases (Roper and Waring 1955) and a high frequency of positive tuberculin skin tests (Antoniskis et al. 1990). In contrast, patients with advanced pulmonary tuberculosis have an ineffective immune response. They often develop progressive and life-threatening disease, and 20%–30% of patients have negative tuberculin skin tests.

3 Immune Defenses Against Tuberculosis

Host defenses against *M. tuberculosis* depend on cell-mediated rather than humoral immunity. Persons with defective cell-mediated immunity, such as those with HIV infection and chronic renal failure, are at markedly increased risk for tuberculosis (AMERICAN THORACIC SOCIETY 1986), whereas persons with defective humoral immunity, such as those with sickle cell disease and multiple myeloma, show no increased predisposition to tuberculosis. Experimental evidence indicates that antimycobacterial immune defenses are mediated primarily by T lymphocytes and macrophages, and adoptive transfer of resistance against tuberculosis in animal models is mediated by T cells (ORME and COLLINS 1983). Although neutrophils and natural killer cells can exhibit mycobacteriostatic effects in vitro (MAY and SPAGNUOLO 1987; BERMUDEZ and YOUNG 1991) and eosinophils can ingest mycobacteria (CASTRO et al. 1991), there is insufficient experimental or clinical evidence to determine if these cell populations contribute significantly to immune defenses in vivo.

4 The Role of Macrophages

When *M. tuberculosis* organisms are inhaled into the lung, they are engulfed by alveolar macrophages, which perform three important functions. First, they produce proteolytic enzymes and other metabolites which exhibit mycobactericidal effects, (detailed elsewhere in this volume). Second, macrophages process and present mycobacterial antigens to T lymphocytes, including $CD4^+$ and $CD8^+$ T lymphocytes, which are central to acquired resistance to *M. tuberculosis*. Third, macrophages produce a characteristic pattern of soluble mediators (cytokines) in response to *M. tuberculosis* that have the potential to exert potent immunoregulatory effects and to mediate many of the clinical manifestations of tuberculosis (VALONE et al. 1988; TOOSSI et al. 1991; BARNES et al. 1992a; ZHANG et al. 1993).

4.1 Interleukin-1

Interleukin-1 consists of two structurally related polypeptides IL-1α and IL-1 β, both of which have a similar spectrum of biologic activities. IL-1α lacks a signal peptide and therefore remains largely in the cell cytosol, whereas IL-1 β can be transported out of the cell and exerts paracrine effects. IL-1 is produced upon stimulation of human monocytes with *M. tuberculosis*, lipoarabinomannan (a heteropolysaccharide embedded in the mycobacterial cell membrane) and mycobacterial proteins with molecular masses 20 and 46 kDa (WALLIS et al. 1990; BARNES et al. 1992a; ZHANG et al. 1993). IL-1 is an endogenous pyrogen and may contribute to the fever that is characteristic of tuberculosis (DINARELLO 1984). In addition, IL-1 may enhance the inflammatory response by inducing macrophages to produce IL-6 and tumor necrosis factor-α (TNF) and by stimulating T cell proliferation through up-regulation of T cell expression of IL-2 receptors and IL-2 production (PLATANIAS and VOGELSANG 1990). However, IL-1 production has also been associated with immunosuppression. In tuberculosis patients with depressed lymphocyte proliferative responses to *M. tuberculosis* antigens and negative tuberculin skin tests, peripheral blood monocytes produce large amounts of IL-1 upon stimulation with purified protein derivative (PPD) of *M.*

tuberculosis compared to monocytes from healthy tuberculin reactors (Ellner 1978; Fujiwara et al. 1986). Although the mechanism of this immunosuppressive activity remains uncertain, one possibility is that excessive production of IL-1 results in expression of IL-2 receptors by monocytes (Toossi et al. 1990), which may inhibit T cell proliferation through consumption of IL-2.

4.2 Tumor Necrosis Factor-α

Human mononuclear cells and alveolar macrophages produce large quantities of TNF in response to *M. tuberculosis* (Valone et al. 1988; Barnes et al. 1992) and specific mycobacterial components such as lipoarabinomannan and proteins with molecular masses of 20, 44, 58 and 65 kDa (Moreno et al. 1989; Barnes et al. 1992a; Wallis et al. 1990, 1993a; Zhang et al. 1993). Lipoarabinomannan has been studied the most extensively and consists of a phosphatidylinositol end attached to a polysaccharide composed of a mannan core and arabinan side chains (Hunter and Brennan 1990). Removal of the polysaccharide components do not affect TNF release, whereas deacylation of the phosphatidylinositol end abrogates TNF production, indicating that the acyl groups are essential for this effect (Barnes et al. 1992a). Induction of TNF is also dependent on the macrophage CD14 molecule (Zhang et al. 1993). Several investigators have demonstrated that lipoarabinomannan from virulent *M. tuberculosis* Erdman strain induces much less TNF production than does lipoarabinomannan from avirulent *M. tuberculosis* H37Ra strain (Chatterjee et al. 1992; Roach et al. 1993; Zhang et al. 1993). It has therefore been suggested that lipoarabinomannan may facilitate intracellular survival of virulent *M. tuberculosis* by failing to induce production of TNF, which has antimycobacterial activity in vitro. However, the H37Ra strain used to generate lipoarabinomannan in these studies was later found to be a nonpathogenic mycobacterium, in contrast to *M. tuberculosis* (Delphi Chatterjee, Ph.D., personal communication). The relationship between virulence and the capacity of lipoarabinomannan to induce TNF production therefore remains uncertain.

Clinical and experimental data in animals and humans suggest that TNF contributes both to protection against tuberculosis and to immunopathology. In favor of a protective role, addition to TNF in vitro to human and murine macrophages enhances antimycobacterial activity (Bermudez and Young 1988; Flesch and Kaufmann 1990) and growth inhibition of *M. tuberculosis* in human alveolar macrophages is associated with enhanced TNF production (Hirsch et al. 1994). Granuloma formation in mice infected with *M. bovis* BCG coincides with local TNF synthesis and injection of anti-TNF antibodies interferes with development of granulomas and elimination of mycobacteria (Kindler et al. 1989). Disruption of the gene for the 55 kDa TNF receptor markedly reduces survival of mice infected with *M. tuberculosis* (Flynn et al. 1995). Furthermore, TNF mRNA and protein are concentrated at the site of disease in patients with tuberculosis pleuritis who mount a resistant immune response to infection (Barnes et al. 1990). Finally, peripheral blood monocytes from patients with chronic refractory pulmonary tuberculosis produce significantly lower amounts of TNF that do monocytes from patients with newly diagnosed tuberculosis (Takashima et al. 1993).

Other findings suggest that TNF plays a role in immunopathology. Administration of TNF to animals results in fever and wasting (Beutler and Cerami 1987). In addition, peripheral blood mononuclear cells (PBMCs) from tuberculosis patients with cachexia and high fever produce more TNF than do PBMCs from tuberculosis patients without these findings (Cadranel et al. 1990).

Because TNF can exhibit both protective and pathologic effects in *M. tuberculosis* infection, it is intriguing to speculate that production of physiologic concentrations of TNF are important to antimycobacterial immune defenses and that local release of TNF at the site of disease contributes to granuloma formation, control of infection and mycobacterial elimination. Excessive local production of TNF may cause marked tissue necrosis that is characteristic of progressive tuberculosis and may result in TNF release into the circulation, contributing to systemic manifestations of tuberculosis such as fever and cachexia.

4.3 Interlelukin-6

Interleukin-6 is a potent B cell growth and differentiation factor that induces immunoglobulin production by activated B cells and which is thought to mediate polyclonal B cell expansion and immunoglobulin production in infectious and neoplastic diseases (Hirano et al. 1990). IL-6 may therefore mediate the hyperglobulinemia that is characteristic of tuberculosis. Limited experimental evidence suggests that IL-6 does not enhance mycobacterial clearance. IL-6 reduces binding of TNF to murine macrophages and antagonizes the antimycobacterial activity of TNF in macrophages infected with *M. avium* (Bermudez et al. 1992). Addition of IL-6 to human monocytes also enhances intracellular and extracellular mycobacterial growth (Shiratsuchi et al. 1991).

4.4 Interleukin-10

Interleukin-10 is an antiinflammatory cytokine that is produced by human and murine macrophages exposed to *M. tuberculosis* in vitro (Barnes et al. 1992a; Orme et al. 1993). IL-10 inhibits cytokine synthesis by monocytes (Fiorentino et al. 1991; De Waal Malefyt et al. 1991a), inhibits the microbicidal activity of murine macrophages (Bogdan et al. 1991; Oswald et al. 1992), and suppresses antigen-specific T cell proliferation by down-regulation of macrophage class II MHC expression (De Waal Malefyt et al. 1991b). Infection of mice with *M. avium* induces IL-10 production, and neutralizing anti-IL 10 antibodies markedly reduce the bacterial burden (Bermudez and Champsi 1993). IL-10 also reverses the mycobactericidal effects of TNF. In tuberculosis patients, neutralizing antibodies to IL-10 enhance *M. tuberculosis*-induced interferon (IFN)-γ production by

PBMC, by enhancing IL-12 production by monocytes (Gong et al. 1996). These results suggest that IL-10 may play a role in inhibiting the immune response to *M. tuberculosis* in humans, and may contribute to the anergy and failure of lymphocytes and proliferate in response to *M. tuberculosis*. Alternatively, IL-10 may prevent excessive inflammation and tissue damage form an uncontrolled inflammatory response.

4.5 Transforming Growth Factor-β

Transforming growth factor (TGF-β) inhibits cytokine synthesis by macrophages and down-regulates class II MHC expression (Espenik et al. 1987; Czarniecki et al. 1988). TGF-β also inhibits IL-2 dependent T cell proliferation and IL-2 receptor expression (Kehrl et al. 1986; Ortaldo et al. 1991). TGF-β is produced constitutively by monocytes from tuberculosis patients, and production is increased in response to PPD. Langhans giant cells and epithelioid cells in tuberculous granulomas also express mRNA for TGF-β (Toossi et al. 1995), suggesting that local production of TGF-β may result in deactivation of macrophages and immunopathology. In human macrophages infected with *M. avium*, production of TGF-β is highest in macrophages infected with the most virulent strains, and addition of recombinant TGF-β inhibits the antimycobacterial effects of TNF (Bermudez 1993). Furthermore, IFN-γ enhances antimycobacterial activity of macrophages only in the presence of neutralizing antibodies to TGF-β. These results indicate that TGF-β inhibits antimycobacterial immune defenses and facilitates mycobacterial survival.

5 CD4⁺ T Lymphocytes

5.1 CD4⁺ Cells in Immune Defenses Against Tuberculosis

The bulk of experimental and clinical data suggest a central role for CD4⁺ cells in immune defenses against tuberculosis. Mice depleted of CD4⁺ cells prior to infection with *M. bovis* are unable to control mycobacterial growth (Pedrazzini et al. 1987; Muller et al. 1987), and adoptive transfer of CD4⁺ cells from sensitized animals confers protection against tuberculosis (Orme 1987, 1988a). In humans, CD4⁺ T cells are selectively expanded at the site of disease in patients with tuberculous pleuritis and a resistant immune response (Barnes et al. 1989), and depletion of CD4⁺ cells by HIV infection markedly increases susceptibility to primary and reactivation tuberculosis (Barnes et al. 1991). Furthermore, in HIV-infected tuberculosis patients, clinical indicators of severe disease, such as extrapulmonary involvement, mycobacteremia and positive acid-fast smears, become progressively more common as the CD4 cell count declines (Jones et al. 1993). For example, the frequency of mycobacteremia rises from 4% in patients with greater than 200 CD4 cells μ to 49% in those with 100 or fewer CD4 cells μ (Jones et al. 1993).

5.2 Th1 and Th2 Cells in Murine Models of Infectious Disease

Murine CD4⁺ T cells comprise two functionally distinct subpopulations which differ in patterns of cytokine production. Th1 cells produce IFN-γ, IL-2 and lymphotoxin, enhance microbicidal activity of macrophages and augment delayed-type hypersensitivity responses. Th2 cells produce IL-4, IL-5, IL-6 and IL-10 support B cell growth and differentiation, and augment humoral immune responses. Both Th1 and Th2 cells produce IL-3, granulocyte macrophage-colony stimulating factor (GM-CSF) and TNF. In the course of an immune response to a specific pathogen, Th1 and Th2 cells exert cross-regulatory influences that favor predominance of one subpopulation. IFN-γ produced by Th1 cells inhibits proliferation by Th2 cells, and IL-10 produced by Th2 cells inhibits cytokine synthesis by Th1 cells. Lymphocytes from mice immunized with purified protein derivative of *M. tuberculosis* produce IFN-γ, but not IL-4 or IL-5 (PEARLMAN et al. 1993). However, immunization with a microfilarial extract prior to immunization with PPD markedly enhances IL-4 and IL-5 production, and this effect is abrogated by anti-IL-4 antibody, demonstrating that ongoing Th2 responses to helminth antigens modulate the Th1 response to mycobacterial antigens through an IL-4-dependent mechanism (PEARLMAN et al. 1993).

Predominance of Th1 or Th2 cells has striking effects on the manifestations of infection by intracellular pathogens. In murine leishmaniasis, Th1 cells mediate immunologic resistance to infection through production of IFN-γ, whereas Th2 cells exacerbate disease through production of IL-4 (SCOTT et al. 1988; HEINZEL et al. (1989)). Immunologic resistance to mycobacterial infection in mice is probably also mediated by Th1 cells. Lymphocytes from mice with immune resistance to *M. bovis* produce high concentrations of IFN-γ and IL-2, but low levels of IL-4, whereas susceptible mice produce high levels of IL-4 and low levels of IFN-γ and IL-2 (HUYGEN et al. 1992). Production of IFN-γ is a functional marker of murine T cells that confer adoptive immunity against *M. tuberculosis* (ORME et al. 1992, 1993). Furthermore, mice with a disrupted IFN-γ gene are unable to control infection with *M. bovis* and are killed with a normally sublethal dose (COOPER et al. 1993; FLYNN et al. 1993).

5.3 Th1 and Th2 Cells in Human Mycobacterial Infection

Human T cells can exhibit dichotomous patterns of cytokine production similar to those of murine Th1 and Th2 cells. However, unlike the case in mice, IL-10 is produced by both human Th1 and Th2 clones in vitro, and inhibits proliferation and cytokine production by both subpopulations (YSSEL et al, 1992; DEL PRETE et al. 1993). In patients with leprosy, the Th1 cytokines IFN-γ and IL-2 predominate in the skin lesions of tuberculoid leprosy patients who mount a resistant immune response to *M. leprae,* whereas the Th2 cytokines IL-4 and IL-10 are prominent in lepromatous leprosy patients with ineffective immunity and enormous bacillary burdens (YAMAMURA et al. 1991).

Studies of cytokine production by human T cells in response to *M. tuberculosis* have focused on evaluation of T cell clones, PBMC and cells from the site of disease, with conflicting results. Some authors have noted that most $CD4^+$ *M. tuberculosis* -reactive T cell clones propagated in vitro are Th1-like, producing high concentrations of IFN-γ, but low concentrations of IL-4 and IL-5 (Del Prete et al. 1991; Hannen et al. 1991). In contrast, others have reported that human *M. tuberculosis*-reactive T cells secrete a broad spectrum of cytokines, including IFN-γ, IL-2, IL-4, IL-5 and IL-10 (Boom et al. 1991; Barnes et al. 1993a). In evaluating PBMCs from tuberculosis patients and healthy tuberculin reactors stimulated with *M. tuberculosis* in vitro, production and mRNA expression of IL-4 and IL-10 were similar in both groups, but PBMC from tuberculosis patients had reduced production and mRNA expression of IFN-γ and IL-2, *compared to healthy* tuberculin reactors (M. Zhang et al. 1995), suggesting that susceptibility to tuberculosis disease is associated with a depressed Th1 response but not an enhanced Th2 response. In contrast to our findings, other investigators have reported that the frequency of IL-4 producing cells is higher in *M. tuberculosis*-stimulated PBMCs from tuberculosis patients than in those from healthy tuberculin reactors, whereas the frequency of IFN-γ-producing cells was similar in both groups (Surcel et al. 1994). These disparate results may result from different experimental methods and the effects of culture conditions in vitro and may not reflect events in vivo. It is therefore critical to evaluate the cytokine response at the site of mycobacterial disease. In patients with tuberculous pleuritis, expression of mRNA for the Th1 cytokines IFN-γ and IL-2 is greater in pleural fluid than in blood, and concentrations of IFN-γ are 15-fold higher than those in serum (Barnes et al. 1993b). In contrast, expression of mRNA for the Th2 cytokine IL-4 is lower in pleural fluid than in blood. Pleural fluid lymphocytes stimulated with *M. tuberculosis* produce more IFN-γ and IL-2 than do peripheral blood lymphocytes. These results provide strong evidence for selective concentration of Th-1 like cells at the site of disease in persons with a resistant immune response and suggest that Th1 cells play an important role in human antimycobacterial defenses.

Although the capacity of IFN-γ to augment mycobacterial killing in human macrophages remains controversial (Douvas et al. 1985; Shiratsuchi et al. 1990), IFN-γ stimulates human macrophages to produce TNF and 1,25- dihydroxyvitamin D, both of which facilitate mycobacterial elimination (Rook et al. 1986; Bermudez and Young 1988), and administration of this cytokine yielded clinical improvement in patients with disseminated *M. avium* complex disease (Holland et al. 1994), and reduced the bacillary burden in patients with lepromatous leprosy (Nathan et al. 1986). IL-2 causes T cell division and is likely to expand populations of antigen-reactive T cells, increasing the local concentration of macrophage-activating factors secreted by T cells. Administration of recombinant IL-2 limits the replication of *M. bovis* BCG in mice (Jeevan and Asherton 1988) and reduces the bacillary burden in lepromatous leprosy (Kaplan et al. 1991). In contrast to the effects of Th1 cytokines, IL-4 deactivates macrophages (Lehn et al. 1989; HO et al. 1992) and blocks T cell proliferation by down-regulation of IL-2 receptor expression (Martinez et al. 1990) and inhibition of transcription of the IL-2 gene (SCHWARZ et al. 1993). IL-4 therefore has the capacity to inhibit the immune response to *M. tuberculosis.* Further studies are needed to determine if excessive production of IL-4 contributes to suppression of the immune response in tuberculosis patients.

5.4 Interleukin-12 and the T Cell Cytokine Response

A central issue in immunology is the elucidation of mechanisms which promote Th1 and Th2 cytokine responses. For many infectious diseases, the cytokine milieu is thought to influence the immune response towards production of a particular T cell cytokine pattern. Of particular interest is IL-12, a macrophage-derived cytokine which augments Th1 responses (HEINZEL et al. 1993; MANETTI et al. 1993; SYPEK et al. 1993; HSIEH et al. 1993). IL-12 is a disulfide-linked heterodimeric cytokine composed of two subunits with molecular masses of 40 kDa (p40) and 35 kDa (p35), both of which are required for bioactivity (KOBAYASHI et al. 1989; STERN et al. 1990; GUBLER et al. 1991; WOLF et al. 1991). IL-12 may play a critical role in favoring a Th1 response, as it induces differentiation of Th1 cells in vitro and in murine leishmaniasis in vivo (HEINZEL et al. 1993; MANETTI et al. 1993; SYPEK et al. 1993; HSIEH et al. 1993; SCOTT 1993). IL-12 mRNA and protein are concentrated at the site of disease in patients with tuberculosis pleuritis, and IL-12 is produced by pleural fluid cells in response to *M. tuberculosis* (ZHANG et al. 1994). These findings suggest that IL-12 participates in a resistant immune response to *M. tuberculosis.*

Production of IL-12 is enhanced in lesions of leprosy patients who manifest a Th1 rather than a Th2 response (SIELING et al. 1994). IL-12 also induces the proliferation and expansion of mycobacterial-specific Th1 but not Th2 cells (SIELING et al. 1994) and enhances cytotoxicity and augments proliferation by antigen-specific cytolytic T cells and natural killer (NK) cells (ROBERTSON et al. 1992; GATELY et al. 1992; BERTAGNOLLI et al. 1992). An intriguing property of IL-12 is its capacity to induce proliferation by cytolytic T cells only upon costimulation of the T cell receptor with antigen or anti-CD3 (BERTAGNOLLI et al. 1992). In contrast, cytokines such IL-2 and IL-7 cause proliferation in the absence or presence of antigen (BERTAGNOLLI and HERRMANN 1990). IL-12 may thus be a candidate to control the cytolytic arm of the initial immune response to microbial pathogens by inducing expansion of cytolytic T cells only when antigen is encountered. Preliminary studies indicate that IL-12 enhances cytotoxicity by $CD4^+$ T cells against human macrophages pulsed with *M. tuberculosis* (BOOM et al. 1992b). IL-12 may also contribute to lymphocyte recognition of *M. tuberculosis* antigens, as anti-IL-12 antibodies inhibit *M. tuberculosis*-induced lymphocyte proliferation (ZHANG et al. 1994). IL-12 may serve as growth factor during the early phase of T cell expansion in response to mycobacterial antigens, as it synergizes with suboptimal amounts of IL-2 to enhance lymphocyte proliferation, resulting in the generation of Th1 responses to the pathogen.

The ability of IL-12 to preferentially augment mycobacterial-specific human $CD4^+$ Th1 responses indicates promise as an immunotherapeutic adjuvant for the treatment of mycobacterial diseases, as has been shown for *Leishmania* infection in mice (SYPEK et al. 1993; HEINZEL et al. 1993). This potential is also suggested by studies of HIV infection in which IL-12 augmented HIV envelope-specific T cell proliferation and Th1 cytokine production (CLERICI et al. 1993). However, in some experimental models, IL-12 may only be effective if given prior to the establishment of the infection (SYPEK et al. 1993). Therefore, the therapeutic potential of IL-12 in combatting mycobacterial disease remains to be determined.

5.5 Cytolytic Activity of CD4⁺ Cells

An alternative mechanism by which T cells may contribute to immune defense is through direct cytolysis of macrophages and nonphagocytic cells infected with *M. tuberculosis*. Human *M. tuberculosis*-specific cytolytic T cells cultured in vitro are CD4⁺ (OTTENHOFF and MUTIS 1990; LORGAT et al. 1992), and *M. tuberculosis*-specific cytolytic activity of CD4⁺ cells at the site of disease is greatly enhanced compared to that of peripheral blood cells (LORGAT et al. 1992). Kaufmann has suggested that many macrophages infected with *M. tuberculosis* have low antimycobacterial potential, allowing the bacilli to evade host defenses (KAUFMANN 1988). Cytolytic T cells that specifically recognize mycobacterial antigens can lyse these macrophages, releasing bacilli to be engulfed and killed by macrophages with greater antimycobacterial activity. Alternatively, cytolytic T cells may play a scavenger role by lysing dead macrophages containing large numbers of dead bacilli, so that they can be catabolized by surrounding mononuclear cells (ORME et al. 1992). Another hypothesis is that cytolytic T cells cause immunopathology by destroying infected macrophages, which in turn release toxic products that result in caseous necrosis (DANNENBERG 1991).

6 CD8⁺ T Lymphocytes

CD8⁺ T cells constitute the major cytolytic T cell population in defenses against many intracellular pathogens in animal models of infection, and there is strong evidence implicating CD8⁺ T cells in protective immunity to tuberculosis. Adoptive transfer and cell depletion studies in vivo have demonstrated that CD8⁺ T cells are involved in controlling *M. tuberculosis* infection (ORME and COLLINS 1984; ORME 1987; MULLER et al. 1987) and are required for immunologic memory (HUBBARD et al. 1991). CD8⁺ T cell lines and clones lyse *M. tuberculosis*-primed macrophages in an antigen-specific manner and restrict the growth of *M. tuberculosis* in macrophages (DELIBERO et al. 1988).

The most striking data to implicate CD8⁺ T cell in protective immunity against tuberculosis come from studies of mice with disrupted β2-microglobulin genes, with resultant failure to express functional MHC class I molecules and a paucity of $CD8^+$ T cells. Infection with *M. tuberculosis* resulted in death of 70% of β2-microglobulin-deficient mice; whereas all the control mice survived. In β2-microglobulin-deficient mice; granuloma formation was intact, but there were tenfold more acid-fast bacilli in the infected tissues compared to controls (FLYNN et al. 1992). These studies suggest that $CD4^+$ cytolytic cells and NK cells are not sufficient for protection. The most likely explanation for these data is that $CD8^+$ T cells function as MHC class I-restricted cytolytic T cells and are required for protection against tuberculosis. Antigens recognized by $CD8^+$ T cells are probably presented by the H-2D locus (APT et al. 1993). The 65 kDa mycobacterial heat shock protein can be presented via the MHC class I pathway, since it can induce CD8 T cell responses and engender protective immunity (SILVA and LOWRIE 1994).

Despite the importance of $CD8^+$ cells in murine models of tuberculosis, their role in human antimycobacterial defenses remains uncertain. Human *M. tuberculosis*-specific cytolytic T cells evaluated to date are not $CD8^+$ (OTTENHOFF and MUTIS 1990; LORGAT et al. 1992), $CD8^+$ T cells are not selectively concentrated at the site of disease in tuberculosis patients (BARNES et al. 1989), and the severity of tuberculosis in HIV-infected patients is unaffected by the CD8 cell count (JONES et al. 1993). However, the lack of correlation between positive tuberculin skin tests and protection against tuberculosis could be explained by the failure of tuberculin skin tests to evaluate $CD8^+$ cytotoxic activity. Furthermore, $CD8^+$ T cell clones derived from the pleural effusion of a tuberculosis patient responded to the *M. tuberculosis* 71 kDa antigen and were restricted by HLA-B8 (REES et al. 1988). More comprehensive studies are needed to assess the role of $CD8^+$ cells in human antituberculosis defenses.

7 γδ T Lymphocytes

Several lines of evidence suggest that γδ T cells play a role in the initial immune response to *M. tuberculosis*. The percentage of γδ T cells is markedly elevated in draining lymph nodes and lungs of mice after primary infection with *M. tuberculosis* (JANIS et al. 1989; AUGUSTIN et al. 1989), and culture of human peripheral blood lymphocytes with live *M. tuberculosis* results in selective expansion of γδ T cells (HAVLIR et al. 1991a; BOOM et al. 1992a). γδ T cells from tuberculin-negative persons and from newborns proliferate in response to *M. tuberculosis*, indicating that human γδ T cells have an innate capacity to recognize mycobacterial antigens without prior exposure to them (O'BRIEN et al. 1989; KABELITZ et al. 1990; TSUYUGUCHI et al. 1991). Rechallenge with *M. tuberculosis* does not expand γδ T cells, suggesting that they do not contribute to the anamnestic response (JANIS et al. 1989; GRIFFIN et al. 1991). The percentage of γδ T cells is not increased in blood of healthy tuberculin reactors or of tuberculosis patients (TAZI et al. 1992; BARNES et al. 1992b) nor at the site of disease in pleural fluid or lymph nodes of tuberculosis patients (TAZI et al. 1991; OHMEN et al. 1991). However, this does not exclude a role for γδ T cells during the early phases of the immune response, such as in the lungs and lymph nodes of persons recently infected with *M. tuberculosis*. Expansion of *M. tuberculosis*-reactive γδ T cells is greater in healthy tuberculin reactors and in patients with tuberculous pleuritis than in those with advanced pulmonary and miliary tuberculosis, suggesting that γδ T cells contribute to immune resistance (BARNES et al. 1992b).

A central issue is the elucidation of the functional role of γδ T cells in the immune response for mycobacteria. Freshly sorted γδ T cells secreted IL-2 and IFN-γ but not IL-4 (FOLLOWS et al. 1992). *M. tuberculosis*-reactive γδ T-cells produce less IL-4 but more IFN-γ than do αβ T-cells (TSUICAGUCH et al. 1995). *M. tuberculosis*-reactive γδ T cell clones consistently produced IFN-γ and TNF (BARNES et al. 1993a), which are likely to contribute to macrophage activation, granuloma formation and elimination of mycobacteria. Production of IL-2, IL-4, IL-5 and IL-10 was variable, indicating some heterogeneity in the cytokine profile of these cells. Since mycobacterium-reactive αβ T cells from the same individuals secreted the identical pattern of cytokines, the specific contribution of γδ T cells to antituberculous defenses may not reside in their capacity to produce a distinct pattern of cytokines.

In addition to their capacity to produce cytokines, γδ T cells can exhibit cytolytic activity against targets pulsed with *M. tuberculosis* and other bacteria (MUNK et al. 1990, TSUKAGUCHI et al. 1995) and recognize antigens distinct from those recognized by αβ T cells. Initial evidence suggested that γδ T cells recognized heat shock or stress proteins, production of which increases in response to stimuli such as high temperature, hypoxia and metabolite deprivation (BORN et al. 1990). More recently, γδ T cells have been shown to respond to mycobacterial phosphate-containing glycolipid antigens, as well as a 10- to 14-kDa protein (TSUYUGUCHI et al. 1991; PFEFFER et al. 1990; BOOM et al. 1994 CONSTANT et al. 1994; TANAKA et al. 1995).

8 CD4⁻ CD8⁻ T Lymphocytes

CD1 is a nonpolymorphic class I molecule that can act as an antigen-presenting molecule for human T cells in vitro. CD1c was specifically recognized by a CD4⁻CD8⁻ "double negative" cytolytic T cell line bearing the γδ T cell receptor (PORCELLI et al. 1989). In contrast, lysis mediated by double-negative cytolytic T cells expressing the αβ T cell receptor was dependent on CD1a. The specificity of these clones for different CD1 isoforms indicates that functional diversity within the CD1 gene family is likely.

The analysis of human T cell clones that responded to *M. tuberculosis*- and *M. leprae*-derived antigens demonstrated the ability of CD1b to present non-protein mycobacterial antigens, lipoarabinomannan and mycolic acid, to, αβ double negative T cell lines, independent of classical class I and class II MHC proteins (PORCELLI et al. 1992, BECKMAN et al. 1994, SIELINE et al. 1995). These findings suggest that CDI-restricted T-cells may play a unique role in immune defenses by recognizing lipid antigens that are not recognized by classical MHC-restricted αβ T-cells.

Double negative T cells have recently been derived from lesions and blood of tuberculoid leprosy patients, and these cells lyse and proliferate in response to target cells pulsed with *M. leprae* (SIELING et al. 1995). Antigen recognition by double negative T cells was restricted by CD1b and CD1c. Four of five double negative T cell lines produced IFN-γ in tenfold excess over IL-4 upon stimulation with anti-CD3 antibodies, characteristic of Th1 cells. Immunohistologic analysis of patient lesions revealed that $CD1a^+$, $CD1b^+$ and $CD1c^+$ cells were tenfold more abundant in lesions of immunologically responsive tuberculoid patients than in unresponsive lepromatous patients. These data suggest that double negative T cells promote cell-mediated immunity at the site of mycobacterial infection.

9 Mycobacterial Antigens Recognized by T Lymphocytes

Mycobacterium tuberculosis is a complex organism with a wide variety of protein antigens, which can be divided into structural and secreted antigens. Immunization of animals with live mycobacteria confers protective immunity against tuberculosis whereas immunization with killed organisms does not (ORME 1988b). Because similar structural antigens are present in both live and killed bacilli, secreted antigens produced only by live organisms are likely to be the most important targets of T cells that confer protective immunity. In support of this hypothesis, immunization of guinea pigs with secreted proteins of *M. tuberculosis* confers substantial protection against respiratory challenge with *M. tuberculosis* (PAL and HORWITZ 1992), and murine T-cells that confer protective immunity against tuberculosis recognize secreted antigens of molecular weights 6 and 30 kDa (ANDERSEN et al. 1995).

Of the wide variety of proteins secreted by *M. tuberculosis*, three are of greatest interest from the standpoint potentially eliciting protective immunity in humans. Two such proteins are the 30 and 32 kDa members of the BCG 85 complex that are secreted in large quantities by rapidly growing mycobacteria. As these proteins can bind fibronectin, a major component of human extracellular matrix, they may mediate adhesion of bacilli to mucosal surfaces and perhaps subsequent intracellular invasion through macrophage fibronectin receptors (ABOU-ZEID et al. 1988a). These proteins may therefore be critical virulence factors that are important targets of the immune response. The 30 kDa antigen (also referred to as BCG 85B, α antigen and antigen 6), elicits proliferation of lymphocytes from healthy tuberculin reactors but not of those from tuberculosis patients, raising the intriguing possibility that failure to recognize this antigen, perhaps a genetically determined characteristic, predisposes patients to development of tuberculosis (HAVLIR et al. 1991b). Alternatively, the 30 kDa antigen may activate suppressive circuits that inhibit the proliferative response in tuberculosis patients. The 32 kDa antigen (also known as BCG 85A or P32) elicits greater proliferation and IFN-γ production by lymphocytes from healthy tuberculin reactors than by those from tuberculosis patients (HUYGEN et al. 1988) and induces development of *M. tuberculosis*-specific cytotoxic T cells (MUNK et al. 1994).

The 10 kDa antigen, also referred to as BCG-a, is a secreted antigen that is associated with the cell wall of *M. tuberculosis* (ABOU-ZEID et al. 1988b; BARNES et al. 1992c). This antigen elicits more proliferation and IFN-γ production by lymphocytes from healthy tuberculin reactors than do other culture filtrate antigens (BARNES et al. 1992c). In addition, proliferative responses to the 10 kDa antigen of pleural fluid lymphocytes from tuberculous pleuritis patients are significantly greater than those of peripheral blood lymphocytes, indicating that this antigen is an important target for T cells at the site of disease.

The mycobacterial antigen which has been studied most intensively is the 65 kDa heat shock protein, a highly conserved molecule that has both bacterial and human homologues. Heat shock proteins are generated in large quantities by cells under stressful conditions and are thought to be produced by mycobacteria in the harsh intracellular environment of the macrophage. The 65 kDa protein is essential for many cellular processes, playing a major role in folding, unfolding and translocation of polypeptides. It is a potent immunogen for generation of antibody by murine B cells and is recognized by a high proportion of *M. tuberculosis*-reactive murine T cells (KAUFMANN 1990). Despite initial hopes that the 65 kDa protein would elicit protective immunity in humans, that is now believed to be unlikely. First, the 65 kDa antigen is not a secreted antigen and is present in high concentration in killed mycobacterial preparations which do not elicit protective immunity. Second, murine T cells that confer protective immunity against tuberculosis do not recognize the 65 kDa antigen (ORME et al. 1992, 1993). Third, the 65 kDa antigen does not elicit proliferation by lymphocytes from most healthy tuberculin reactors (HAVLIR et al. 1991b; BARNES et al. 1992c).

10 Expansion of Specific T Lymphocyte Subpopulations by *Mycobacterium tuberculosis* Antigens

Unlike nominal antigens which stimulate a small fraction of T cells, superantigens stimulate the majority of T cells bearing a T cell receptor encoded by a specific V region polypeptide chain resulting in systemic illness (KAPPLER et al. 1989; CHOI et al. 1990). The clinical manifestations of tuberculosis indicate a systemic inflammatory response characterized by fever, weight loss and tissue necrosis. To identify T cell populations that may be involved in this systemic response and to characterize the antigens activating these T cells, V region gene usage of the T cell receptor β chain was evaluated in patients with tuberculous pleuritis. Analysis by RT-PCR and flow cytometry indicated an expansion of $V\beta8^+$ T cells at the site of disease in some donors, suggesting that *M. tuberculosis* contains a superantigen. *M. tuberculosis* induced strong T cell proliferative responses in healthy tuberculin-negative donors in vitro, with preferential expansion of $V\beta8^+$ T cells, independent of the CDR3 region. T cell stimulation was MHC class II-dependent and did not require antigen processing, consistent with the presence of a superantigen in *M. tuberculosis* (Ohmen et al. 1994).

The mycobacterial superantigen may contribute to disease pathogenesis by inducing local release of inflammatory cytokines from all T cells bearing Vβ8-encoded T cell receptors. These cytokines may contribute not only to macrophage activation and mycobacterial elimination, but also to tissue damage which may further the infection. The initial burst of cytokines induced by a mycobacterial superantigen may also contribute to the powerful adjuvant activity of *M. tuberculosis* by expanding T cells required for antibody production and cell-mediated immunity. Further analysis will be required to elucidate the precise molecules responsible for this superantigen activity.

11 Interaction Between *Mycobacterium tuberculosis* and HIV

Clinical and epidemiologic data indicate that persons coinfected with HIV and *M. tuberculosis* are at increased risk for progressive disease from both pathogens. HIV-infected patients are at markedly increased risk for development of primary and reactivation tuberculosis (Selwyn et al. 1989; Small et al. 1994), and the manifestations of tuberculosis are more severe and life-threatening in HIV-infected persons (Barnes et al. 1991). HIV-infected tuberculosis patients have shortened survival times compared to HIV-infected controls without tuberculosis, matched for CD4 cell counts, and die from complications of progressive HIV infection rather than from tuberculosis (Whalen et al. 1995).

Recent experimental data also indicate that coinfection with HIV and *M. tuberculosis* results in significant alterations in the cellular immune response to both pathogens. Upon stimulation with *M. tuberculosis* in vitro, PBMCs from HIV-infected tuberculosis patients had reduced proliferative responses and diminished production of IFN-γ and IL-2 compared to those of PBMCs from tuberculosis patients with HIV infection (Zhang et al. 1994). Production of IL-4 and IL-10 were comparable in tuberculosis patients with or without HIV infection, suggesting that coinfection with HIV in tuberculosis patients results in a diminished Th1 but not an enhanced Th2 response. The reduced production of Th1 cytokines in HIV-infected patients resulted from the depressed CD4 cell counts in this population. In HIV-infected tuberculosis patients, *M. tuberculosis* induced proliferative responses and IFN-γ production were significantly enhanced by neutralizing antibodies to IL-10 but not by antibodies to IL-4 (ZHANG et al. 1994), suggesting that the diminished proliferative and Th1 responses are mediated at least in part by IL-10

Experimental data also suggest that tuberculosis accelerates the course of HIV infection. Activation of T cells by exposure to *M. tuberculosis* results in production of cytokines, many of which can enhance HIV replication in vitro. For example, PBMCs from dually infected patients produce more TNF in response to PPD than do PBMCs from patients with tuberculosis of HIV infection alone (WALLIS et al. 1993b), and TNF can promote HIV replication through activation of NFkB (MATSUYAMA et al. 1991). Other cytokines produced in response to *M. tuberculosis*, such as IL-1, IL-6 and IL-2, also up-regulate HIV expression and replication in vitro, and IL-10, TGF- β and IFN-γ can enhance or inhibit HIV replication, depending on experimental conditions (FAUCI 1993). Furthermore, mycobacteria and their soluble products activate HIV replication in the latently infected line U1 (LEDERMAN et al. 1994) and enhance replication of HIV human monocytic cell. Lines by increasing transcriptional activation at the HIV-1 long terminal repeat (Y. ZHANG et al. 1995). This increase in transcriptional activation was inhibited by antibodies to TNF-α and IL-1 Monocytes from tuberculosis patients show enhanced susceptibility to productive HIV infection in vitro (TOOSSI et al. 1993). If tuberculosis facilitates progression of HIV disease, preventive therapy of tuberculosis assumes critical importance, and inhibitors of TNF production such as pentoxifylline and thalidomide may be important adjuncts to chemotherapy.

12 Conclusion

The past decade has witnessed significant advances in our understanding of the human immune response to *M. tuberculosis* infection. Nevertheless, more progress is urgently needed, as development of antituberculosis vaccines, adjunctive immunotherapy for drug-resistant tuberculosis, and approaches to reduce immunopathology will hinge on a comprehensive understanding of the interaction between *M. tuberculosis* and the human immune system.

Acknowledgements. Supported in part by grants from the National Institutes of Health (AI27285, AI31066, AI22553, AI36069) and the World Health Organization (IMMTUB).

References

Abou-Zeid C, Ratliff TL, Wiker HG, Harboe M, Bennedson J, Rook GAW (1988a) Characterization of fibronectin-binding antigens released by *Mycobacterium tuberculosis* and *Mycobacterium bovis* BCG. Infect Immun 56: 3046–3051

Abou-Zeid C, Smith I, Grange JM, Ratliff TL, Steele J, Rook GAW (1988b) The secreted antigens of *Mycobacterium tuberculosis* and their relationship to those recognized by the available antibodies. J Gen Microbiol 134: 531–533

American Thoracic Society (1986) Treatment of tuberculosis and tuberculosis infection in adults and children. Am Rev Respir Dis 134: 355–363

Andersen P, Andersen AB, Sørensen AL, Nagai S (1995) Recall of long-lived immunity to *Mycobacterium tuberculosis* infection in mice. J Immunol 154: 3359–3372

Antoniskis D, Amin K, Barnes PF (1990) Pleuritis as a manifestation of reactivation tuberculosis. Am J Med 89: 447–450

Apt AS, Avdienko VG, Nikonenko BV, Kramnik IB, Moroz AM, Skamene E (1993) Distinct H-2 complex control of mortality and immune responses to tuberculosis infection in virgin and BCG-vaccinated mice. Clin Exp Immunol 94: 322–329

Augustin A, Kubo RT, Sim GK (1989) Resident pulmonary lymphocytes expressing the T-cell receptor. Nature 340: 239–241

Barnes PF, Mistry SD, Cooper CL, Pirmez C, Rea TH, Modlin RL (1989) Compartmentalization of a $CD4^+$ T lymphocyte subpopulation in tuberculous pleuritis. J Immunol 142: 1114–1119

Barnes PF, Fong S-J, Brennan PJ, Twomey PE, Mazumder A, Modlin RL (1990) Local production of tumor necrosis factor and interferon-γ in tuberculous pleuritis. J Immunol 145: 149–154

Barnes PF, Bloch AB, Davidson PT, Snider DE Jr (1991) Tuberculosis in patients with human immunodeficiency virus infection. N Engl J Med 324: 1644–1650

Barnes PF, Chatterjee D, Abrams JS, Lu S, Wang E, Yamamura M, Brennan PJ, Modlin RL (1992a) Cytokine production induced by *Mycobacterium tuberculosis* lipoarabinomannan. Relationship to chemical structure. J Immunol 149: 541–547

Barnes PF, Grisso CL, Abrams JS, Band H, Rea TH, Modlin RL (1992b) γf T lymphocytes in human tuberculosis. J infect Dis 165: 506–512

Barnes PF, Mehra V, Rivoire B, Fong S-J, Brennan PJ, Voegtline MS, Minden P, Houghten RA, Bloom BR, Modlin RL (1992c) Immunoreactivity of a 10 kD antigen of *Mycobacterium tuberculosis*. J Immunol 148: 1835–1840

Barnes PF, Abrams JS, Lu S, Sieling PA, Rea TH, Modlin RL (1993a) Patterns of cytokine production by mycobacterium-reactive human T cell clones. Infect Immun 61: 197–203

Barnes PF, Lu S, Abrams JS, Wang E, Yamamura M, Modlin RL (1993b) (Cytokine production at the site of disease in human tuberculosis. Infect Immun 61: 3482–3489

Beckman EM, Porcelli SA, Morita CT, Behar SM, Furlong ST, Brenner MB (1994) Recognition of a lipid antigen by CDI-restricted $\alpha\beta^+$ T cells. Nature 372: 691–694

Bermudez LE (1993) Production of transforming growth factor- β by *Mycobacterium avium*- infected human macrophages is associated with unresponsiveness of IFN-γ. J Immunol 150: 1838–1845

Bermudez LE, Champsi J (1993) Infection with *Mycobacterium avium* induces production of interleukin-10 (IL-10), and administration of anti-IL-10 antibody is associated with enhanced resistance to infection in mice. Infect Immun 61: 3093–3097

Bermudez LEM, Young LS (1988) Tumor necrosis factor, alone or in combination with IL-2, but not IFN-γ, is associated with macrophage killing of *Mycobacterium avium* complex. J Immunol 140: 3006–3013

Bermudez LEM, Young LS (1991) Natural killer cell-dependent mycobacteriostatic and mycobactericidal activity in human macrophages. J Immunol 146(1): 265–270

Bermudez LE, Wu M, Petrofsky M, Young LS (1992) Interleukin-6 antagonizes tumor necrosis factor-mediated mycobacteriostatic and mycobactericidal activities in macrophages. Infect Immun 60: 4245–4552

Bertagnolli M, Herrmann S (1990) IL-7 supports the generation of cytotoxic T lymphocytes from thymocytes. Multiple lymphokines required for proliferation and cytotoxicity. J Immunol 145: 1706–1712

Bertagnolli MM, Lin B-Y, Young D, Herrmann SH (1992) IL-12 augments antigen-dependent proliferation of activated T lymphocytes. J Immunol 149: 3778–3783

Beutler B, Cerami A (1987) Cachectin: more than a tumor necrosis factor. N Engl J Med 316: 379–385

Bogdan C, Vodovotz Y, Nathan C (1991) Macrophage deactivation by interleukin 10. J Exp Med 174: 1549–1555

Boom WH, Wallis RS, Chervenak KA (1991) Human *Mycobacterium tuberculosis* -reactive $CD4^+$ T-cell clones: heterogeneity in antigen recognition, cytokine production, and cytotoxicity for mononuclear phagocytes. Infect Immun 59: 2737–2743

Boom WH, Chervenak KA, Mincek MA, Ellner JJ (1992a) Role of the mononuclear phagocyte as an antigen-presenting cell for human T cells activated by live *Mycobacterium tuberculosis*. Infect Immun 60: 3480–3488

Boom WH, Toossi Z, Wolf SF, Chervenak KA (1992b) The modulation by IL-2, NKSF (IL-12/CLMF) and TGF- β of $CD4^+$ T-cell-mediated cytotoxicity for macrophages. Proceedings of the 27th joint conference on tuberculosis and leprosy, pp 163–167 (abstract)

Boom WH, Balaji KN, Nayak R, Tsu Kaguchi K, Chervenak KA (1994) Characterization of a 10- to 14-kilodaltan protease-sensitive *Mycobacterium tuberculosis* H 37Ra antigen that stimulates human $\gamma\delta$ T cells. Infect Immun 62: 5511–5518

Born W, Happ MP, Dallas A, Reardon C, Kubo R, Shinnick T, Brennan P, O'Brien R (1990) Recognition of heat shock proteins and T cell function. Immunol Today 11: 40–43

Cadranel J, Philippe C, Perez J, Milleron B, Akoun G, Ardaillou R, Baud L (1990) In vitro production of tumour necrosis factor and prostaglandin E_2 by peripheral blood mononuclear cells from tuberculosis patients. Clin Exp Immunol 81: 319–324

Castro AG, Esaguy N, Macedo PM, Aguas AP, Silva MT (1991) Live but not heat-killed mycobacteria cause rapid chemotaxis of large numbers of eosinophils in vivo and are ingested by the attracted granulocytes. Infect Immun 59(9): 3009–3014

Chatterjee D, Roberts AD, Lowell K, Brennan PJ, Orme IM (1992) Structural basis of capacity of lipoarabinomannan to induce secretion of tumor necrosis factor. Infect Immun 60: 1249–1253

Choi Y, Lafferty JA, Clements JR, Todd JK, Gelfand EW, Kappler J, Marrack P, Kotzin BL (1990) Selective expansion of T cells expressing Vβ2 in toxic shock syndrome. J Exp Med 172: 981–984

Clerici M, Lucey DR, Berzovsky JA, Pinto LAA, Wynn TA, Blatt SP, Dolan MJ, Hendrix CW, Wolf SF, Shearer GM (1993) Restoration of HIV-specific cell-mediated immune responses by interleukin-12 in vitro. Science 262: 1721–1724

Constant P, Davodeau F, Peyrat M-A, Poquet Y, Puzo G, Bonneville M, Fournie J-J (1994) Stimulation of human T cells by nonpeptidic mycobacterial ligands. Science 264: 267–270

Cooper AM, Dalton DK, Stewart TA, Griffin JP, Russell DG, Orme IM (1993) Disseminated tuberculosis in interferon-γ gene-disrupted mice. J Exp Med 178: 2243–2247

Czarniecki CW, Chiu HH, Wong GHW, McCabe SM, Palladino MA (1988) Transforming growth factor β modulates the expression of class II histocompatibility antigens on human cells. J Immunol 140: 4217–4223

Dannenberg AM (1991) Delayed-type hypersensitivity and cell-mediated immunity in the pathogenesis of tuberculosis. Immunol Today 12: 228–233

de Waal Malefyt R, Abrams J, Bennett B, Figdor C, De Vries JE (1991a) IL-10 inhibits cytokine synthesis by human monocytes: an autoregulatory role of IL-10 produced by monocytes. J Exp Med 174: 1209–1220

de Waal Malefyt R, Haanen J, Spits H, Roncarlo M-G, te Velde A, Figdor C, Johnson K, Kastelein R, Yssel H, De Vries JE (1991b) Interleukin-10 (IL-10) and viral IL-10 strongly reduce antigen-specific human T cell proliferation by diminishing the antigen-presenting capacity of monocytes via downregulation of class II major histocompatibility complex expression. J Exp Med 174: 915–924

Del Prete GF, De Carli M, Mastromauro C, Biagiotti R, Macchia D, Falagiani P, Ricci M, Romagnani S (1991) Purified protein derivative of *Mycobacterium tuberculosis* and excretory-secretory antigen(s) of *Toxocara canis* expand in vitro human T cells with stable and opposite (type 1 T helper or type 2 T helper) profile of cytokine production. J Clin Invest 88: 346–350

Del Prete G, De Carli M, Almerigogna F, Giudizi MG, Biagotti R, Romagnani S (1993) Human IL-10 is produced by both type 1 helper (Th1) and type 2 helper (Th2) T cell clones and inhibits their antigen-specific proliferation and cytokine production. J Immunol 150: 353–360

DeLibero G, Flesch I, Kaufmann SHE (1988) Mycobacteria-reactive Lyt-2+ T cell lines. Eur J Immunol 18: 59–66

Dinarello C (1984) Interleukin-1 and the pathogenesis of the acute-phase response. N Engl J Med 311: 1413–1418

Douvas GS, Looker DL, Vatter AE, Crowle AJ (1985) Gamma interferon activates human macrophages to become tumoricidal and leishmanicidal but enhances replication of macrophage-associated mycobacteria. Infect Immun 50: 1–8

Ellner JJ (1978) Suppressor adherent cells in human tuberculosis. J Immunol 121: 2573–2579

Espenik T, Figari IS, Shalaby MR, Lewis GD, Shepard HM, Palladino MS (1987) Inhibition of cytokine production by cyclosporin A and transforming growth factor β. J Exp Med 166: 571–576

Fauci AS (1993) Multifactorial nature of human immunodeficiency virus disease: implications for therapy. Science 262: 1011–1018

Fiorentino DF, Zlotnik A, Mosmann TR, Howard M, O'Garra A (1991) IL-10 inhibits cytokine production by activated macrophages. J Immunol 147: 3815–3822

Flesch IEA, Kaufmann SHE (1990) Activation of tuberculostatic macrophage functions by gamma interferon, interleukin-4, and tumor necrosis factor. Infect Immun 58: 2675–2677

Flynn JL, Goldstein MM, Triebold KJ, Koller B, Bloom BR (1992) Major histocompatibility complex class I-restricted T cells are required for resistance to *Mycobacterium tuberculosis* infection. Proc Natl Acad Aci USA 89: 12013–12017

Flynn JL, Chan J, Triebold KJ, Dalton DK, Stewart TA, Bloom BR (1993) An essential role for interferon-γ in resistance to *Mycobacterium tuberculosis* infection. J Exp Med 178: 2249–2254

Flynn JL, Goldstein MM, Chan J, Triebold KJ, Pfeffer K, Lowenstein CJ, Screiber R, Mak JW, Bloom BR (1995) Tumor necrosis factor-α is required in the protective immune response against *Mycobacterium tuberculosis* in mice. Immunity 2: 561–572

Follows GA, Munk ME, Gatrill AJ, Conradt P, Kaufmann SHE (1992) Gamma interferon and interleukin 2, but not interleukin 4, are detectable in gamma/delta T-cell cultures after activation with bacteria. Infect Immun 60: 1229–1231

Fujiwara H, Kleinhenz ME, Wallis RS, Ellner JJ (1986) Increased interleukin-1 production and monocyte suppressor cell activity associated with human tuberculosis. Am Rev Respir Dis 133: 73–77

Gately MK, Wolitzky AG, Quinn PM, Chizzonite R (1992) Regulation of human cytolytic lymphocyte responses by interleukin-12. Cell Immunol 143: 127–142

Gong J-H, Zhang M, Modlin RL, Linsley PS, Iyer D, Lin Y, Barnes PF (1996) Interleukin-10 down regulates *Mycobacterium tuberculosis*-induced Th1 responses and CTLA-4 expression. Infect Immun 64: 913–918

Griffin JP, Harshan KV, Born WK, Orme IM (1991) Kinetics of accumulation of gamma/delta receptor-bearing T lymphocytes in mice infected with live mycobacteria. Infect Immun 59: 4263–4265

Gubler U, Chua AO, Schoenhaut DS, Dwyer CM, McComas W, Motyka R, Nabavi N, Wolitzky AG, Quinn PM, Familletti PC, Gately MK (1991) Coexpresssion of two distinct genes is required to generate secreted bioactive cytotoxic lymphocyte maturation factor. Proc Natl Acad Aci USA 88: 4143–4147

Haanen JBAG, de Waal Malefijt R, Res PCM, Kraakman EM, Ottenhoff THM, de Vries RRP, Spits H (1991) Selection of a human T helper type 1-like T cell subset by mycobacteria. J Exp Med 174: 583–592

Havlir DV, Ellner JJ, Chervenak KA, Boom WH (1991a) Selective expansion of human T cells by monocytes infected with live *Mycobacterium tuberculosis*. J Clin Invest 87: 729–733

Havlir DV, Wallis RS, Boom WH, Daniel TM, Chervenak K, Ellner JJ (1991b) Human immune response to *Mycobacterium tuberculosis* antigens. Infect Immun 59: 665–670

Heinzel FP, Sadick MD, Holaday BJ, Coffman RL, Locksley RM (1989) Reciprocal expression on interferon γ or interleukin 4 during the resolution or progression of murine leishmaniasis. Evidence for expansion of distinct helper T cell subsets. J Exp Med 169: 59–72

Heinzel FP, Schoenhaut DS, Rerko M, Rosser LE, Gately MK (1993) Recombinant interleukin 12 cures mice infected with *Leishmania major*. J Exp Med 177: 1505–1509

Hirano T, Akira S, Taga T, Kishimoto T (1990) Biological and clinical aspects of interleukin 6. Immunol Today 11: 443-449

Hirsch CS, Ellner JJ, Russell DG, Rich EA (1994) Complement receptor-mediated uptake and tumor necrosis factor-α-mediated growth inhibition of *Mycobacterium tuberculosis* by human alveolar macrophages. J Immunol 152: 743–753

Ho JL, He SH, Rios MJC, Wick EA (1992) Interleukin-4 inhibits human macrophage activation by tumor necrosis factor, granulocyte-monocyte colony-stimulating factor, and interleukin-3 for antileishmanial activity and oxidative burst capacity. J Infect Dis 165: 344–351

Holland SM, Eisenstein EM, Kuhns DB, Turner ML, Fleisher TA, Strober W, Gallin JI (1994) Treatment of refractory disseminated nontuberculous mycobacterial infection with interferon gamma. A preliminary report. N Engl J Med 330: 1348–1355

Hsieh C-S, Macatonia SE, Wolf SF, O'Garra A, Murphy KM (1993) Development of T_H1 $CD4^+$ T cells through IL-12 produced by Listeria-induced macrophages. Science 260: 547–549

Hubbard RD, Flory CM, Collins FM (1991) Memory T cell-mediated resistance to *Mycobacterium tuberculosis* infection in innately susceptible and resistant mice. Infect Immun 59: 2012–2016

Hunter SW, Brennan PJ (1990) Evidence for the presence of a phosphatidylinositol anchor on the lipoarabinomannan and lipomannan of *Mycobacterium tuberculosis*. J Biol Chem 265: 9272–9279

Huygen K, Van Vooren JP, Turneer M, Bosmans R, Dierckx P, De Bruyn J (1988) Specific lymphoproliferation, gamma interferon production, and serum immunoglobulin G directed against a purified 32 kDa mycobacterial protein antigen (P32) in patients with active tuberculosis. Scand J Immunol 27: 187–194

Huygen K, Abramowicz D, Vandenbussche P, Jacobs F, DeBruyn J, Kentos A, Drowart A, Van Vooren JP, Goldman M (1992) Spleen cell cytokine secretion in *Mycobacterium bovis* BCG-infected mice. Infect Immun 60: 2880–2886

Janis EM, Kaufmann SH, Schwartz RH, Pardoll DM (1989) Activation of T cells in the primary immune response to *Mycobacterium tuberculosis*. Science 244: 713–716

Jeevan A, Asherton GL (1988) Recombinant interleukin-2 limits the replication of *Mycobacterium lepraemurium* and *Mycobacterium bovis* BCG in mice. Infect Immun 56: 660–664

Jones BE, Young SMM, Antoniskis D, Davidson PT, Kramer F, Barnes PF (1993) Relationship of the manifestation of tuberculosis to CD4 cell counts in patients with human immunodeficiency virus infection. Am Rev Respir Dis 148: 1292–1297

Kabelitz D, Bender A, Schondelmaier S, Schoel B, Kaufmann SHE (1990) A large fraction of human peripheral blood gamma/delta^{+} T cells is activated by *Mycobacterium tuberculosis* but not by its 65-kD heat shock protein. J Exp Med 171: 667–679

Kaplan G, Britton WJ, Hancock GE, Theuvenet WJ, Smith KA, Job CK, Roche PW, Molloy A, Burkhardt R, Barker J, Pradhan HM, Cohn ZA (1991) The systemic influence of recombinant interleukin 2 on the manifestation of lepromatous leprosy. J Exp Med 173: 993–1006

Kappler J, Kotzin B, Herron L, Gelfand EW, Bigler RD, Boylston A, Carrel S, Posnett DN, Choi Y, Marrack P (1989) Vβ-specific stimulation of human T cells by staphylococcal toxins. Science 244: 811–813

Kaufmann SHE (1988) CD8+ T lymphocytes in intracellular microbial infections. Immunol Today 9: 168–174

Kaufmann SHE (1990) Heat shock proteins and the immune response. Immunol Today 11(4): 129–136

Kehrl JH, Wakefield LM, Roberts AB, Jakowiew S, Alvarez-Mon M, DeRynck R, Sporn MB, Fauci AS (1986) Production of transforming growth factor β by human T lymphocytes and its potential role in the regulation of T cell growth. J Exp Med 163: 1037–1050

Kindler V, Sappino A-P, Grau GE, Piguet P-F, Vassalli P (1989) The inducing role of tumor necrosis factor in the development of bactericidal granulomas during BCG infection. Cell 56: 731–740

Kobayashi M, Fitz L, Ryan M, Hewick RM, Clark SC, Chan S, Loudon R, Sherman F, Perussia B, Trinchieri G (1989) Identification and purification of natural killer cell stimulatory factor (NKSF), cytokine with multiple biologic effects on human lymphocytes. J Exp Med 170: 827–845

Lederman MM, Georges DL, Kusner DJ, Mudido P, Giam C-Z, Toossi Z (1994) *Mycobacterium tuberculosis* and its purified protein derivative activate expression of the human immunodeficiency virus. J AIDS 7: 727–733

Lehn M, Weiser WY, Engelhorn S, Gillis S, Remold HG (1989) IL-4 inhibits H_2O_2 production and antileishmanial capacity of human cultured monocytes mediated by IFN-γ. J. Immunol 143: 3020–3024

Lorgat F, Keraan MM, Lukey PT, Ress SR (1992) Evidence for in vivo generation of cytotoxic T cells. PPD-stimulated lymphocytes from tuberculous pleural effusions demonstrate enhanced cytotoxicity with accelerated kinetics of induction. Am Rev Respir Dis 145: 418–423

Manetti R, Parronchi P, Giudizi MG, Piccinni MP, Maggi E, Trinchieri G, Romagnani S (1993) Natural killer cell stimulatory factor (interleukin 12 [IL-12]) induces T helper type 1 (Th1)-specific immune responses and inhibits the development of IL-4-producing Th cells. J Exp Med 177: 1199–1204

Martinez OM, Gibbons RS, Garovoy MR, Aronson FR (1990) IL-4 inhibits IL-2 receptor expression and IL-2-dependent proliferation of human T cells. J Immunol 144: 2211–2215

Matsuyama T, Kobayashi N, Yamamoto N (1991) Cytokines and HIV infection: Is AIDS a tumor necrosis factor disease? AIDS 5: 1405–1417

May ME, Spagnuolo PJ (1987) Evidence for activation of a respiratory bursts in the interaction of human neutrophils with *Mycobacterium tuberculosis*. Infect Immun 55: 2304–2307

Moreno C, Taverne J, Mehlert A, Bate CAW, Brealey RJ, Meager A, Rook GAW, Playfair JHL (1989) Lipoarabinomannan from *Mycobacterium tuberculosis* induces the production of tumour necrosis factor from human and murine macrophages. Clin Exp Immunol 76: 240–245

Muller I, Cobbold SP, Waldmann H, Kaufmann SHE (1987) Impaired resistance to *Mycobacterium tuberculosis* infection after selective in vivo depletion of L3T4$^+$ and Lyt-2$^+$T cells. Infect Immun 55(9): 2037–2041

Munk ME, Gatrill AJ, Kaufmann SHE (1990) Target cell lysis and IL-2 secretion by T lymphocytes after activation with bacteria. J Immunol 145: 2434–2439

Munk ME, De-Bruyn J, Gras H, Kaufmann SHE (1994) The *Mycobacterium bovis* 32-kilodalton protein antigen induces human cytotoxic T-cell responses. Infect Immun 62(2): 726–728

Nathan CF, Kaplan G, Levis WR, Nusrat A, Witmer MD, Sherwin SA, Job CK, Horowitz CR, Steinman RM, Cohn ZA (1986) Local and systemic effects of intradermal recombinant interferon-γ in patients with lepromatous leprosy. N Engl J Med 315: 6–15

O'Brien RL, Happ MP, Dallas A, Palmer E, Kubo R, Born WK (1989) Stimulation of a major subset of lymphocytes expressing T cell receptor gamma delta by an antigen derived from *Mycobacterium tuberculosis*. Cell 57: 667–674

Ohmen JD, Barnes PF, Uyemura K, Lu S, Grisso CL, Modlin RL (1991) The T-cell receptor of human gamma-delta T-cells reactive to *Mycobacterium tuberculosis* are encoded by specific V genes but diverse V-J junctions. J Immunol 147: 3353–3359

Ohmen JD, Barnes PF, Grisso CL, Bloom BR, Modlin RL (1994) Evidence for a superantigen in human tuberculosis. Immunity 1: 35–43

Orme IM (1987) The kinetics of emergence and loss of mediator T lymphocytes acquired in response to infection with *Mycobacterium tuberculosis*. J Immunol 138: 293–298

Orme IM (1988a) Characteristics and specificity of acquired immunologic memory to *Mycobacterium tuberculosis* infection. J Immunol 140: 3589–3593

Orme IM (1988b) Induction of nonspecific acquired resistance and delayed-type hypersensitivity, but not specific acquired resistance, in mice inoculated with killed mycobacterial vaccines. Infect Immun 56: 3310–3312

Orme IM, Collins FM (1983) Protection against *Mycobacterium tuberculosis* infection by adoptive immunotherapy. J Exp Med 158: 74–83

Orme IM, Collins FM (1984) Adoptive protection of the *Mycobacterium tuberculosis*-infected lung. Dissociation between cells that passively transfer protective immunity and those that transfer delayed-type hypersensitivity. Cell Immunol 84: 113–120

Orme IM, Miller ES, Roberts AD, Furney SK, Griffin JP, Dobos KM, Chi D, Rivoire B, Brennan PJ (1992) T lymphocytes mediating protection and cellular cytolysis during the course of *Mycobacterium tuberculosis* infection. J Immunol 148: 189–196

Orme IM, Roberts AD, Griffin JP, Abrams JS (1993) Cytokine secretion by CD4 T lymphocytes acquired in response to *Mycobacterium tuberculosis* infection. J Immunol 151: 518–525

Ortaldo JR, Mason AT, O'Shea JJ, Smyth MJ, Falk LA, Kennedy CS, Longo DL, Ruscetti FW (1991) Mechanistic studies of transforming growth factor β inhibition of IL-2 dependent activation of CD3$^-$ large lymphocyte functions. J Immunol 146: 3791–3798

Oswald IP, Gazzinelli RT, Sher A, James SL (1992) IL-10 synergizes with IL-4 and transforming growth factor- β to inhibit macrophage cytotoxic activity. J Immunol 148: 3578–3582

Ottenhoff THM, Mutis T (1990) Specific killing of cytotoxic T cells and antigen-presenting cells by CD4$^+$ cytotoxic T cell clones. J Exp Med 171: 2011–2024

Pal PG, Horwitz MA (1992) Immunization with extracellular proteins of *Mycobacterium tuberculosis* induces cell-mediated immune responses and substantial protective immunity in a guinea pig model of pulmonary tuberculosis. Infect Immun 60: 4781–4792

Pearlman E, Kazura JW, Hazlett FE Jr, Boom WH (1993) Modulation of murine cytokine responses to mycobacterial antigens by helminth-induced T helper 2 cell responses. J Immunol 151: 4857–4864

Pedrazzini T, Hug K, Louis JA (1987) Importance of L3T4+ and Lyt-2+ cells in the immunologic control of infection with *Mycobacterium bovis* strain Bacillus Calmette-Guerin in mice. J Immunol 139: 2032–2037

Pfeffer K, Schoel B, Gulle H, Kaufmann SHE, Wagner H (1990) Primary responses of human T cells to mycobacteria: a frequent set of gamma/delta T cells are stimulated by protease-resistant ligands. Eur J Immunol 20: 1175–1179

Platanias LC, Vogelzang NJ (1990) Interleukin-1: biology, pathophysiology, and clinical prospects. Am J Med 89: 621–629

Porcelli S, Brenner MB, Greenstein JL, Balk SP, Terhorst C, Bleiche PA (1989) Recognition of cluster of differentiation 1 antigens by human CD4-CD8- cytolytic T lymphocytes. Nature 341: 447–450
Porcelli S, Morita CT, Brenner MB (1992) CD1b restricts the response of human CD4- CD8- T lymphocytes to a microbial antigen. Nature 360: 593–597
Rees A, Scoging A, Mehlert A, Young DB, Ivanyi J (1988) Specificity of proliferative response of human CD8 clones to mycobacterial antigens. Eur J Immunol 18: 1881–1887
Roach TI, Barton CH, Chatterjee D, Blackwell JM (1993) Macrophage activation: lipoarabinomannan from avirulent and virulent strains of *Mycobacterium tuberculosis* differentially induces the early genes c-fos, KC, JE, and tumor necrosis factor-α. J Immunol 150: 1886–1896
Robertson MJ, Soiffer RJ, Wolf SF, Manley TJ, Donahue C, Young D, Herrmann SH, Ritz J (1992) Response of human natural killer cells to NK cell stimulatory factor (NKSF): Cytolytic activity and proliferation of NK cells are differentially regulated by NKSF. J Exp Med 175: 779–788
Rook GAW, Steele J, Fraher L, Barker S, Karmali R, O'Riordan J (1986) Vitamin D_3, gamma interferon, and control of proliferation of *Mycobacterium tuberculosis* by human monocytes. Immunology 57: 159–163
Roper WH, Waring JJ (1955) Primary serofibrinous pleural effusion in military personnel. Am Rev Tuberc 71: 616–634
Schwarz EM, Salgame P, Bloom BR (1993) Molecular regulation of human interleukin 2 and T- cell function by interleukin 4. Proc Natl Acad Sci USA 90: 7734–7738
Scott P (1993) IL-12: Initiation cytokine for cell-mediated immunity. Science 260: 496–497
Scott P, Natovitz P, Coffman RL, Pearce E, Sher A (1988) Immunoregulation of cutaneous leishmaniasis. T cell lines that transfer protective immunity or exacerbation belong to different T helper subsets and respond to distinct parasite antigens. J Exp Med 168: 1675–1684
Selwyn PA, Hartel D, Lewis VA, Schoenbaum EE, Vermund SH, Klein RS, Walker AT, Friedland GH (1989) A prospective study of the risk of tuberculosis among intravenous drug users with human immunodeficiency virus infection. N Engl J Med 320: 545–550
Shiratsuchi H, Johnson JL, Toba H, Ellner JJ (1990) Strain- and donor-related differences in the interaction of *Mycobacterium avium* with human monocytes and its modulation by interferon-gamma. J Infect Dis 162: 932–938
Shiratsuchi H, Johnson JL, Ellner JJ (1991) Bidirectional effects of cytokines on the growth of *Mycobacterium avium* within human macrophages. J Immunol 146: 3165–3170
Sieling PA, Wang X-H, Gately MK, Oliveros JL, McHugh T, Barnes PF, Wolf SF, Yamamura M, Yogi Y, Uyemura K, Rea TH, Modlin RL (1994) Interleukin-12 regulates T-helper type 1 cytokine responses in human infectious disease. J Immunol 153: 3639–3647
Sieling PA, Chatterjee D, Porcelli SA, Prigozy TI, Mazzaccaro RJ, Soriano T, Bloom BR, Brenner MB, Kronenberg M, Brennan PJ, Modlin RL (1995) CDI-restricted T cell recognition of microbial lipoglycan antigens. Science 269: 227–230
Silva CL, Lowrie DB (1994) A single mycobacterial protein (hsp 65) expressed by a transgenic antigen-presenting cell vaccinates mice against tuberculosis. Immunology 82: 244–248
Small PM, Hopewell PC, Singh SP, Paz A, Parsonnet J, Ruston DC, Schecter GF, Daley CL, Schoolnik GK (1994) The epidemiology of tuberculosis in San Francisco. A population-based study using conventional and molecular methods. N Engl J Med 330: 1703–1709
Stern AS, Podlaski FJ, Hulmes JD, Pan YE, Quinn PM, Wolitzky AG, Familletti PC, Stremlo DL, Truitt T, Chizzonite R, Gately MK (1990) Purification to homogeneity and partial characterization of cytotoxic lymphocyte maturation factor from human B-lymphoblastoid cells. Proc Natl Acad Aci USA 87(9): 6808–6812
Surcel H-M, Troye-Blomberg M, Paulie S, Andersson G, Moreno C, Pasvol G, Ivanyi J (1994) Th1/Th2 profiles in tuberculosis, based on the proliferation and cytokine responses of blood lymphocytes to mycobacterial antigens. Immunology 81: 171–176
Sypek JP, Chung CL, Mayor SEH, Subramanyam JM, Goldman SJ, Sieburth DS, Wolf SF, Schaub RG (1993) Resolution of cutaneous leishmaniasis: interleukin 12 initiates a protective T helper type 1 immune response. J Exp Med 177: 1797–1802
Takashima T, Ueta C, Tsuyuguchi I, Kishimoto S (1990) Production of tumor necrosis factor alpha by monocytes from patients with pulmonary tuberculosis. Infect Immun 58: 3286–3292
Tanaka Y, Morita CT, Tanaka Y, Nieves E, Brenner MB, Bloom BR (1995) Natural and synthetic non-peptide antigens recognized by human γδ T-cells. Nature 375: 155–158
Tazi A, Fajac I, Soler P, Valeyre D, Battesti JP, Hance AJ (1991) γδ T- lymphocytes are not increased in number in granulomatous lesions of patients with tuberculosis or sarcoidosis. Am Rev Respir Dis 144: 1373–1375

Tazi A, Bouchonnet F, Valeyre D, Cadranel J, Battesti JP, Hance AJ (1992) Characterization of T-lymphocytes in the peripheral blood of patients with active tuberculosis. A comparison with normal subjects and patients with sarcoidosis. Am Rev Respir Dis 146: 1216–1221

Toossi Z, Sedor JR, Lapurge JP, Ondash RJ, Ellner JJ (1990) Expression of functional interleukin 2 receptors by peripheral blood monocytes from patients with active pulmonary tuberculosis. J Clin Invest 85: 1777–1784

Toossi Z, Gogate P, Shiratsuch H, Young T, Ellner JJ (1995) Enhanced production of transforming growth factor-β (TGFβ) from patients with active tuberculosis and presence of TGF-β in tuberculous granulomatous lung lesions. J Immunol 154: 465–473

Toossi Z, Sierra-Madero JG, Blinkhorn RA, Mettler MA, Rich EA (1993) Enhanced susceptibility of blood monocytes from patients with pulmonary tuberculosis to productive infection with human immunodeficiency virus type 1. J Exp Med 177: 1511–1516

Tsukaguchi K, Balaji KN, Boom WH (1995) $CD4^+$ αβ T cell and γδ T cell responses to *Mycobacterium tuberculosis* similarities and differences in Ag recognition, cytotoxic effector function, and cytokine production. J Immunol 154: 1786–1796

Tsuyuguchi I, Kawasumi H, Ueta C, Yano I, Kishimoto S (1991) Increase of T-cell receptor gamma/delta-bearing T cells in cord blood of newborn babies obtained by in vitro stimulation with mycobacterial cord factor. Infect Immun 59: 3053–3059

Valone SE, Rich EA, Walllis RS, Ellner JJ (1988) Expression of tumor necrosis factor in vitro by human mononuclear phagocytes stimulated with whole *Mycobacterium bovis* BCG and mycobacterial antigens. Infect Immun 56: 3313–3315

Wallis RS, Amir-Tahmasseb M, Ellner JJ (1990) Induction of interleukin 1 and tumor necrosis factor by mycobacterial proteins: the monocyte Western blot. Proc Natl Acad Sci USA 87: 3348–3352

Wallis RS, Paranjape R, Phillips M (1993a) Identification by two-dimensional gel electrophoresis of a 58-kilodalton tumor necrosis factor-inducing protein of *Mycobacterium tuberculosis*. Infect Immun 61: 627–632

Wallis RS, Vjecha M, Amir-Tahmasseb M, Okwera A, Byekwaso F, Nyole S, Kabengera S, Mugerwa RD, Ellner JJ (1993b) Influence of tuberculosis on human immunodeficiency virus (HIV-1): Enhanced cytokine expression and elevated β_2-microglobulin in HIV-1- associated tuberculosis. J Infect Dis 167: 43–48

Whalen C, Horsburgh CR, Hom D, Labart C, Simberkoff M, Ellner JJ (1995) Accelerated course of human immunodeficiency virus infection after tuberculosis. AM J Respir Crit Care Med 151: 129–135

Wolf SF, Temple PA, Kobayashi M, Young D, Dicig M, Lowe J, Dzialo R, Fitz L, Ferenz C, Azzoni L, Chan SH, Trinchieri G, Perussia B, Hewick RM, Kelleher K, Herrmann SH, Clark SC (1991) Cloning of cDNA for natural killer cell stimulatory factor, a heterodimeric cytokine with multiple biologic effects on T and natural killer cells. J Immunol 146: 3074–3081

Yamamura M, Uyemura K, Deans RJ, Weinberg K, Rea TH, Bloom BR, Modlin RL (1991) Defining protective responses to pathogens: cytokine profiles in leprosy lesions. Science 254: 277–279

Yssel H, de Waal Malefyt R, Roncarolo MG, Abrams JS, Lahesmaa R, Spits H, deVries JE (1992) IL-10 is produced by subsets of human CD4 T cell clones and peripheral blood T cells. J Immunol 149: 2378–2384

Zhang M, Gately MK, Wang E, Wolf SF, Lu S, Modlin RL, Barnes PF (1994) Interleukin-12 at the site of disease in tuberculosis. J Clin Invest 93: 1733–1739

Zhang M, Gong J, Iyer DV, Jones BE, Modlin RL, Barnes PF (1994) T-cell cytokine responses in persons with tuberculosis and human immunodeficiency virus infection. J Clin Invest 94: 2435–2442

Zhang M, Lin Y, Iyer DV, Gong J, Abrams JS, Barnes PF (1995) T-cell cytokine responses in human infection with *Mycobacterium tuberculosis*. Infect Immun 63: 3231–3234

Zhang Y, Doerfler M, Lee TC, Guillemin B, Rom WN (1993) Mechanisms of stimulation of interleukin-1β and tumor necrosis factor-α by *Mycobacterium tuberculosis* components. J Clin Invest 91: 2076–2083

Zhang Y, Nakata K, Weiden M, Rom WN (1995) *Mycobacterium tuberculosis* enhances human immuno deficiency virus-1 replication by transcriptional activation at the long terminal repeat. J Clin Invest 95: 2324–2331

Mechanisms of Anergy in Tuberculosis

Z. Toossi and J.J. Ellner

1 Introduction

Immunological hyporesponsiveness (anergy) is a concomitant of diseases due to intracellular pathogens and well-demonstrated in tuberculosis. The expression of anergy in tuberculosis encompasses depression of both delayed-type hypersensitivity (DTH), manifest as depressed tuberculin skin test reaction, and in vitro lymphocyte responses to antigens of *Mycobacterium tuberculosis*. Evidence for immunosuppression, which appears to be mostly limited to mycobacterial antigens, has been documented in different populations with active tuberculosis worldwide, regardless of prior BCG vaccination status. Whether attenuation of the host responses to antigens of *M. tuberculosis* that stimulate or maintain protective immunity is a factor in the development of tuberculosis or contributes to the chronicity of the disease is not known. However, it appears that anergy in active tuberculosis is mediated largely through monocytes that display signs of activation, possibly by exposure to mycobacterial antigens or cytokines in situ. *M. tuberculosis* and its constituents are potent in stimulation of

Department of Medicine, Case Western Reserve University, University Hospitals of Cleveland and Veterans Administration Medical Center, Cleveland, 10701 East Boulevard, Cleveland, OH 44106, USA monocytes to produce an array of cytokines, notably transforming growth factor β (TGF-β), which is suppressive to lymphocytes, deactivates macrophage effector functions, and promotes fibrosis (WAHL 1992). Immunosuppressive circuits are activated in situ as newly recruited monocytes encounter *M. tuberculosis* and its products; at the site of infection these circuits may contribute to immunopathogenesis, while at local sites of exposure to antigens (skin test) they may modulate the expression of DTH.

2 Spectrum of Anergy in Tuberculosis

2.1 Expression of Anergy

The frequency of the two components of anergy, i.e., depressed skin test tuberculin reaction and in vitro lymphocyte responses to mycobacterial antigens, correlates with the spectrum of clinical disease. In pulmonary tuberculosis, the clinically predominant form of disease, from 17% to 25% of patients are unresponsive to PPD skin testing (DANIEL et al. 1981). Low blastogenic responses to PPD are present in 40%–60% of patients, which include almost all skin test negative and some test positive tuberculosis patients (KLEINHENZ and ELLNER 1987; TOOSSI et al. 1986). In disseminated (miliary) tuberculosis 50%–75% of patients are skin test nonresponsive; although blastogenic responses have been studied to a more limited extent, low responses appear to be common (BHATNAGAR et al. 1977). Both skin test reactivity and blastogenic responses are preserved in localized forms of tuberculosis, such as lymphatic disease, in which 100% of patients are skin test reactive. A high percentage of cutaneous reactions (40%) ulcerate in patients with tuberculous adenitis, indicating the intensity of the DTH response (HUSSAIN et al., submitted).

Both skin test anergy and lymphocyte responses to PPD vary according to the mycobacterial load and/or extent of pathology at the disease site (Toossi et al. 1986; Hussain et al., submitted). For example, among patients with pulmonary tuberculosis, those with far advanced disease not only have a higher frequency of negative skin tests (up to 50%), the size of their reaction is significantly smaller than in patients with moderate or minimal pulmonary tuberculosis. This group also has the lowest in vitro blastogenic responses to PPD. In contrast, 100% of patients with minimal disease are PPD skin test positive and demonstrate robust reactions. Lymphocyte responses tend to be preserved in this group.

2.2 Correlation of Delayed-Type Hypersensitivity with Protective Immunity

The successful containment of a primary tuberculous infection in healthy individuals is characterized by the development of a positive tuberculin reaction and strong T cell proliferative responses to PPD in vitro. These responses are long-lasting; for example, the reversion rate for positive tuberculin reactions is about 5% each year (GRZYBOWSKI and ALLEN 1964). Preservation of in vitro lymphocyte responses to PPD over a prolonged period (19 years) has been recently shown (HAVLIR et al. 1991a). Development of lymphocyte responses and DTH in the naive individual is indicative of the establishment of a memory immune response to *M. tuberculosis*, known to be mediated by CD4 lymphocytes. Evidence that tuberculin responses correlate with protective immunity to *M. tuberculosis* is based on epidemiologic observations that individuals are protected against exogenous reinfection with the organism (STEAD 1981). Furthermore, since over 90% of immunocompetent subjects remain free of tuberculosis during their lifetime subsequent to a primary infection, at least some degree of immunologic protection against an active *M. tuberculosis* infection is maintained. Observations in HIV infection, in fact, indicate that active immunologic surveillance is necessary to maintain latent tuberculosis foci as quiescent. Tuberculosis occurs relatively early in HIV infection; 50%–67% of cases in the USA occurred 6–9 months prior to development of another AIDS defining illness (ELLNER 1990). Also, CD4 lymphocyte counts are relatively preserved in HIV-infected patients with pulmonary tuberculosis (THEUER et al. 1990). Collectively these data indicate that subtle defects, which may well include a selective loss or dysfunction of memory T cell early in the progressive immune dysfunction of HIV infection, may be sufficient for reactivation of *M. tuberculosis*.

Recent studies have established that protective immunity and DTH are separable in the murine model of *M. tuberculosis* infection, based on the fact that each can be transferred to naive animals with different subpopulations of T cells (ORME and COLLINS 1984). Live organisms induce protective immunity, whereas DTH can be induced by live or dead organisms. The generation of protective immunity in mice is mediated by CD4 cells and is directed towards secreted proteins of *M. tuberculosis*, thereby requiring actively metabolizing organisms (ORME et al. 1992). DTH responses, by contrast, are also mediated by CD4 cells but can be elicited towards nonviable mycobacteria (ORME 1988). Whether the same holds true in human disease is not known. However, development of positive PPD skin test reactions after vaccination with BCG does not correlate with protection from tuberculosis (COMSTOCK et al. 1974). Nonetheless, PPD reactive nurses exposed to patients with active tuberculosis had a significantly (fourfold) lower incidence of disease than those who were tuberculin negative at entry (HEIMBECK 1928). It is important to note that, historically, PPD skin testing has been the only studied immunologic correlate in humans that may relate to protective immunity. In humans, components of the protective immune response and DTH to *M. tuberculosis* may in fact overlap and/or be similarly influenced by cytokine regulatory circuits during tuberculosis.

2.3 Purified Protein Derivative Skin Test Reaction

The tuberculin reaction, manifest as an induration palpable 48–72 h after the intradermal injection of PPD of *M. tuberculosis,* is a prototype DTH response. There is a strong correlation between skin test reactivity and lymphocyte responses, as demonstrated in a 19-year follow-up of 22 tuberculin reactors; only three subjects showed a skin test reversion to negative, accompanied by loss of blastogenic responses to PPD (HAVLIR et al. 1991a). Whereas even reactions of 1–4 mm of induration correlate with positive lymphocyte proliferative responses in vitro (MILLER and JONES 1973), depending on epidemiologic features, an induration of 5–15 mm is considered as positive. However, due to cross-reactivity between mycobacterial antigens, at least in areas endemic for nontuberculous mycobacteria, larger reactions (15 mm) are more likely to indicate a true *M. tuberculosis* infection. Reactions due to nontuberculous mycobacteria are also more likely to revert to negative over time. Other factors associated with increased frequency of reversion to a negative tuberculin reactivity include a lack of exposure to family with tuberculosis, age <20 or >65, and an initial response of <15 mm.

Histologically, in healthy subjects, tuberculin response is characterized by an early infiltration of CD4 lymphocytes and monocytes and expression of cellular markers of activation, such as HLA-DR and IL-2 receptors (PLATT et al. 1983). A preferential activation of CD4 T cells displaying a Th1 cytokine profile, such as expression of interferon-γ (IFNγ) and interleukin-2 (IL-2), has more recently been shown in these lesions (TSICOPOULOS et al. 1992). In contrast, allergen-induced skin reactions in atopic subjects express IL-4 and IL-5, markers of a Th2 response (KAY et al. 1991). It appears that, histologically, the dermal response to PPD studied in a small group of patients with pulmonary tuberculosis was similar to that of healthy tuberculin responders except for a more intense infiltration of OK-M1 positive monocytes, which represented 50% of the cells (FULLMER et al. 1987). Functional analysis of these monocytes in comparison to those obtained from PPD skin reactions of healthy responders may be informative as regards mechanisms of anergy.

Conditions that are suppressive to the host immunity, including tuberculosis itself, decrease the size of the tuberculin reaction. Old age, cachexia, malnutrition, lymphoproliferative malignancies, cytotoxic and immunosuppressive therapies, and overwhelming viral and bacterial infections may lead to cutaneous anergy (PESANTI 1994). Overall 10% of patients with active tuberculosis fail to react to PPD, despite a high variability reported in different studies. Of more that 5000 patients with all forms of tuberculosis, some of whom had received antituberculous medications for some time, only 4% did not demonstrate a measurable response to PPD (PALMER and EDWARDS 1966). In contrast, STEAD (1969) reports a 20% tuberculin negativity in patients with newly diagnosed, sputum smear positive tuberculosis. Even when cut-offs of 4–5 mm were chosen, 17%–21% of patients with active pulmonary tuberculosis were skin test nonresponsive (HOLDEN et al. 1971; ROONEY et al. 1976). Most nonresponders became tuberculin positive after 2 weeks of specific antituberculous therapy (ROONEY et al. 1976). Overall, half of the tuberculous patients who are tuberculin negative do not respond to other antigens. Therefore, in active tuberculosis use of nonmycobacterial skin tests (i.e., anergy panel) may not be helpful in clarification of false negative tuberculin reactions. Among HIV-infected persons, tuberculin nonresponsiveness is common and in part reflects a nonspecific loss of memory T cell responses secondary to T cell immunodeficiency. In one study, 43% of 65 HIV-seronegative as compared to only 9% of 43 HIV-seropositive prisoners were tuberculin reactive (CANESSA et al. 1989). Similarly, the relative risk of tuberculin nonreactivity in HIV-infected compared with that in noninfected pregnant women in Uganda was 2.89 (CDC 1990). HIV-infected subjects also were six times more likely to not respond to a panel of antigens (*Candida*, mumps, tetanus toxoid, histoplasmin, coccidioidin) than uninfected subjects (GRAHAM et al. 1992). Tuberculin nonreactivity in HIV disease clearly correlates with the degree of T cell immunodeficiency. In HIV-infected patients with active tuberculosis, PPD skin tests are positive in 60%–80% in the absence of other AIDS-related opportunistic diseases. Skin test reactivity drops to 20%–40% once another AIDS-defining diagnosis has occurred (CDC 1991). In geographic areas such as Uganda, where *M. tuberculosis* and HIV infection both are common, 93% of patients with active tuberculosis had a positive skin test (NSUBUGA et al. 1994).

2.4 T Cell Responses in Tuberculosis

A number of studies have investigated the in vitro responses of T cells during active tuberculosis. Depression of lymphocyte blastogenic responses to PPD and several purified and recombinant antigens of *M. tuberculosis* has been acccompanied in most studies by intact responses to mitogens. In some studies low PPD responses are concomitant with low responses to other microbial antigens, however with differing underlying mechanisms (KLEINHENZ and ELLNER 1985). Other studies indicate intact responses to nonmycobacterial antigens (TOOSSI et al. 1986) and thus a more specific T cell unresponsiveness. Early studies demonstrated an inverse correlation between T cell responses and circulating levels of immunoglobulins. Furthermore, with clinical improvement upon antituberculous chemotherapy, immunologic abnormalities reverted to normal, suggesting that the abnormalities correlated with disease activity. Even in the patients with preserved lymphocyte responses during active disease, the progressive enhancement of T cell reactivity during the course of chemotherapy is consistent with earlier transient immunosuppression. Based on multiple studies from Japan (SHIRATSUCHI et al. 1987), Brazil (ANDRADE-ARZABE et al. 1991), and

Pakistan (Hussain et al., submitted), it is now established that neither repeated exposure to *M. tuberculosis* in areas with high prevalence of tuberculosis nor prior vaccination with BCG alters the pattern of T cell hyporesponsiveness in active tuberculosis. However genetic factors may be linked to low responsiveness to *M. tuberculosis*; an association of the MHC class II molecule HLA-B14 (DR1) with low responses to PPD in Mexican American patients with tuberculosis has been demonstrated (Cox et al. 1988).

The basis of the lower T cell responsiveness during active tuberculosis is a selective defect in the production of IL-2 and expression of IL-2 receptors(Rs) (Toossi et al. 1986). The relative numbers of CD4 and CD8 T cells are not affected in tuberculosis (Kleinhenz and Ellner 1985). This observation was recently confirmed in a study involving 25 HIV seronegative, newly diagnosed tuberculous patients (Vanham et al., submitted). Normal distribution of CD4 and CD8 cells was associated with an increase in expression of HLA-DR on both subsets and increased serum levels of the activation markers tumor necrosis factor α (TNFα), β-2 microglobulin, and neoptrin. Expression of CD45RO (memory T cell) was not lower on either CD4 or CD8 cells. Furthermore, the frequency of PPD-reactive T cells is not significantly lower than that of healthy skin test reactors (Fujiwara and Tsuyuguchi 1984). Reversal of the low blastogenic responses to PPD in mononuclear cells from patients with tuberculosis in the presence of exogenous IL-2 was not observed in one of the two studies examining this issue (Toossi et al. 1986; Shiratsuchi et al. 1987). Whereas this observation may reflect low IL-2R expression by lymphocytes (Toossi et al. 1986), the possibility of interference, presumably by other cytokines, with the interaction of IL-2 and its receptor needs to be considered. Several studies have documented a defect in mononuclear cell production of IFNγ in patients with active pulmonary tuberculosis (Onwubalili et al. 1985; Vilcek et al. 1986; Huygen et al. 1988). Importantly, depression of both T cell blastogenesis and IFNγ production is transient and tends to normalize with therapy in patients with pulmonary tuberculosis (Ellner 1978a; Huygen et al. 1988). In tuberculous pleuritis, which is both paucibacillary and self-healing, the pleural mononuclear cells abundantly produce the macrophage activating molecules IFNγ and TNFα (Barnes et al. 1990). Recently, pleural mononuclear cells have also been shown to produce Ih-12 (Zhang et al. 1994), which may be a critical cytokine in the development of a Th1 response (Wu et al. 1993).

T cell hyporesponsiveness in subjects with tuberculosis is displayed toward a wide spectrum of both crude antigenic preparations and moleculary well-characterized antigens of *M. tuberculosis*. This most probably reflects the well documented enormous heterogeneity in the human immune response to the organism (Havlir et al. 1991b). A number of studies have shown low responses to PPD, culture filtrate and sonicates of *M. tuberculosis*. Low responses to the highly conserved heat shock antigens 71 kDa of *M. tuberculosis* and 65 kDa of *M. bovis* have been shown, however, presently these responses are of unknown significance. Lowered responses to a 10 kDa heat shock antigen of

M. tuberculosis /BCG, analogous to *E. coli* GroES heat shock protein, has been shown in blood mononuclear cells from patients with pulmonary tuberculosis, but not in the pleural mononuclear cells from tuberculous pleuritis. This antigen induces T cell proliferation and IFNγ production in T cells from healthy subjects (BARNES et al. 1992a).

Falla et al. demonstrated that, despite comparable responses to sonic extracts of virulent *M. tuberculosis* (H37Rv), only about 50% of patients and 84% of healthy household contacts responded to peptides of the MTP40 (14 kDa) antigen (FALLA et al. 1991). Furthermore, patients with active tuberculosis have reduced responses to several fractions of the A60 antigen complex prior to chemotherapy, which increases significantly after several weeks of treatment (CARLUCCI et al.1993). The A60 complex is comprised of heat resistant cytoplasmic antigens in growing mycobacteria which accumulate in the cell wall of stationary organisms and are released when mycobacteria are in the declining phase of growth in culture. In healthy individuals these antigens induce both IFNγ production and T cell proliferation. However, VORDERMEIER et al. (1992) have demonstrated a selective anergy to an immunodominant epitope of 38 kDa, also known as antigen 5 and antigen 78, in patients with pulmonary but not lymphatic tuberculosis.

Recently, attention has focused on the secreted proteins of *M. tuberculosis* since live but not heat-killed organisms and culture filtrate have been shown to confer immunity in animals (PAL and HOROWITZ 1992). Two of the three major antigens secreted by growing cultures of *M. tuberculosis* (the 85 complex), the 30 kDa antigen (85 B; also known as α antigen), and the 32 kDa antigen (85A), induce both T cell proliferation and cytokine production in healthy PPD skin test reactors (HAVLIR et al. 1991b; HUYGEN et al. 1988). Also, the 30 kDa antigen elicits DTH in mice (SALATA et al. 1991). Studies in the USA (HAVLIR et al. 1991b), Mexico (TORRES et al. 1994), and Pakistan (HIRSCH et al.,submitted) indicated that the majority of patients with active tuberculosis failed to respond to the 30 kDa antigen, whereas healthy tuberculin positive household contacts of patients did. Interestingly, the response to sonic extracts of *M. tuberculosis* was comparable between the patients and their contacts in the study from Mexico. Also, serum antibody to the 30 kDa antigen was significantly higher in patients. Whether in humans lack of T cell responses to the 30 kDa antigen is a marker of susceptibility to infection, or indicates activation of antigen-specific immunosuppressive circuits by this molecule remains to be seen.

3 Mechanisms of Anergy

Over the last few decades several mechanisms have been postulated for the observed immune hyporesponsiveness during tuberculosis.

3.1 Compartmentalization of Antigen Responsive Cells

As a chronic inflammatory disease, the recruitment of mononuclear cells to sites of infection is characteristic of tuberculosis in the immunocompetent host. However, the degree of cellular infiltrate seen pathologically may predominantly reflect an in situ expansion of antigen reactive T cells in addition to recruitment of monocytes and T cells at sites of disease. Evidence that the basis of anergy during active tuberculosis is due to expansion of mycobacterial reactive T cells, comes from studies of tuberculous pleuritis and meningitis. The number of PPD-reactive T cells is expanded in the pleural space as compared to peripheral blood in patients with tuberculous pleuritis (Fugiwara et al. 1984). Also, pleural fluid mononuclear cells are enriched for CD4+CDw29+cells (memory cells) that produce IFNγ and proliferate in response to *M. tuberculosis* (Barnes et al. 1989). Expression of IL-2 and IFNγ is higher in pleural mononuclear cells than in peripheral blood mononuclear cells (PBMCs). In contrast, expression of mRNA for IL-4, a Th2 cytokine which inhibits Th1 responses, is lower in pleural than in blood mononuclear cells. Adequate antigen-reactive cells present in the peripheral blood of patients with tuberculous pleuritis (Fugiwara et al. 1985) argues for mechanisms other than or in combination with T cell compartmentalization as underlying the hyporesponsiveness in tuberculosis. In tuberculous pleuritis, blood monocyte-mediated immunosuppression (Ellner 1978b) (discussed below) may be operative, in addition to a peripheral overexpression of Th2 responses. However, there also is evidence for local expression of immunosuppressive cytokines such as IL-10 (Barnes et al. 1993) and TGFβ (Maeda et al. 1993) in tuberculous pleuritis, which may limit T cell responses at the disease site.

3.2 Adherent Cell Suppression

Monocytosis is a characteristic of active tuberculosis. DNA labeling and cytochemical studies indicate that monocytes from tuberculous patients are immature (Schmitt et al. 1977). Also characteristic of active disease is that circulating monocytes appear to be "activated" in vivo to produce cytokines and suppress lymphocyte functions. The possibility of a direct participation of *M. tuberculosis* and its protein and polysaccharide constituents in activation of monocytes has been raised by recent studies. Furthermore, as well be discussed, the immaturity of monocytes may be key in suppression of T cell responses during disease.

Numerous studies indicate that monocytes from tuberculous patients differ from those of healthy individuals and those from patients with other mycobacterial infections both phenotypically and functionally. Phenotypically, monocytes from patients with active tuberculosis are OKM1 positive and are radiosensitive in that they lose their suppressive capacity with γ-irradiation (Kleinhenz and Ellner 1985). Furthermore, monocytes display an altered pattern of expression of HLA-DR antigens in that, as compared to healthy individuals, initial surface DR is significantly lower and is followed by a sharp rise in DR expression after in vitro culture for short periods (TWEARDY et al. 1984). Cell mixing experiments demonstrated increased suppressive activity of DR monocytes. Initial low DR expression may be marker of immaturity of circulating monocytes in tuberculosis. Alternatively, exposure to cytokines that downmodulate DR on monocytes, such as TGFβ, IL-4, or IL-10, may explain low surface DR. Expression of CD16 is increased on monocytes from patients with tuberculosis (Vanham et al., submitted). This may reflect excess in vivo TGFβ activity, as TGFβ induces CD16 expression on monocytes in vitro (WELCH et al. 1990).

Functionally, monocytes from tuberculous subjects demonstrate enhanced adherence to plastic surfaces, an increase in hexose monophosphate shunt activity (ELLNER et al. 1981), and increased killing of *Schistosoma*, *staphylococcus*, and *Listeria monocytogenes* (ELLNER et al. 1990). Tumor killing and both spontaneous and lipopolysaccharide-induced production of prostaglandins are not altered. Furthermore, monocytes from patients with tuberculosis have an increased ability to produce the proinflammatory cytokines TNFα, IL-6 (TAKASHIMA et al. 1990; OGAWA et al. 1991) and IL-1 (FUJIWARA et al. 1986) upon stimulation in vitro. Recently, monocytes from tuberculous subjects have been shown to have an enhanced susceptibility to in vitro infection with HIV; this may directly relate to the increase in the synthetic capacity of certain of the above cytokines (TOOSSI et al. 1993).

The best documented characteristic of blood monocytes during active tuberculosis, however is their suppression of lymphocyte functions in vitro (ELLNER 1978a). In fact, both T cell blastogenesis and production of IL-2 in response to PPD increase upon removal of adherent monocytes (ELLNER 1978a; TOOSSI et al. 1986). Monocyte-mediated suppression of T cell blastogensis was observed specifically with PPD but not with nonmycobacterial antigens (KLENHENZ and ELLNER 1985; HIRSCH et al. 1994). Furthermore, small numbers of monocytes (2% of T cell) reconstitute suppression of IL-2 production in response to PPD (TOOSSI et al.1989) (Fig.1). These data indicate that mycobacterial-specific T cell anergy is often superimposed on nonspecific anergy, however mechanistically separable from it. Recent observations have demonstrated that, analogous to lipopolysaccharide (LPS)-activated monocytes from healthy subjects, freshly isolated monocytes from patients with pulmonary tuberculosis spontaneously express IL-2R mRNA, display IL-2Rs on their surface, and shed the molecule upon in vitro culture (TOOSSI et al. 1990). Also, sera of patients with pulmonary tuberculosis have elevated levels of soluble IL-2Rs that are sustained for up to 3 months after initiation of therapy (BROWN et al. 1989). Despite the low numbers of high affinity IL-2Rs on monocytes from tuberculous subjects, as shown by Scatchard plot analysis, these cells efficiently remove IL-2 when cultured with the cytokine. Besides being indicative of in vivo activation, expression of IL-2R by monocytes during tuberculosis may be a possible mechanism for lowered T cell responses. However, lack of reconstitution of mononuclear cell responses to PPD in the

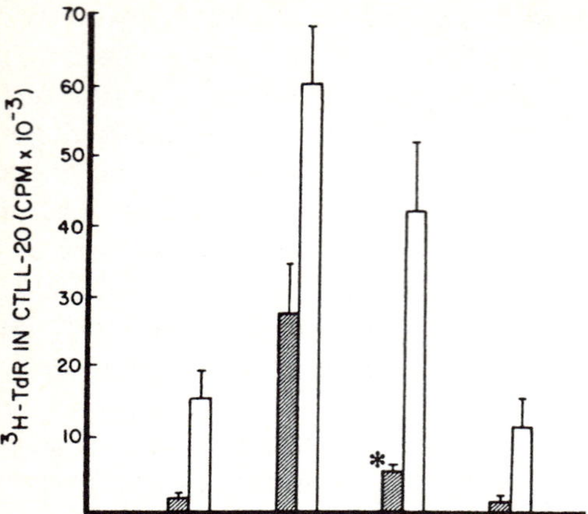

Fig. 1. Suppression of T cell interleukin-2 (IL-2) production by monocytes from patients with tuberculosis and healthy individuals. Peripheral blood mononuclear cells (PBMCs), T cells, or T cells with 2% or 25% adherent monocytes (ADHs) were cultured with PPD. After 48 h, cell-free supernatants were assessed for IL-2 activity using the IL-2 dependent murine T cell line CTLL-20. Data show mean ± SE for 10 patients with tuberculosis(filled columns) and 10 tuberculin-reactive healthy subjects (open columns). Significant suppression (*) of IL-2 by 2% adherent monocytes of patients when cocultured with T cells as compared to T cells cultured alone. (From Toossi et al. 1989. Copyright 1989 by The University of Chicago Press)

presence of exogenous IL-2 indicates other mechanisms, such as lower expression of IL-2Rs on T cells (Toossi et al. 1986) or interference with the interaction of IL-2 and IL-2R on T cells, as by cytokines (such as TGFβ).

The direct monocyte stimulatory properties of *M. tuberculosis* and its constituents may underlie the in vivo activation and the associated immunosuppression of monocytes in tuberculosis. Interestingly, it may also offer an explanation for the apparent antigen specificity of suppression during disease. PPD is a potent stimulant of monocyte production of the proinflammatory cytokines IL-1 (Wallis et al. 1986) and TNFα (Valone et al. 1988). *M. tuberculosis* and its PPD transcriptionally activate the expression of TGFβ mRNA and induce TGFβ production in monocytes (Hirsch et al. 1994; Toossi et al. 1995a). PPD and the 30 kDa antigen of *M. tuberculosis* also induce increased TGFβ production in monocytes from tuberculous patients (Hirsch et al. submitted). Whether the other proteins of *M. tuberculosis*, such as the 58 kDa antigen that induces TNFα in monocytes from healthy subjects (Wallis et al. 1993), stimulate the production of TGFβ or other suppressive mediators is not known. Lipoarabinomannan (LAM) of virulent *M. tuberculosis* induces TNFα, IL-1, IL-10, and IL-6 poorly, yet stimulates TGFβ production by monocytes (Dahl et al. 1996). Of note, LAM constitutes 5 mg/g of mycobacterial weight (Hunter et al. 1986) and therefore its contribution to the cytokine milieu may be substantial. Other constituents of *M. tuberculosis* that are associated with suppression of lymphocyte responses in vitro, such as arabinogalactan, may operate through induction of prostaglandins in monocytes from healthy subjects (Kleinhenz et al. 1981). LPS- induced prostaglandin release of monocytes from patients with tuberculosis, however, was found to be similar to that of healthy individuals (Ellner et al. 1981).

Recently, spontaneous production of TGFβ by monocytes from patients with active pulmonary tuberculosis has been shown in both the USA (Toossi et al. 1995b) and Pakistan (Hirsch et al. 1994). Furthermore, enhancement of the constitutive expression of TGFβ mRNA was present in three of five patients studied. Also, TGFβ activity was seen in multinucleated giants cells and epithelioid cells of lung lesions in samples from three patients with untreated tuberculosis by immunohistochemical staining (Toossi et al. 1995b). TGFβ is a potent suppressor of both IL-2 (Brabletz et al. 1993) and IFNγ (Espevik et al. 1987) production and inhibits IL-2 mediated T cell activation (Ahuja et al. 1993); therefore its excess production may prove to be the basis of lymphocyte hyporesponsiveness seen in active tuberculosis. In fact, neutralization of TGFβ by specific antibody reversed low T cell blastogenesis and production of IFNγ in patients with pulmonary tuberculosis (Hirsch et al.,submitted). Another immunosuppressive cytokine that needs to be considered is IL-10. However, in contrast to TGFβ, PPD-stimulated release of IL-10 in unseparated mononuclear cells and monocytes from patients was similar to that of healthy subjects.

3.3 T Cells in Tuberculosis

Whereas adherent monocyte suppressive pathways appear to be universal among patients with active pulmonary tuberculosis, a number of patients with far advanced disease do not recover T cell responses to mycobacterial antigens after depletion of monocytes (Toossi et al. 1986). As discussed above, the distribution of subsets of T cells is not altered significantly during tuberculosis, but T cells display markers of activation such as DR and transferrin receptor reactivity. Other studies indicate that FcγR+ and CD16 reactive lymphocytes are expanded during active disease and differentially affect lymphocyte responses; FcγR+ lymphocytes contribute to low blastogenic responses directly (Kleinhenz and Ellner 1987), whereas CD16 cells collaborate with monocytes in suppression of IL-2 production (Toossi et al. 1989). CD4 and CD8 cells are not suppressive of lymphocyte responses in tuberculosis (Kleinhenz and Ellner 1987).

Another population of circulating lymphocytes examined recently for their role in human tuberculosis is the γδ T cells. These cells proliferate in response to live mycobacteria (Havlir et al. 1991c) even in tuberculin-negative individuals and in newborns (Tsuyuguchi et al. 1991). Whereas the frequency of γδ T cells is not altered in tuberculosis (Tazi et al. 1992; Barnes et al. 1992b), their proliferation in patients with advanced tuberculosis is lower than in both healthy tuberculin reactors and patients with tuberculous pleuritis (Barnes et al. 1992b). However, functionally, γδ cells likely contribute to the immune response of primary infection, and do not show suppressive activity in tuberculosis. Lack of expansion of γδ cells may therefore be a marker of the generalized hyporesponsiveness of tuberculosis.

Recently, the cytokine profile of T cells has been used to functionally distinguish two groups of CD4 cells: those with a Th1 cytokine profile (IL-2 and IFNγ), and those with a Th2 cytokine pattern (IL-4, IL-5, IL-10) (STREET and MOSSMAN 1991). Evidence for expansion of CD4 cells with Th1 profile has been documented at sites of disease in tuberculoid leprosy (YAMAMURA et al. 1991) and in tuberculous pleuritis (BARNES et al. 1993). By contrast, lepromatous patients have an expansion of CD4 cells with a Th2 profile (STELING et al. 1993). Th2 cytokines, such as IL-4 and IL-10, are cross-modulatory in that they reduce Th1 responses and increase antibody production. Whether a Th2 profile is expressed at the disease site in advanced pulmonary or miliary tuberculosis or at sites of tuberculin injection in patients with skin test anergy is presently unknown. An increase in the frequency of IL-4 producing T cells in response to several mycobacterial antigens in tuberculosis has recently been shown (SURCEL et al. 1994); however, it did not correlate with either low T cell blastogenesis or production of IFNγ. Other studies indicate that IL-4 mRNA expression is increased in the blood mononuclear cells of patients with tuberculosis (BARNES et al. 1993). In our studies IL-4 mRNA in PBMCs from tuberculous patients and their healthy contacts was similarly expressed (HIRSCH et al. 1994).

Recent immunogenetic studies have demonstrated an association between low PPD-induced blastogenic responses and the haplotype HLA-DRI (B14) in patients with tuberculosis (COX et al. 1988). Interestingly, steroid 21-hydroxylase deficiency is associated with the same haplotype (DAVIS et al. 1987). This trait is associated with altered expresssion of the MHC class II molecule, HLA-DR1 and a failure to activate alloreactive T cell clones or provision of accessory function to MHC-restricted T cell clones.

4 Transforming Growth Factor-β-Mediated Immunopathogenic Circuits in Tuberculosis

As noted, there is now substantial evidence for immunologic hyporesponsiveness in active tuberculosis, which includes both tuberculin anergy and reduced lymphocyte responses to mycobacterial antigens. Two features that need to be underscored are the antigen specificity and the transient nature of this hyporesponsiveness. Mechanistically, in vitro studies have implicated blood monocytes to be activated, possibly a result of heavy exposure to mycobacterial constituents and/or cytokines in vivo, and suppressive of T cell functions. Recent studies have shown an excess production of TGFβ by monocytes and tissue macrophages during tuberculosis. Therefore this molecule is probably a mediator of anergy. However, TGFβ has strong effects on many parameters of the immune response and therefore likely affects expression of the disease systemically and locally.

The basis for enhanced TGFβ production in tuberculosis presumably may be dependent upon two features of circulating monocytes. First, in active tuberculous infection there is an early release of bone marrow cells including monocytes, many of which are not fully mature. These moncytes in turn display a heightened capacity to produce and respond to TGFβ as they circulate through the infected tissues in vivo. Of note, monocytes but not macrophages display high affinity receptors for TGFβ (McCartney-Francis and Wahl 1994) and therefore are subject to autoinduction to produce more TGFβ (Kim et al. 1990). Our recent observations indicate lower production of TGFβ by alveolar macrophages than by autologous monocytes in response to LPS, phorbol esters, and TGFβ itself (Toossi et al., 1996). Second, *M tuberculosis* and its constituents are strong inducers of TGFβ production. Freshly recruited monocytes at sites of active infection are exposed to the *M. tuberculosis,* and its secretory and structural moieties (e.g., 30 kDa antigen and LAM), and produce excess TGFβ. Furthermore, favorable conditions for release of biologically active TGFβ from its latent form are more likely at sites of infection. For example, enhanced sialidase activity at sites of *M. tuberculosis* infection, as seen in alveolar macrophages of BCG-infected rabbits (Pilatte et al. 1987), may contribute to activation of latent TGFβ (Miyazono and Heldin 1989). Also, autoinduction of TGFβ may be key in sustaining high local levels of TGFβ and thus in maintenance of immunopathogenic circuits (Fig. 2). Since TGFβ is a potent chemoattractant for monocytes, their predominance in situ is favored. Local overexpression of TGFβ under this scenario undermines macrophage effector functions and limits T cell responses to mycobacteria and its products. Recent data show that the intracellular growth of *M. tuberculosis* is enhanced by TGFβ, and macrophage activation by TNFα and IFNγ are antagonized by TFGβ (Hirsch et al. 1994). TGFβ strongly deactivates macrophages (Tsunawki et al. 1988). Inhibition of production of IFNγ by TGFβ in situ may further decrease macrophage effector function. Low production and response to IL-2 may limit expansion of antigen responsive cells. In addition, by counteracting IL-3, IFNγ, and TNFα, cytokines likely to be important in formation of granulomas (Kindler et al. 1989; Enelow et al. 1992), excess TGFβ may result in ineffective granuloma formation. TGFβ inhibits T cell, in particular CD4 cell, cytotoxicity (Fontana et al. 1989; Wahl 1992) possibly further dampening host responses to *M. tuberculosis*. Furthermore, since TGFβ preferentially down-regulates CD4 cells, the cytokine profile of CD8 cells may predominate locally, possibly contributing to tissue damage. However, TGFβ enhances extracellular matrix deposition and induces fibroblast proliferation, thereby promoting fibrosis (Wahl 1992). Excess TGFβ may lead to exuberant scarring of tissues, as implicated in the pathogenesis of other diseases associated with fibrosis such as glomerulonephritis (Border and Rusolahti 1992) and which is characteristic of tuberculosis. Inhibitors of TGFβ may be important future agents as adjuvants in the treatment of tuberculosis.

At the site of tuberculin injection recruited monocytes are exposed to PPD; production of TGFβ and amplification of its activity may ensue. TGFβ, mainly by

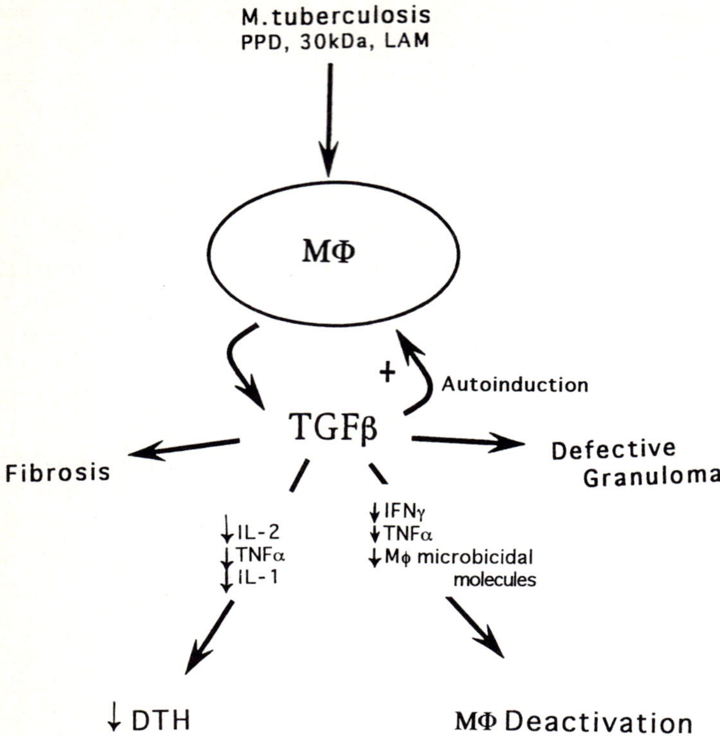

Fig. 2. Central role of transforming growth factor-β (TGFβ) in modulation of anergy in tuberculosis. Production of TGFβ by mononuclear phagocytes upon exposure to *M. tuberculosis* and its antigens initiates autocrine and paracrine effects that may eventually culminate in suppression of cellular responses, deactivation of macrophage effector functions, and fibrosis at sites of infection, and to depressed delayed-type hypersensitivity (DTH) at site of tuberculin injection

limiting T cell proliferation and counteracting proinflammatory cytokines may limit the size of a DTH response. Excess TGFβ may explain the apparent paradox of skin test anergy in tuberculosis and its correlation with low T cell blastogenic responses. As the cytokine profile of the newly recruited monocytes at sites of tuberculin injection will be determined by their state of activation and maturation, tuberculin anergy therefore will directly correlate with the mycobacterial load during active tuberculosis.

References

Ahuja SS, Paliogianni F, Yamada H, Balow JE, Boumpas DT (1993) Effect of transforming growth factor-β on early and late activation events in human T cells. J Immunol 150: 3109–3118

Andrade-Arzabe R, Machado IV, Fernndez B, Blanca I, Ramirez R, Bianco NE (1991) Cellular immunity in current active pulmonary tuberculosis. Am Rev Respir Dis 143: 496–500

Barnes PF, Mistry SD, Cooper CL, Cooper CL, Pirmez C, Rea TH, Modlin RL (1989) Compartmentalization of a CD4+ T lymphocyte subpopulation in tuberculous pleuritis. J Immunol 142: 1114–1119
Barnes PF, Fong F-J, Brennan PJ, Twomey PE, Mazumuder A, Modlin RL (1990) Local production of TNFα and IFNγ in tuberculous pleuritis. J Immunol 145: 149–154
Barnes PF, Mehra V, Rivoire B, Fong SJ, Brennan PJ, Voegtline PM, Houghten RA, Bloom BR, Modlin RL (1992a) Immunoreactivity of a 10 kD antigen of *Mycobacterium tuberculosis*. J Immunol 148: 1835–1840
Barnes PF, Grisso CL, Abram JF, Band H, Rea TM, Modlin RL (1992b) γδ T-lymphocytes in human tuberculosis. J Infect Dis 165: 506–512
Barnes PF, Lu S, Abrams JS, Wang E, Yamamura M, Modlin RL (1993) Cytokine production at the site of disease in human tuberculosis. Infect Immun 61: 3482–3489
Bhatnagar R, Malaviya AN, Naray Anan P, Ragopalan P, Kumar R, Bharadwaj OP (1977) Spectrum of immune response abnormalities in different clinical forms of tuberculosis. Am Rev Respir Dis 115: 207–212
Border WA, Rusolahti ER (1992) Transforming growth factor β in disease: the dark side of tissue repair. J Clin Invest 90: 1–7
Brabletz T, Pfeuffer I, Schorr E, Siebelt F, Wirth T, Serfling E (1993) Transforming growth factor β and cyclosporin A inhibit the inducible activity of the interleukin-2 gene in T cells through a non canonical octamer-binding site. Mol Cell Bio 13: 1155–1162
Brown AE, Reicker TK, Webster HK (1989) Prolonged elevated soluble interleukin-2 receptors in tuberculosis. Am Rev Respir D s 139: 1036–1038
Canessa PA, Fasano L, Lavecchia MA, Torraca A, Schiattone ML (1989) Tuberculin skin test in asymptomatic HIV seropositive carriers. Chest 96: 1215–1218
Carlucci S, Beschin A, Tuosto L, Ameglio F, Gandolfo GM, Cocito C, Fiorucci F, Saltini C, Piccolella E (1993) Mycobacterial antigen complex A60-specific T cell response during the course of pulmonary tuberculosis. Infect Immun 61(2): 439–447
Centers for Disease Control (1990) Tuberculin reaction in apparently health HIV seropositive and HIV-seronegative women: Uganda. MMWR 39: 638–646
Centers for Disease Control (1991) Purified protein derivative (PPD)-tuberculin anergy and HIV infection: guidelines for anergy testing and management of anergic persons at risk of tuberculosis. MMWR 40: 27–38
Comstock GW, Livesay VT, Woolpert SF (1974) The prognosis of a positive tuberculin reaction in childhood and adolescence. Am J Epidemiol 99: 131–138
Cox RA, Downs M, Neimes RE, Ognibene AJ, Yamashita TS, Ellner JJ (1988) Immunogenic analysis of human tuberculosis. J Infect Dis 158: 1302–1308
Dahl KE, Shiratsuchi H, Hamilton BD, Ellner JJ, Toossi Z (1996) Selective induction of TGFβ in human monocytes by lipoarabinomannan of *M. tuberculosis.* Infect Immun 64: 399-405
Daniel TM, Oxtoby MJ, Pinto E, Moreno S (1981) the immune spectrum in patients with pulmonary tuberculosis. Am Rev Respir Dis 123: 556–559
Davis J, Rich RR, Van M, Le MV, Pollach MS, Cook RG (1987) Defective antigen presentation and noval structural properties of DR1 from an HLA haplotype associated with 21-hydroxylase deficiency. J Clin Invest 80: 898–904
Ellner JJ (1978a) Suppressor adherent cells in human tuberculosis. J Immunol 121: 2573–2578
Ellner JJ (1978b) Pleural fluid and peripheral blood lymphocyte function in tuberculosis. Ann Intern Med 89: 932–933
Ellner JJ (1990) Tuberculosis in the times of AIDS. The facts and the message. Chest 98: 1051–1052
Ellner JJ, Spagnuolo PJ, Schacter BZ (1981) Augmentation of selective monocyte functions in tuberculosis. J Infect Dis 144: 391–398
Ellner JJ, Boom WB, Edmonds KL, Rich EA, Toossi Z, Wallis RS (1990) Regulation of the immune response to *Mycobacterium tuberculosis*. In: Ayoub EM et al. (eds) Microbial determinants of virulence and host response. American Society for Microbiology, Washington DC, pp 77–91
Enelow RI, Sullivan GW, Carper HT, Mandell G (1992) Induction of MGC formation from in vitro culture of human monocytes with IL-3 and IFNγ. Comparison with other simulating factors. Am J Respir Cell Mol Biol 6: 57–62
Espevik T, Figari IS, Shalaby MR, Lackides GA, Lewis GD, Shepard HM, Pallandino MA Jr (1987) Inhibition of cytokine production by cyclosporin A and transforming growth factor β.J Exp Med 166: 571–576

Falla JC, Para CA, Mendoza M, Franco LC, Guzmaan F, Orozco O, Patarroyo ME (1991) Indentification of B and T cell epitopes within the MTP 40 protein of *M. tuberculosis* and their correlation with the disease course. Infect Immun 59: 2265–2273

Fontana A, Frei K, Bodmer S, Hoefer E, Scheier MH, Pallandino MA, Zingkernagel M (1989) Transforming growth factor-β inhibits the generation of cytotoxic T cells in virus infected mice. J Immunol 143: 3220–3234

Fujiwara H, Tsuyuguchi I (1984) Frequency of tuberculin-reactive T-lymphocytes in pleural fluid and blood from patients with tuberculous pleuritis. Chest 89: 530–535

Fujiwara H, Kleinhenz ME, Wallis RS, Ellner JJ (1986) Increased interleukin-1 production and monocyte suppressor cell activity associated with human tuberculosis. Am Rev Respir Dis 133: 73–77

Fullmer MA, Shen J-Y, Modlin RL, Rea TH (1987) Immunohistochemical evidence of lymphokine production and lymphocyte activation antigens in tuberculin reaction. Clin Exp Immunol 67: 383–390

Graham NMH, Nelson KE, Solomon L, Bonds M, Rizzo RT, Scavotto J, Astemborski J, Vlahov D (1992) Prevalence of tuberculin positivity and skin test anergy in HIV-1 seropositve and seronegative intravenous drug users. JAMA 267: 369–372

Grzybowski S, Allen EA (1964) The challenge of tuberculosis in decline: a study based on the epidemiology of tuberculosis in Ontario, Canada. Am Rev Respir Dis 90: 707–720

Havlir DV, van der Kupy F, Duffy E, Marshall R, Hom D, Ellner JJ (1991a) A 19 year follow-up of tuberculin reactors: assessment of skin test reactivity and *in vitro* lymphocyte responses. Chest 99: 1172–1176

Havlir DV, Wallis RS, Boom WH, Daniel TM, Chevernak K, Ellner JJ (1991b) Human immune response to *Mycobacterium tuberculosis* antigens. Infect Immun 59: 665–670

Havlir DV, Ellner JJ, Chevernak KA, Boom WH (1991c) Selective expansion of human γδ T-cells by monocytes infected with live *Mycobacterium tuberculosis* J Clin Invest 87: 729–733

Heimbeck (1928) Immunity to tuberculosis. Arch Intern Med 41: 336–342

Hirsch CS, Yoneda T, Ellner JJ, Averill LE, Toossi Z (1994) Enhancement of intracellular growth of *M. tuberculosis* in human monocytes by transforming growth factor beta. J Infect Dis 170: 1229–1237

Holden M, Dubin MR, Diamond PH (1971) Frequency of negative intermediate-strength tuberculin sensitivity in patients with active tuberculosis. N Engl J Med 285: 1506–1509

Hunter SW, Gaylord H, Brennan PJ (1986) Structure and antigenicity of the phosophorylated antigens from leprosy and tubercle bacilli. J Biol Chem 261: 12345–12351

Hussain R, Dawood G, Obaid M, Toossi Z, Wallis RS, Minai A, Dojki M, Sturm AW, Ellner JJ (submitted) Depressed cellular and augmented humoral responses in patients with active tuberculosis from Pakistan.

Huygen K, van Vooren JP, Turneer M, Bosmans R, Dierckx P, De Bruyn J (1988) Specific lymphoproliferation, gamma interferon production, and serum immunoglobulin G directed against a purified 32 kDa mycobacterial protein antigen (P32) in patients with active tuberculosis. Scand J Immunol 27: 187–194

Kay AB, Ying S, Varney V, Durham SR, Moqbel R, Wardlaw AJ, Hamid Q (1991) Messenger RNA expression of the cytokine gene cluster, IL-3, IL-4, IL-5, and GM-CSF in allergen-induced late-phase reactions in atopic subjects. J Exp Med 173: 775–779

Kim SJ, Angel P, Lafyatis R, Hattori K, Kim KY, Spron MB, Karin M, Roberts AB (1990) Autoinduction of transforming growth factor β1 is mediated by the AP-1 complex. Mol Cell Biol 10: 1492–1496

Kindler V, Syepino AP, Gran GE et al (1989) The inducing role of tumor necrosis factor in the development of bactericidal granulomas during BCG infection. Cell 56: 731–740

Kleinhenz ME, Ellner JJ (1985) Immunoregulatory adherent cells in human tuberculosis: radiation-sensitive antigen-specific suppression by monocytes. J Infect Dis 152: 171–176

Kleinhenz ME, Ellner JJ (1987) Antigen responsiveness during tuberculosis: regulatory interaction of T-cells subpopulations and adherent cell. J Lab Clin Med 110: 31–40

Klienhenz ME, Ellner JJ, Spagnulo PJ, Daniel TM (1981) Suppression of lymphocyte responses by tuberculous plasma and mycobacteril arabinogalctan: monocyte dependence and indomethecin reversibility.68: 153–158

Maeda J, Ueki N, Ohkawa T, Iwahashi N, Nakano T, Hada T, Higashino K (1993) Local production and localization of transforming growth factor beta in tuberculous pleurisy. Clin Exp Immunol 92: 32–38

McCartney-Francis NL, Wahl SM (1994) Transforming growth factor β: a matter of life and death. J Leukoc Biol 55: 401–409

Miller SD, Jones HE (1973) Correlation of lymphocyte transformation with tuberculin skin-test sensitivity. Am Rev respir Dis 107: 530–538

Miyazono K, Heldin CH (1989) Role for carbohydrate structures in TGF-β 1 latency. Nature 338: 158–160

Nsubuga P, Whalen C, Johenson JL, Byekwaso F, Okwera A, Mugerwa R, Ellner JJ (1994) Preserved PPD reactivity and frequent cavitary disease as initial clinical manifestations of pulmonary tuberculosis in HIV-infected Ugandans. International conference on AIDS 1994, 7–12; Aug, Vol 10(2), p 28 (abstact no 406B)

Ogawa T, Uchida H, Kusumoto Y, Mori Y, Yamamura Y, Hamada S (1991) Increase in tumor necrosis factor alpha and interluekin-6 secreting cells in peripheral blood mononuclear cells from subjects infected with *Mycobacterium tuberculosis*. Infect Immun 59: 3021–3025

Onwubalili JK, Scott GM, Robinson JA (1985) Deficient immune interferon production in tuberculosis. Clin Exp Immunol 59: 405–413

Orme IM (1988) Induction of non specific acquired resistance and delayed type hypersensitivity but not specific acquired resistance in mice inoculated with killed mycobacterial vaccines. Infect Immun 56: 3310–3312

Orme IM, Collins FM (1984) Adoptive protection of the *Mycobacterium tuberculosis*-infected lung. Dissociation between cells that passively transfer protective immunity and those that transfer delayed-type hypersensitivity. Cell Immunol 84: 113–120

Orme IM, Miller ES, Roberts AD (1992) T lymphocytes mediating protection and cellular cytolysis during the course of *Mycobacterirum tuberculosis* infection. J Immunol 148: 189–196

Pal PG, Horowitz MA (1992) Immunization with extracellular proteins of *M. tuberculosis* induces cell-mediated immune responses with substantial protective immunity in a guinea pig model of pulmonary tuberculosis. Infect Immun 60(11): 4782–4792

Palmer CE, Edwards LB (1966) The tuberculin test: in retrospect and prospect. The Baker Lecture, presented at the University of Michigan School of Public Health, Ann Arbor, MI

Pesanti EL (1994) The negative tuberculin test. Am J Resp Crit Care Med 149: 1699–1709

Pilatte Y, Bignon J, Lambre CR (1987) Lysosomal and cytosolic sialidases in rabbit alveolar macrophages, demonstration of increased lysosomal activity after *in vivo* activation with BCG. Biochem Biophys Acta 923: 150–155

Platt JL, Grant BW, Eddy AA, Michael AF (1983) Immune cell populations in cutaneous delayed-type hypersensitivity. J Exp Med 158: 1227–1242

Rooney JJ, Crocco JA, Kramer S, Lyons HA(1976) Further observations on tuberculin reactions in tuberculosis. Am J Med 60: 517–522

Salata RA, Sanson AJ, Malhotra IJ, Wiker HG, Harboe M, Philips NB, Daniel TM (1991) Purification and characterization of the 30,000 dalton native antigen of *Mycobacterium tuberculosis* and characterization of six monoclonal antibodies reactive with a major epitope of this antigen. J Lab Clin Med 118: 589

Schmitt E, Meuret G, Stix L (1977) Monocyte recruitment in tuberculosis and sarcoidosis. Br J Haematol 35: 11–17

Shiratsuchi H, Okuda Y, Tsuyuguchi I (1987) Recombinant human IL-2 reverses *in vitro* deficient cell-mediated immune responses to tuberculin purified derivative by lymphocytes of tuberculous patients. Infect Immun 55: 2126–2131

Sieling PA, Abrams JS, Yamamura M, Salgame P, Bloom BR, Rea TH, Modlin RL (1993) Immunosuppressive roles for IL-10 and IL-4 in human infection. J Immunol 150: 5501–5510

Stead WW (1969) The new face of tuberculosis. Hosp Prac 4: 62

Stead WW (1981) Tuberculosis among elderly persons: an outbreak in a nursing home. Ann Intern Med 94: 606–610

Street NE, Mossman TR (1991) Functional diversity of T lymphocytes due to secretion of different cytokine patterns. FASEB J 5: 171–175

Surcel HM, Tory-Blomberg M, Paulie S, Anderson G, Moreno C, Pasvol G, Ivanyi J (1994) TH1/TH2 profiles in tuberculosis, based on the proliferation and cytokine response of blood lymphocytes to mycobacterial antigens. Immunology 81: 171–176

Takashima T, Ueta C, Tsuyuguchi I, Kishimoto S (1990) Production of tumor necrosis factor by monocytes from patients with pulmonary tuberculosis. Infect. Immun 58: 3286–3292

Tazi A, Bouchonnet F, Valeyre O, Battesti JP, Hance AJ (1992) Characterization of γδ T lymphocytes in the peripheral blood of patients with active tuberculosis. Am Rev Respir Dis 146: 1216–1221

Theuer CP, Hopewell PC, Elias D, Schecter GF, Rutherford GW, Chaisson RF (1990) Human immunodeficiency virus infection in tuberculosis patients. J Infect Dis 162: 8–12

Toossi Z, Kleinhenz ME, Ellner JJ (1986) Defective interleukin-2 production and responsiveness in human pulmonary tuberculosis. J Exp Med 163: 1162–1172

Toossi Z, Edmonds KE, Tomford WJ, Ellner JJ (1989) Suppression of PPD-induced interleukin-2 production by interaction of CD 16 lymphocytes and adherent mononuclear cells in tuberculosis. J Infect Dis 159: 352–356

Toossi Z, Lapurga JP, Ondash R, Sedor JR, Ellner JJ (1990) Expression of functional interleukin 2 receptors by peripheral blood monocytes from patients with active pulmonary tuberculosis. J Clin Invest 85: 1777–1784

Toossi Z, Sierra-Madero JG, Blinkhorn RA, Mettler MA, Rich EA (1993) Enhanced susceptibility of blood monocytes from patients with pulmonary tuberculosis to productive infection with human immunodeficiency virus. J Exp Med 177: 1511–1517

Toossi Z, Young TG, Averill LE, Hamilton BD, Shiratsuchi H, Ellner JJ (1995a) Induction of Transforming growth factor-β (TGF-β) by purified protein derivative (PPD) of mycobacterium tuberculosis. Infect Immun 63: 224–228

Toossi Z, Gogate P, Shiratsuchi H, Young T, Ellner JJ (1995b) Enhanced production of transforming growth factor-β (TGF-β) by blood monocytes from patients with active tuberculosis and presence of TGFβ in tuberculous granlomatous lung lesions. J Immunol 154: 465–473

Toossi Z, Hirsch CS, Hamilton BD, Knuth CK, Friedlander MA, Rich EA (1996) Decreased production of transforming growth factor β1 (TGF-β1) in human alveolar macrophages. J Immunol (in press)

Torres M, Mendez-Sampiero P, Jimenez-Zamudio L, Teran L, Camerena A, Quezada R, Ramos E, Sada E (1994) Comparison of the immune response against Mycobacterium tuberculosis antigens between a group of patients with active pulmonary tuberculosis and healthy household contacts. Clin Exp Immunol 96: 75–78

Tsicopoulos A, Hamid Q, Varney V, Ying V, Moqbel R, Durham SR, Kay AB (1992) Preferential messenger RNA expression of Th1-type cells (IFN gamma+, IL-2+) in classical delayed-type (tuberculin) hypersensitivity reactions in human skin. J Immunol 148: 2058–2061

Tsunawki S, Spron M, Nathan C (1988) Deactivation of macrophages by TGFβ. Nature 334: 260–264

Tsuyuguchi I, Kawasumi H, Ueta C, Yano I, Kishimoto S (1991) Increase of T-Cell Receptor gamma/delta-bearing T cells in cord blood of newborn babies obtained by *in vitro* stimulation with mycobacterial cord factor. Infect Immun 59: 3053–3059

Tweardy DJ, Schacter BZ, Ellner JJ (1984) Association of altered dynamics of monocyte surface expression of human leukocyte antigen-DR with immunosuppression in tuberculosis. J Infect Dis 149: 31–37

Valone SE, Rich EA, Wallis RS, Ellner J (1988) Expression of Tumor Necrosis Factor *In Vitro* by human mononuclear phagocytes stimulated with whole mycobacterium bovis BCG and mycobacterial antigens. Infect Immun 56: 3313–3315

Vanham G, Edmonds KE, Qing L, Hom D, Toossi Z, Joness B, Daley C, Huebner R, Kestens L, Gigase P, Ellner JJ (submitted) Global immune activation during *M. tuberculosis* infection in humans.

Vilcek J, Klion A, Henriksen-DeStefano D, Zemtsov A, Davidson DM, Davidson M, Friedman-Kien A (1986) Defective gamma-interferon production in peripheral blood leukocytes of patients with acute tuberculosis. J Clin Immunol 6: 146–151

Vordemeier HM, Harris DP, Friscia G, Roman E, Surcel HM, Moreno C, Pasvol G, Ivanyi J (1992) T cell repertoire in tuberculosis: Selective anergy to an immunodominant epitope of the 380kDa antigen in patients with active disease. Eur J Immunol 22: 2631–2637

Wahl S (1992) Transforming growth factor Beta: a cause and a cure. J Clin Immunol 2: 61–71

Wallis RS, Fujiwara H, Ellner JJ (1986) Direct stimulation of monocyte release of interleukin 1 by mycobacterial protein antigens. J Immunol 136: 193–196

Wallis RS, Paranjape R, Phillips M (1993) Identification by two-dimensional gel electrophoresis of a 58-kilodalton tumor necrosis factor-inducing protein of *Mycobacterium tuberculosis.* Infect Immun 61: 627–632

Welch GR, Wong HL, Wahl SM (1990) Selective induction of Fc gamma RIII on human monocytes by transforming growth factor β. J Immunol 144: 3444–3448

Wu CY, Demeure C, Kiniwa M, Gately M, Delepess G (1993) IL-2 induces the production of IFN-γ by neonatal human CD4 T cells. J Immunol 151: 1938–1949

Yamamura M, Uyemura K, Deans RJ, Weinberg K, Rea TH, Bloom BR, Modlin RL (1991) Defining protective responses to pathogens: cytokine profiles in leprosy lesions. Science 254: 277–279

Zhang M, Gately MK, Modlin RL, Barnes PF (1994) Interleukin 12 at the site of disease in tuberculosis. J Clin Invest 93: 1733–1739

The Koch Phenomenon and the Immunopathology of Tuberculosis

G.A.W. Rook and J.L. Stanford

Department of Bacteriology, University College London Medical School, Windeyer Building, 46 Cleveland Street, London W1P 6DB, UK

1 Introduction

Tuberculosis kills more than three million people every year, and the problem is rapidly increasing. This is paradoxical when we know that the immune response is perfectly capable of coping with this infection. Before the advent of HIV, only 5% of infected individuals developed disease, and perhaps another 5% did so when T cell function was compromised by old age, stress or protein malnutrition. Therefore 90% of the population was resistant under normal circumstances. Moreover the incidence in the suceptible 10% could be reduced by up to 80% by BCG vaccination in those countries in which the vaccine worked. Why then are we failing to control the rapid increase in tuberculosis?

Our failure is partly attributable to the fact that conventional treatment takes at least 6 months, with consequent problems of cost, compliance and drug resistance. These are two reasons for this: First, although chemotherapy kills the majority of the organisms within a week or two, there is a subpopulation of organisms, known as persisters, that may be dormant or perhaps in true "stationary" phase (GRANGE 1992; SIEGELE and KOLTER 1992). These organisms are poorly susceptible to chemotherabpy and it is not certain that drugs can be devised that will kill them. Chemotherapy must be prolonged so that every dormant organism has time to enter a phase of metabolic activity while the drugs are still present.

Second, the cell-mediated response in tuberculosis patients does little or nothing to assist the physician's efforts to sterilize the lesions. The patients' response causes tissue necrosis and has some limited ability to wall off the organisms, but it has little bactericidal activity. Clinically this is obvious. For instance, chemotherapy is as effective in patients whose tuberculosis has become active because T cell function has been compromised by HIV infection as it is in non-HIV-infected tuberculosis patients. This implies that even in the latter case, the T cell-mediated response fails to contribute to sterilization of the lesions. Similarly, the fact that tuberculosis patients show a high relapse rate if chemotherapy is stopped at 4 months, when few live bacilli are present, implies that the pattern of response that exists even as long as 4 months after the bacterial load has been drastically lowered is incapable of destroying *Mycobacterium tuberculosis*.

These well-known observations tell us something fundamental. It seems that although protective immunity is perfectly possible and even common, in humans, conventional chemotherapy does not convert the response from the disease-associated tissue-damaging pattern to the more efficiently protective mode.

The fundamental issues therefore are the relationship between the tissue-damaging response and protection, and whether the physician can do anything to switch the patient's response from one to the other, thereby accelerating cure.

2 The Koch Phenomenon

In the 1890s Koch noted that 4–6 weeks after establishment of infection in guinea pigs, intradermal challenge with whole organisms or culture filtrate resulted in necrosis both locally and in the original tuberculous lesion (KOCH 1891). Similar phenomena occur in humans. The tuberculin test is frequently necrotic in subjects who are, or have been, tuberculous. This is not an inevitable consequence of the delayed hypersensitivity response to tuberculin because necrosis does not occur when the same test is performed in normal BCG recipients or in tuberculoid leprosy patients. Similarly it does not occur when tuberculosis patients are skin-tested with sonicates derived from environmental saprophytic mycobacteria, in spite of the fact that there are large numbers of shared epitopes (KARDJITO et al. 1986).

This necrotizing reaction, now known as the "Koch phenomenon," protected guinea-pigs against intradermal challenge with live organisms because the local necrosis caused sloughing of the tissue containing them. However, similar necrosis in deep sites, or in the lungs, failed to eliminate the bacteria. Thus, if guinea-pigs were preimmunized with protocols that gave rise to necrotizing skin-test reactivity equivalent to the local necrosis elicited by KOCH, they were rendered more, rather than less, susceptible to infection by *intramuscular* injection of a small number of virulent organisms. In contrast, immunisation protocols priming small tuberculin reactions were protective (WILSON et al. 1940) These observations led to endless confusion. There are two particularly crucial questions:

1) Is the Koch phenomenon as exaggerated version of the tissue–damaging process seen routinely in tuberculosis lesions, or is it a quite separate curiosity attributable to an effect of sudden injection of large quantities of bacterial products?
2) What is the relationship between this tissue-damaging response and protection? Are tissue damage and protection "excessive" and "regulated" manifestations of similar pathways, or are they the result of qualitatively different immunological mechanisms?

Without knowing the answer to these questions, which are discussed in detail below, KOCH sought to exploit the necrotizing phenomenon for the treatment of human tuberculosis and found that injection of larger quantities of culture filtrate (old tuberculin) subcutaneously into tuberculosis patients would evoke necrosis in established tuberculous lesions at distant sites (ANDERSON 1891). This resulted in necrosis and sloughing of the lesions of skin tuberculosis (lupus vulgaris, usually caused by bovine strains), but when similar necrosis was evoked in deep lesions in the spine or lungs, the results were disastrous, and merely provided further necrotic tissue in which the bacteria could proliferate. This treatment was therefore abandoned.

The induction of tissue damage both locally and at distant inflammatory sites by an injection of microbial components suggests involvement of cytokines, and this aspect of the immunopathology of tuberculosis is explored below.

2.1 Cytokine Release in Tuberculosis

Can Koch's obervations, or the necrosis routinely present in tuberculous lesions, be attributed in part to tumour necrosis factor-α (TNFα)? Evidence for release of TNFα tuberculosis is strong. The bacteria produce potent triggers of cytokine release (MORENO et al. 1989; ROOK et al. 1987a; SILVA and FACCIOLI 1988; VALONE et al. 1988). Blood monocytes (TAKASHIMA et al. 1990) and alveolar macrophages (ROOK and AL ATTIYAH 1991) from tuberculosis patients release TNFα "spontaneously" in large quantities, and the cytokine is present lesions (BARNES et al. 1990). In view of the weight loss seen in humans, it is interesting that cytokine-induced wasting can be evoked by injecting tiny quantities of trehalose dimycolate (cord factor) dissolved in oil into the peritoneal cavities of mice (SILVA and FACCIOLI 1988). Circulating level of TNFα inhibitors are also high in the serum of tuberculosis patients (FOLEY et al. 1990). These are extracellular domains of receptors shed largely in response to TNFα release, and they make it difficult to assay TNFα in the sera of tuberculosis patients. Thus TNFα is certainly released in the human disease.

2.2 Is Cytokine Release Involved in Tissue Damage?

Is release of TNFα and of other synergistic cytokines responsible in part for the necrosis in tuberculosis lesions in humans? There is little doubt that in mice TNFα plays a critical protective role. In vitro it mediates additional activation of macrophages previously activated by interferon-γ (IFNγ) (CHAN et al. 1992). This may contribute to mycobactericidal activity by causing production of nitric oxide (NO). Similarly neutralization of TNFα in vivo using antibodies has revealed that it contributes to granuloma formation (KINDLER et al. 1989) and to protection. However the relevance of these findings to humans is uncertain. TNFα does not trigger significant NO release from human macrophages which seem to be deficient in tetrahydrobiopterin an essential cofactor (STUEHR et al. 1991). This does not rule out a protective role for TNFα in humans because it may act through other pathways, and there is some evidence that it can increase the antimycobacterial efficacy of human alveolar macrophages (HIRSCH et al. 1994). Nevertheless there are good reasons for suggesting that TNFα plays a central role in the pathogenesis of tuberculosis. First, release of TNFα could be excessive, and secondly, the tuberculous lesions could be particularly sensitive to the necrotizing action of the cytokine. There is evidence for both of these possibilities, which are obviously interrelated.

2.3 Excessive Release of Tumor Necrosis Factor-α

Thalidomide, which is routinely used for the treatment of erythema nodosum leprosum (ENL) and of graft-vs-host disease following bone marrow transplantation, probably works in these conditions by reducing TNFα levels. Thalidomide appears to shorten the half-life of TNFα mRNA (Moreira et al. 1993). Thus it reduces but does not eliminate TNFα. When administered to tuberculosis patients it results in remarkable weight gain and symptomatic relief (Prof. G. Kaplan, personal communication). This could imply that release of TNFα is excessive in relation to the release of TNFα inhibitors (free receptors). This does not appear to be the case in the peripheral blood, where, as already pointed out, the inhibitors are usually present at a considerable excess (Foley et al. 1990) and free TNFα is rarely detectable. Nevertheless it is perfectly possible that such an imbalance occurs in the lesions.

2.4 Parallels Between the Koch Phenomenon and the Shwartzman Reaction: Release of Tumor Necrosis Factor-α into Cytokine-Sensitive Sites

The alternative hypothesis is that in tuberculosis the lesions themselves are excessively sensitive to TNFα. TNFα is only toxic under certain circumstances, and in certain types of inflammatory site. Shwartzman observed that a site primed by an injection gram-negative bacteria (though endotoxin will substitute) undergoes necrosis if a second dose of gram-negative organisms (or endotoxin) is injected intravenously 24 h later (Shwartzman 1937). It is thought that the "prepared" inflammatory site is abnormally susceptible to circulating cytokines and activated cells resulting from the second challenge injection. This view is supported by the finding that direct injection of cytokines, particularly TNFα, into such sites will cause similar necrosis (Rothstein and Schreiber 1988).

Evidence that this is relevant to mycobacterial pathology is suggestive and has a long history. Several early workers demonstrated that mycobacterial lesions, like LPS-prepared sites, will undergo necrosis if the animal is subsequently challenged intravenously or subcutaneously with endotoxin-rich bacteria (Bordet 1931), endotoxin (Shands and Senterfitt 1972), or muramyl dipeptide (Nagao and Tanaka 1985), and this necrosis is accompanied by massive systemic release of TNFα (Carswell et al. 1975). Is there direct evidence that TNFα and other synergistic cytokines such as interleukin (IL)-1 will cause damage in mycobacterial lesions? In other words, can we demonstrate by direct inejection of cytokines into mycobacterial lesions that they are prepared sites in the sense used by Shwartzman?

2.5 Direct Induction of Tissue Damage by Tumor Necrosis Factor-α in Mycobacterial lesions

We have found that the injection of mycobacterial components (if genuinely endotoxin-free) will not prepare a site for TNFα-mediated necrosis in nonimmune mice. However if CD4+T cell reactivity has previously been primed and a delayed-type hypersensitivity (DTH) response is elicited, the site containing the DTH response can be exquisitely sensitive to TNFα (AL ATTIYAH et al. 1992a,b). It is therefore theoretically possible that the Koch phenomenon represents a "T cell-dependent" Shwartzman reaction. However it then emerged that not all DTH reactions to mycobacterial antigen are sensitive to TNFα (ROOK and AL ATTIYAH 1991), and recent work suggests that this may be related to the balance of Th1 to Th2 cytokines. There is evidence from other systems that inflammatory lesions mediated by mixed Th1+Th2 (or Th0?) T cell activity are susceptible to necrosis. For instance in murine Schistosomiasis there is initially a relatively "pure" Th1 pattern of response, with priming for release of IFNγ. Subsequently (when the ova are produced at about day 42) a Th2 response becomes superimposed on this Th1 pattern and release of IL-4 and IL-5 can be demonstrated (GRZYCH et al. 1991). At this point tissue damage begins to occur in the granulomata, and further tissue damage can be evoked if systemic cytokine (TNFα) release is triggered by endotoxin (CARSWELL et al. 1975; FERLUGA 1979). Then from day 84 the Th2 component starts to decline, the Th1 component returns and the granulomata promptly become non-necrotizing.

Subsequent studies have confirmed this correlation in a model that uses mycobacterial antigen (HERNANDEZ-PANDO and ROOK 1994a). As mentioned above, direct injection of TNFα into T cell-mediated responses to mycobacterial antigen often, but not always, causes necrosis. The effect is dependent on CD4+ T cells (AL ATTIYAH et al. 1992a). However the sensitivity of a DTH skin-test site to a subsequent injection of TNFα, into the same site depended not on the size of the DTH response but on the immunization schedule(ROOK and AL ATTAIYAH 1991). DTH reaction sites elicited in mice with apparently "pure" Th1 responses (and suppressed IL-4 release) are not affected by a subsequent injection of TNFα, whereas when such DTH sites are elicited in mice with a mixed Th1/Th2 (or perhaps Th0) response, the site is exquisitely sensitive to TNFα (HERNANDEZ-PANDO and ROOK 1994a,b). For this factor to be relevant in tuberculosis we need evidence that there is a Th2 component in the immune response of tuberculosis patients.

2.6 The Balance of Th1 to Th2 Cytokines in Tuberculosis: Disease Progression, Decreased Macrophage Function and Loss of Cytotoxic T Lymphocytes

Patients have clear evidence of a mixed Th1+Th2 (or Th0) response, because they have IgE antibody to *M. tuberculosis* (YONG et al. 1989), peripheral T cells that release IL-4 in response to mycobacterial antigens (SURCEL et al. 1994) and activated IL-4 genes in their peripheral blood T cells (Prof. G. Kaplan, New York, personal communication). This is particularly interesting because all the evidence from animal studies leads to the conclusion that immunity requires a Th1 or "type 1" pattern with Th1 cytokines and CD8+ MHC class 1-restricted cells, presumably cytotoxic T lymphocytes (CTL) capable of lysing cells that contain organisms that have moved into the cytoplasmic compartment (COOPER et al. 1993; FLYNN et al. 1992, 1993; McDONOUGH et al. 1993; ORME et al. 1993). Therefore it is possible that an inappropriate imbalance in Th1/Th2 ratio lies behind susceptibility to tuberculosis and leads to progressive tissue-destructive disease. This would be analogous to what is seen in *Leishmania* infection in Balb/c mice, in leprosy, secondary syphilis and HIV infection progressing to AIDS (FITZGERALD 1992; SALGAME et al. 1991; SALK et al. 1993). Even when the Th1 cytokines continue to be produced their efficacy is reduced if excessive levels of Th2 cytokines are also present. This may be due in part to the ability of IL-4 and IL-10 to oppose macrophage-activating effects of IFNγ and TNFα (SIELING et al. 1993). It may also result from the inability of the immune system to develop CD8+ cytotoxic T cell activity when the underlying Th1 response is undergoing suppression from a superimposed Th2 response. For instance mice in this phase of Schistosomiasis cannot develop a CTL response to a virus infection (ACTOR et al. 1993). An obvious example in human disease is the hepatitis B carrier state, inwhich a Th1→Th2 shift results in massive antibody production, eosinophil infiltration, and a lack of protective CTL. (It is also of interest that the livers of these patients, like those of patients with mycobacterial lesions, and schistosome granulomata, are sensitive to cytokine-induced hemorrhagic necrosis).

2.7 Direct Toxicity *Mycobacterium tuberculosis* and Enhanced Susceptibility of Infected Cells to Tumor Necrosis Factor-α

The necrotic component is greater in tuberculosis than in other chronic infections in which a Th1 to Th2 shift can be detected. There are several likely reasons for this. The excessive release of cytokines attributable to mycobacterial components such as lipoarabinomannan (LAM) has already been discussed. It is also clear that *M. tuberculosis* has some inherent toxicity for cells in vitro. This is particularly noticeable with monocytes, which often die if they take up more than about five bacilli. Conversely, monocytes or macrophages can take up very large numbers of *M. avium* without obvious toxic effects. This toxicity of *M. tuberculsosis* may relate to the fact that it escapes from the phagolysosome (MYRVIK et al. 1984). More recent studies show that it appears to bud or extrude from the fused phagolysosome to form a unique vesicle, with the organisms enclosed by a very tightly apposed membrane (McDONOUGH et al. 1993). Over time, fusion fails to occur in these organism-containing vesicles, and much of the multiplication of *M. tuberculosis* occurred within them Between 4 and 7 days, only the virulent strain of *M. tuberculosis*, H37Rv, escaped from these tightly apposed membrane vesicles and entered the cytoplasm (McDonough et al. 1993), while BCG did not appear to be able to do this. *M. tuberculosis* that enters the cytoplasm of macrophages in this way may exert direct toxic effects on the cells, or may be responsible for the increased susceptibility of infected cells to TNFα that is discussed below.

M. tuberculosis is readily taken up by a wide variety of non-macrophage cell types in vitro (Filley et al. 1992; Filley and Rook 1991; Shepard 1958), and such cells are much less susceptible to the toxicity of the organism. This has been shown for several cell lines and for normal human endothelial cells and fibroblasts. The uptake is paradoxical, because unlike *M. leprae, M. tuberculosis* is not seen inside such cells in vivo. One possiblity is that in vivo these cells are killed quickly so that parenchymal cells infected with bacilli are rarely seen in histological sections of tissues. An alternative answer may lie in the observation that cells containing *M. tuberculosis* are rendered exquisitely sensitive to killing by TNFα (Filley et al. 1992; Filley and Rook 1991). Therefore, macrophages infected in vitro may be killed by their own production of TNFα, while non-macrophage cells survive in vitro in the absence of TNFα, but are rapidly killed in vivo since TNFα is probably abundant in lesions, as discussed below. The ability to increase sensitivity of TNFα was prominent in virulent strains of *M. tuberculosis* but weak in H37Ra and virutally absent from *M. avium* and from BCG strains (Filley and Rook 1991). Finally, if CTL are engaged, they would have the capability of lysing parenchymal cells expressing mycobacterial antigens in association with MHC class I.

These effects of *M. tuberculosis* of cell lines (Filley et al. 1992; Filley and Rook 1991) and on macrophages (McDonough et al. 1993) are not shared by BCG and are less pronounced in avirulent strains such as H37Ra, Thus, they may represent the first useful correlates of virulence to be described in vitro.

3 Other Inflammatory Mediators and Immunopathology

It is also possible that some of the necrosis is due to excessive release of oxygen reduction products such as superoxide anion, hydrogen peroxide and hydroxyl radicals, and release of NO. All of these substances are toxic to host cells. We also cannot rule out a role for excessive activity of cytotoxic T lymphocytes. They are clearly part of the protective pathway, but as in all models in which CTL are involved (such as lymphocytic choriomeningitis virus infection in mice), there is a point beyond which the destruction of infected autologous cells becomes counterproductive. At present there is no evidence for or against these views.

4 Mycobacteria and Autoimmunity

There is another type of immunological phenomenon which may contribute to the immunopathology of tuberculosis. Tuberculosis patients have a spectrum of autoantibodies remarkably similar to that seen in rheumatoid arthritis patients (SHOENFELD and ISENBERG 1988). Tlhey also have the striking change in the glycosylation of the IgG heavy chain (ROOK et al. 1994). This change is also characteristic of rheumatioid arthritis, Crohn's disease, and a subset of patients with sarcoidosis (RADEMACHER et al. 1988; ROOK et al. 1993a) and is proving to be a useful and robust marker of immunological recovery in tuberculosis, as explained later (ROOK et al. 1994). Moreover all mycobacterial diseases can also be accompanied by arthropathy (Reviewed and referenced in MORELAND and KOOPMAN 1991; ROOK et al. 1993a). Such findings, together with the known capacity of mycobacterium-containing adjuvants (Freund's complete adjuvant) to facilitate experimental induction of autoimmunity, have reawakened speculation that some "autoimmune"syndromes may be cryptic infections or triggered by past encounters with mycobacterium-like organisms (ROOK et al. 1993a; VAN-EDEN et al. 1988). The other side of the coin is that the immunopathology of tuberculosis itself could involve tissue damage mediated via recognition of autologous antigens such as the heat shock proteins (hsp). In tuberculosis, such autoimmune reactions could contribute to both caseous necrosis and liquefaction, but there is at present no direct evidence for this.

This argument is relevant to the point made later, that cell-mediated protective immunity to slowly growing intracellular bacteria such as mycobacteria is often mediated via recognition of common epitopes rather than of species-specific ones. It has been suggested that by concentrating its repertoire on molecules such as hsps which are enormously conserved throughout all life forms, the immune system maximizes its chances of rapidly recognizing any new pathogen (COHEN and YOUNG 1991). The possibility that bacteria will provoke cross-reactive autoimmunity to similar epitopes in the host's own hsps may be a small price to pay. In fact it may be a mechanism that is actually exploited to facilitate recognition of stressed, putatively infected autologous cells. CTL in humans have been reported to kill macrophages pulsed with the mycobacterial 65 KDa hsp (Ottenhoff et el. 1988). More importantly CTL responding to the mycobacterial hsp65 have also been claimed to kill stressed autologous cells in the absence of mycobacterial antigen, presumably through cross-reactive recognition of the autologous hsp (KOGA et al.1989). If this is true, then recognition of autologous and of microbial hsps may be an important normal part of the immune response, particularly to slow-growing intracellular infections.

5 Th1 and Th2 cells: Selection of Functional T Cell Subsets, "Locking In" vs Disease Progression

If, as suggested earlier, tuberculosis is yet another disease characterized by an in-appropriate balance of Th1 to Th2 cytokines, we must ask why this imbalance occurs and whether we can do anything about it. This is an area in which clinical observation is in disagreement with immunological dogma. Current dogma states that the cytokines released by Th1 cells enhance Th1 activity and inhibit Th2 and vice versa. Therefore Th1 responses should be stable. Why therefore does the Th1→Th2 shift occur? There is no doubt that it does so in several chronic infections such as *Leishmania* infection in Balb/c mice, leprosy, secondary syphilis and HIV infection progressing to AIDS (Bretscher et al. 1992; Fitzgerald 1992; Salgame et al. 1991; Salk et al. 1993). To some extent rising antigen dose can be blamed, but it is more complicated than that. Under some circumstances the Th1 response is stable ("locked in"), but at other times it is readily suppressed and replaced by Th2. For instance the DTH primed by low doses of sheep erythrocytes (SRBCs) is suppressed if a subsequent large antibody-inducing dose is given. If the primary immunization is into a BCG-primed animal, the DTH response is stable and is only transiently suppressed by subsequent high dose challenge (Lagrange and Mackaness 1975). This observation can now be reinterpreted as an example of the preservation of the Th1 component when a mycobacterial adjuvant was used. Similar observations have been made in the murine *Leishmania* system. After an initial low dose exposure, subsequent challenge with a large dose merely boosts the Th1 response. It does not evoke a Th2 response as it would in unimmunized animals or lead to progressive disease. This type of observation has led to the concept that the response to an antigen can become "imprinted" or locked in to Th1 (Bretscher et al. 1992). It seems probable that the response to common bacterial epitopes, like those found in conserved components such as hsps, is locked in to Th1 as a result of prolonged exposure to low doses during childhood. If so, this must be a further influence on the Th1/Th2 decision that is made when a new organism is encountered, and this point is expanded later.

Attention has focussed on these concepts because many workers think that it would be desirable to achieve the locked into Th1 state with the antigens of HIV (Clerici and Shearer 1993; Salk et al. 1993). This arises from speculation that the appearance of HIV seropositivity in individuals exposed to HIV represents a detrimental Th1→Th2 shift, since loss of TH1 cytokine release by peripheral blood lymphocytes precedes loss of TH2 cytokine release, and individuals (homosexuals, hemophiliacs, needle-stick health care workers, and babies born to HIV+ mothers) who remain healthy after many years of exposure to HIV tend to have T cells that secrete IL-2 in response to HIV-derived peptides, but no antibody response (Clerici and Shearer 1993; Meyaard et al. 1993). If this view, currently controversial, proves correct, the mechanisms leading to such switching will be of crucial contemporary importance. Moreover if there is a property of live mycobacteria and of *Leishmania* that is capable of inducing the locked in state, it would be useful to understand the mechnisms involved so that we can exploit them more easily.

5.1 Are Immunopathology and Protection Related to Epitope Specificity of the Helper T Cell Response?

How do these Th1 and Th2 components of the response relate to the specificity of the T cells in tuberculosis? Do different epitopes drive the Th1 and the Th2 components of the response? Do different epitopes drive protection and immunopathology? It has long been the dream of clinical mycobacteriologists to separate these two functions, but most workers have approached the probelm by mapping epitopes recognized by T cells, and this approach has so far proved unhelpful. Meanwhile useful and clinically applicable conclusions can be drawn merely by thinking about some of the things that have been known about tuberculosis for many years.

There has been a tacit assumption that protective epitopes will be species-specific. This assumption seems to be an historical consequence of the early successes of species-specific antibody-mediated vaccines such as the toxoids of tetanus and diptheria. Neutralizing antibodies bind to conformational epitopes that are likely to be parts of species specific toxins or receptor molecules. We know that is not how T cells operate. T cells do not neutralize, they merely recognize small linear peptides, and they are not bacterial taxonomists. During its evolution the immune response is likely to have economized by using recognition of conserved microbial components for protection, just as it uses recognition of conserved adjuvant components to direct response mechanisms and to activate natural immunity. The discovery that hsps are protent immunogens in numerous infections highlighted this point and led to a brief period of publicity for this view (COHEN and YOUNG 1991), as pointed out earlier. In fact conserved epitopes are found in many other molecules such as mycobacterial superoxide dismutase, or the 30 kDa fibronectin-binding proteins.

It is particularly clear in the context of mycobacterial disease that protective responses can be mediated by these common epitopes. Conclusive evidence for this assertion has been available for some years. Most mycobacteriologists have failed to see the implications of the fact that BCG protects against leprosy as well as it does against tuberculosis (BROWN et al. 1966; FINE et al. 1986). This tells us that protection can be mediated via recognition of epitopes that are common to *M. tuberculosis* and *M. leprae* and raises questions about the role of species-specific epitopes. It is also clear that small positive skin-test reactions in non-BCG recipients correlate with protection from both tuberculosis and leprosy (FINE 1993; PALMER and LONG 1966). These reactions are caused by contact with environmental mycobacteria. Similar support comes from animal studies. A "common" protein (the 65 kDa hsp) from *M. leprae*, expressed in murine macrophages, has evoked potent protection against tuberculosis in a murine model (C. Silva and D. Lowrie, personal communication). Moreover *M. vaccae*, an intensely immunogenic member of the fast-growing mycobacterial subgenus, will protect birds against *M. avium*, (R. Cromie, personal communication) and mice against *M. tuberculosis* (J. Watson and R. Prestidge, Auckland, presented at International Leprosy Congress in Orlando, 1993). Since protection clearly can be mediated via the common epitopes it is interesting that tuberculosis patients lose their skin-test responses to "common" epitopes (Kardjito et al. 1986), while retaining necrotic skin-test reactivity to *M. tuberculosis* itself.

5.2 Epitopes That Are Targets for the Immunopathological Response

The last point raises the interesting possibility that, far from being protective, species-specific epitopes are the targets of the necrotizing immunopathology in tuberculosis (the Koch phenomenon). The tissue damaging responses are evoked only by tuberculin. Are the species-specific epitopes the targets of the unwanted Th2 component? We suggest that they are, and Fig. 1 provides a theoretical sequence of events. Prolonged exposure to low levels of environmental mycobacteria primes a Th1 response, and this is probably locked in to Th1 because of the adjuvant properties of mycobacteria discussed earlier. Therefore it is protective (Fine 1993; Palmer and Long 1966). Most individuals (90%–95%) encountering *M. tuberculosis* are protected from the disease by Th1 cells recognizing these common epitopes. If this protection fails the organism proliferates and drives a Th2 response to the species-specific epitopes which the immune system is encountering for the first time. The response to these is not locked in to Th1. We propose that this gives rise to the immunopathological situation outlined above, in which a Th2 component is superimposed on a Th1 response, and TNFα provokes damage (Hernandez-Pando and Rook 1994a). At the same time the macrophage-activating function of Th1 cells will be reduced, as will the ability to generate CD8+ CTL.

The sequence of events seen in the murine model of Schistosomiasis can be represented in the same manner, though in this system the equivalent of the common epitopes are those that are those that are shared by the worm and the ovum, while the equivalent of the species-specific epitopes are the Th2-driving components found only in the ovum (Actor et al. 1993; Grzych et al. 1991).

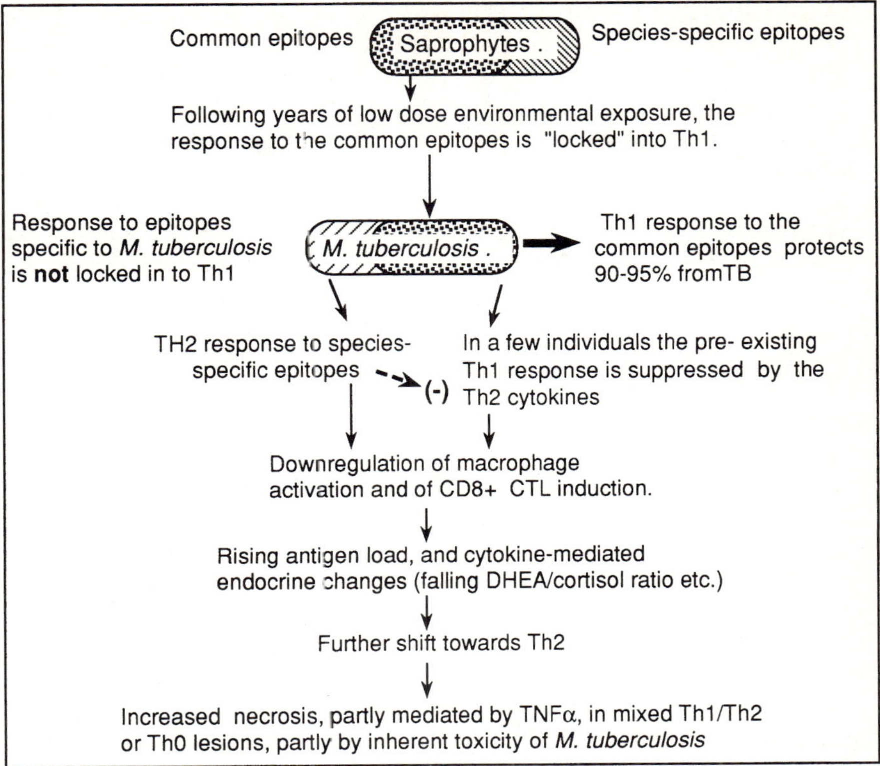

Fig. 1. The possible role of exposure to environmental saprophytes and the different roles of common and of species specific epitopes. The common epitopes, shown as *dots*, are likely to be particularly abundant in conserved components such as heat shock proteins

6 Factors Contributing to the Switch from Th1 to Th2 in Chronic Infections: Endocrine Effects on the Th1/Th2 Balance

Once infection is established, a number of additional factors must play an important role in driving the response further towards Th2. The next sections consider the effects of the interactions between the immune mediators and the endocrine system. In reality some of the mechanisms discussed below, such as formation of calcitriol or dehydroepiandrosterone (DHEA) in draining lymphoid tissue, are also relevant in the early stages of Th1/Th2 selection and of disease development; however, for convenience, all the endocrinological factors are considered together.

6.1 Formation of Calcitriol Mycobacterial Lesions

The macrophages of tuberculosis patients express an active 1α-hydroxylase and rapidly convert 25(OH) cholecalciferol (vitamin D3) to calcitriol (Rook et al. 1986; Rook 1988). Their T cells may also produce this enzyme (Cadranel et al. 1990). Therefore 25(OH) cholecalciferol can be regarded as a prohormone that is converted into its active form in the periphery. This a potent phenomenon, leading occasionally to leakage of calcitriol into the periphery, and to hypercalcaemia, though it has in the past been difficult to understand its role in the disease (Rook 1988). It now seems likely that this is a feedback mechanism that tends to down-regulate Th1 and enhance Th2 responses, because the active vitamin D3 metabolite, $1,25(OH)_2$ cholecalciferol (calcitriol), inhibits production of IFN-γ and IL-2 and increases production of IL-4 and IL-5 (Daynes et al. 1991; Rigby et al. 1987). Intense conversion of 25(OH) D3 to $1,25(OH)_2D3$ (calcitriol) also occurs in the lesions of other chronic T cell-dependent inflammatory disorders such as sarcoidosis (Rook 1988). The enzyme responsible, 1α-hydroxylase, is up-regulated in macrophages exposed to IFNγ or to bacterial components such as LPS. This enzyme is also present in murine lymphoid tissues (G. Rook, M. Hewison and R. Hernandez-Pando, unpublished observations). Calcitriol is best known for its systemic effects on calcium balance, but hypercalcemia, although it can occur, is uncommon in tuberculosis and sarcoidosis (Rook 1988). Therefore calcitriol can act independently in the systemic and lymphoid compartments in which it exerts entirely unrelated functions, with little cross-talk between the two. This is an important fundamental principle which may be relevant to our thinking about the roles of several other hormones. Calcitriol cannot easily be used systemically to obtain this down-regulation of Th1 activity in vivo because at immunologically relevant doses it causes severe hypercalcemia. However new calcitriol analogues such as KH1060 (Leo Pharmaceuticals, Princes Risborough, Bukinghamshire, UK) have increased effects on T cells relative to their effects on calcium balance, so they can be given systemically (J.L. Touraine, Lyons, personal communication). These new analogues rival cyclosporin A in their ability to prolong skin graft survival by inhibiting Th1 activity, while sparing antibody production, and their effects are additive, perhaps synergistic with those of cyclosporin.

One striking new finding may be interpreted as an effect of calcitriol on lymphocyte homing. When calcitriol is administered peripherally with or shortly after antigen it evokes a response which tends to localize to mucosal surfaces (Daynes et al. 1995), resulting in massive IgA titers and mucosal immunity. Thus mucosal immunity can be evoked by merely painting calcitriol onto skin sharing lymphatic drainage with the immunization site. Following exposure to calcitriol, the draining node behaves temporarily like gut-associated lymphoid tissue (Daynes et al. 1995).

6.2 Treatment of Tuberculosis with Calcitriol

These findings emphasize the likely physiological importance of calcitriol formation in tuberculosis. Moreover calcitriol provides a link between the factors causing the Th1→Th2 shift and those causing necrosis. Vitamin D was extensively used during the 1940s as a treatment for chronic skin tuberculosis (lupus vulgaris) (MACRAE 1947). This condition was usually caused by bovine strains and differed from the classical contemporary disease in ways which are unclear. Treatment with calcitriol was sometimes successful because it provoked necrosis and sloughing of the chronic skin lesions (MACRAE 1947). It was also very dangerous because it provoked liquefation of deep lesions and pulmonary foci, resulting in spread via the bronchi, disseminated disease and death (BRINCOURT 1967). These effects are fascinating because they mimic the effects of Koch's own treatment with soluble culture filtrate (ANDERSON 1891). Macrophages incubated with calcitriol have a somewhat increased capacity to inhibit *M. tuberculosis*, but they are also primed for greater release of TNFα (ROOK et al. 1986, 1987b). We suggest that the calcitriol causes a shift towards Th2 accompanied by increased release of TNFα, and that this results in increased necrosis for the reasons outlined earlier.

6.3 Peripheral Conversion of Prohormones and Th1/Th2 Regulation

Apart from 25(OH) cholecalciferol there are several other immunologically important prohormones. One of these, DHEA sulphate (DHEAS), may, like calcitriol, be important in the regulation of the TH1/Th2 balance. DHEAS is the major adrenal steroid in the serum of both men and women. Specific receptors for the free hormone (DHEA) are found in T cells (MEIKLE et al. 1992). A single injection of DHEA before a large dose of dexamethasone can almost completely block the ability of the dexamethasone to cause apoptosis of thymocytes and unresponsiveness of peripheral T cells (BLAUER et al. 1991). This antiglucocorticoid effect is indirectly Th1-enhancing because glucocorticoids suppress Th1 activity and can synergize with Th2 cytokines such as IL-4 (FISCHER and KONIG 1991). Moreover DHEA also directly enhances Th1 T cell activity (DAYNES et al. 1991; SUZUKI et al. 1991). It was found that the balance of Th1 to Th2 cytokines released from normal mouse lymphoid tissue in response to anti-CD3 is related to the DHEA sulphatase activity in each lymphoid compartment (DAYNES et al. 1990). DHEA levels fall to very low levels in old age, and DHEA supplements can correct many of the immunological defects and excessive cytokine release seen in old mice (ARANEO et al. 1993; DAYNES et al. 1993), and cause clear immunological changes in postmenopausal women (CASSON et al. 1993). A particularly striking finding in this study was the increase in NK cell activity, leading to the obvious hypothesis that the Th1-promoting effects are due to enhancement of IL-12 production.

It may therefore be significant that a fall in serum DHEAS heralds the downgrading from HIV+ to AIDS (Rook et al. 1993b; Wisniewski et al. 1993), regarded by many as at least partly attributable to loss of Th1 activity. Tuberculosis patients often have reduced levels of DHEAS (our own unpublished observations). To some extent this occurs in all chronic disease states and its significance is unclear. However DHEAS can be undetectable in the serum of some very severely ill tuberculosis patients and it seems likely that this contributes to the shut down of Th1 function in these individuals.

6.4 Regulation of the Prohormone-Converting Enzymes

We know almost nothing about the regulation of DHEA sulphatase. It is a microsomal enzyme, and it is not clear how DHEAS, which cannot cross membranes, gains access to it (Rose 1982; van Diggeln et al. 1989). Does the enzyme shuttle backwards and farwards to the cell surface? Investigating the regulation of the enzyme has been remarkably difficult. Levels are high in Triton X-100 lysates of the human macrophage lines U937 and MonoMac 6, but absent from murine J774 macrophages, in which we have been unable to induce it (Setloboko and Rook, unpublished observations). When present, the activity of intact cultured cells is less than 1% of that of Triton X-100 lysates. Thus trivial changes in cell viability and permeability cause huge problems for in vitro studies of stimuli that might increase or decrease the activity of intact cells. However the recent discovery of potent inhibitors of the enzyme (Howarth et al. 1994) will facilitate rapid dissection of its immunological role in vivo such that somewhat speculative discussion of the role of DHEA in tuberculosis can now addressed directly.

6.5 Indirect Endocrine Effects on the Adrenal via the Hypothalamic/Pituitary/Adrenal Axis

Another mechanism tending to cause a Th1→Th2 switch in chronic infection is feedback via the endocrine system. It has been clear since the pioneering observations of Besedovsky, Sorkin and colleagues in the 1970s that there is a two-way flow information between the immune system and the hypothalamic/pituitary/adrenal axis (HPA) (Besedovsky and Sorkin 1977) Rats were immunized with sheep erythrocytes, and a peak of antibody-forming cells in the spleen was observed 5–8 days later. However, there was simultaneously a peak in serum cortisol, a trough in thyroxine and a threefold increase in the firing rate of neurones in the ventromedial hypothalamic nuclei. Equally striking was the poor response to a second immunogen given at the time of the cortisol peak. This "antigenic competition" was elimiated by adrenalectomy. The cytokines TNFα, IL-1, and IL-6 have been identified as important mediators of the HPA response to activation of the immune system in vivo (Besedovsky et al. 1991). In general these cytokines act by causing release of conrticotropin-releasing hormone (CRH), leading to ACTH production by the pituitary. However recent studies reveal that there is another as yet unidentified, pathway which can activate the pituitary adrenal axis, because in the chronic phase of adjuvant arthritis, levels of mRNA for CRH and the levels of CRH peptide in hypophyseal portal blood are in fact below normal, but increased production of ACTH and glucocorticoids persists (HARBUZ et al. 1992). Since glucocorticoids enhance Th2 activity and synergize with IL-4, these observations are of fundamental relevance.

An alternative way of activating the pituitary adrenal axis is stress. As little as 5 min of restraint stress causes readily apparent increased expression of mRNA for *c-fos* and CRH in the rat hypothalamus (HARBUZ et al. 1993). Students stressed by their exams show increased titers of antibody of Epstein-Barr virus, implying transient reactivation of the latent infection which is normally controlled by CTLs (reviewed in ZWILLING 1992). This could imply that stress drives a Th1→Th2 switch. A study of the consequences of an intensely stressful training course to which the US Rangers are exposed (Edward W. Bernton, Walter Reed Army Institute of Research, Washington DC, personal communication) showed that thyroid hormone levels fell so dramatically (as in Besedovsky's early studies, in which antigen rather than stress was the stimulus), that the subjects became clinically hypothyroid. Testosterone levels fell to castrate levels. From an immunological point of view it is particularly important to note that the ratio of DHEAS to cortisol fell and that there was loss of DTH responses with relative sparing of humoral immunity. This also suggested that stress drives a shift in the Th1/Th2 balance towards Th2. The stress of chronic tuberculosis is therefore one more factor tending to distort the Th1 response.

6.6 Direct Effects of Cytokines on Adrenal Function

There is a further set of endocrine changes in chronic infectious disease which may contribute to the shift from Th1 to th2 that is often seen in these situations. The changes may be attributable to TNFα and to other cytokines released with it. It begins to appear likely that TNFα plays a part in the fall in DHEAS/cortisol ratio in chronic infections, because in vitro at least, TNFα acts directly on adrenal cells to reduce steroid output (JÄÄTTELÄ et al 1991; STANKOVIC et al. 1994b). In particular, TNFα causes a dose-related reduction of DHEAS production by human fetal adrenals in vitro (STANKOVIC et al. 1994b and Dr. Parker, Birmingham Alabama, personal communication). IL-1 does not have this effect (HARLIN and PARKER 1991), but transforming growth factor-β (TGFβ) does, and like TNFα, it inhibits DHEAS more than cortisol (STANKOVIC et al. 1994a). This may be important because TGFβ is also abundant in tuberculous lesions, (MAEDA et al. 1993) and has been implicated in the suppressive effects of patients' monocytes in vitro (ELLNER 1991).

Is there any evidence for such effects in vivo? For instance does cytokine excess cause reduced adrenal function in tuberculosis? Clincally this is a reasonable suggestion. Other authors have argued that an adrenal deficit can explain "tuberculin shock," a potentially fatal complication sometimes seen after treatment of severe disease with rapidly bactericidal drugs such as rifampicin (SCOTT et al. 1990). It is likely that death of bacilli causes massive release of cytokine-triggering components such as LAM (MORENO et al. 1989). This would indeed be expected to have serious consequences in the absence of an adequate glucocorticoid response from the adrenals. In a minority of cases poor adrenal function may be due to direct infection of the adrenals, but this does not apply to most patients, so a cytokine-mediated effect is possible. Tuberculin shock is most often seen in patients with severe disseminated disease and in patients with protein malnutrition and liver damage. In experimental models liver damage enhances susceptibility to the lethal toxicity of TNFα (FREUDENBERG and GALANOS 1991).

In spite of this suggestive clinical evidence for poor adrenal reserve in tuberculosis, this has remained controversial because serum cortisol levels measured in the moring are usually normal or even slightly raised. The synacthen test has also not resolved the debate, because the percentage of patients deemed to be abnormal depends on the criteria used. This point is discussed in detail by POST et al. (1994). However this dilemma may also be explained by the fact that during severe illness adrenocortical metabolism changes so as to maintain serum cortisol levels (DRUCKER and MCLAUGHLIN 1986; SEMPLE et al. 1987), and in order to get more precise information about adrenal steroid output it is better to study 24 h urine collections. Using HPLC and mass spectrometry to indentify and quantify steroid metabolites in such urine samples we have found that tuberculosis patients' adrenals do indeed have strikingly reduced sterioid output of both glucocorticoids and androgens (G. Rook, H. Honour, R. Davidson and R. Wilkinson, in press). Output of essentially all steroid metabolites was reduced by about 50%.

If TNFα acting directly on the adrenal is partly responsible for these changes in adrenal function, as in vitro studies suggest, the previously reported *M. tuberculosis*-derived TNFα-enhancing activity could also play a role. It is possible that it enhances the adverse effect of TNFα on adrenal function (FILLEY et al. 1992). Some such mechanism might explain the remarkable enlargement of the adrenals seen as an early event in tuberculosis, even in the absence of direct infection of the gland. This has been documented in humans by using CT scans and in the rabbits studied by Max Lurie by weighing the adrenals at post-mortem (LURIE 1964). The degree of englargement was different in rabbit strains with different susceptibility to disease. It is controversial whether this enlargement can be attributed entirely to ACTH.

6.7 Sick Euthyroid Syndrome

Patients with chronic infections such as tuberculosis can also have the "sick euthyroid syndrome," in which there is reduced peripheral conversion of thyroxine (T4) into triiodothyronine (T3) with increased reverse T3 (rT3). Recently

TNFα has been implicated in this change in monodeiodinase activity, because infusion of TNFα into volunteers reproduces the syndrome (VAN-DER-POLL et al. 1990). Subsequent studies suggest that the effect was indirect, due to the IL-6 release triggered by the TNFα, because levels of IL-6 rather than of TNFα correlate with the changes in T3 (BOELEN et al. 1993; HASHIMOTO et al. 1994; REINCKE et al. 1993). It is not clear what effect these changes have on T cell function or immunopathology.

7 Can the Balance Between Protection and Immunopathology Be Clinically Manipulated? Immunotherapy or "Therapeutic Vaccination"

As pointed out in the Introduction, an important aim for the mycobacteriologist is to find a way to switch the response from tissue-damaging mode to protection during the first 2–3 weeks of chemotherapy, because conventional chemotherapy fails to do this, and in the absence of an efficient bactericidal protective response the treatment must be continued for 6–8 months. Tuberculin or BCG cannot be used for immunotherapy because they evoke the Koch phenomenon, as Robert Koch discovered to his cost (ANDERSON 1891). Logically this must be due to a response to species-specific epitopes as outlined above. Clearly it is not realistic to think in terms of engineering a BCG or *M. tuberculosis* variant that contains no such epitopes. At present the only way to achieve a preparation with suitable adjuvant properties, containing a broad spectrum of common, but not of species-specific antigens of *M. tuberculosis*, is to use a related mycobacterial species. Good preliminary results of immunotherapy have been achieved with an autoclaved member of the fast-growing subgenus *M. vaccae*. Single or repeated intradermal injections of *M. vaccae* have been used for this purpose because it is intensely immunogenic when autoclaved and can be regarded as a preparation of common mycobacterial epitopes in a potent Th1 mycobacterial adjuvant (HERNANDEZ-PANDO and ROOK 1994a). The results have been encouraging (COHEN 1994; ETEMADI et al. 1992; ONYEBUJOH and ROOK 1991; STANFORD and GRANGE 1994; STANFORD et al. 1993), and the preparation is now entering phase 3 clinical trials. A further useful property of this organism is the readiness with which foreign genes can be expressed in it (GARBE et al. 1994), so it may be possible to adapt it for treatment or vaccination of other diseases in which a Th1 response is required. A striking piece of evidence that immunotherapy for tuberculosis modulates mechanisms at the heart of the complex networks of cytokine- and endocrine-mediated events described above is the immediate fall in the percentage agalactosyl IgG that follows immunotherapy (ROOK et al. 1994). In contrast, in tuberculosis patients receiving antimicrobial therapy alone, there is a further increase in the percentage of abnormally glycosylated IgG for at least 2 months before a very slow decline in the level is seen (Rook et al. 1994).

In conclusion, surprisingly few mycobacteriologists have worked on the nature of the immunopathology of tuberculosis, yet we consider that it holds the key to novel types of treatment.

References

Actor JK, Shirai M, Kullberg MC, Buller RM, Sher A, Berzofsky JA (1993) Helminth infection results in decreased virus-specific CD8+ cytotxic T-cell and Th1 cytokine responses as well as delayed virus clearance. Proc Natl Acad Sci USA 90: 948–952

Al Attiyah R, Moreno C, Rook GAW (1992a)TNF-α mediated tissue damage in mouse foot-pads primed with mycobacterial preparations. Res Immunol 143: 601–610

Al Attiyah R, Rosen H, Rook GAW (1992b) A model for the investigation of factors influencing haemorrhagic necrosis mediated by tumour necrosis factor in tissue sites primed with mycobacterial antigen preparations. Clin Exp Immunol 88: 537–542

Anderson MC (1981) On Koch's treatment. Lancet i: 651–652

Araneo BA, Woods ML, Daynes RA (1993) Reversal of the immunosenescent phenotype by dehydroepiandrosterone: hormone treatment provides an adjuvant effect on the immunization of aged mice with recombinant hepatitis B surface antigen. J Infect Dis 167: 830–846

Barnes PF, Fong SJ, Brennan PJ, Twomey PE, Mazumder A, Modlin RL (1990) Local production of tumor necrosis factor and IFN-γ in tuberculous pleuritis. J Immunol 145: 149–154

Besedovsky H, Sorkin E (1977) Network of immune-neuroendocrine interactions. Clin Exp Immunol 27: 1–12

Besedovsky HO, del-Rey A, Klusman I, Furukawa H, Monge-Arditi G, Kabiersch A (1991) Cytokines as modulators of the hypothalamus-pituitary-adrenal axis. J Steroid Biochem Mol Biol 40: 613–618

Blauer KL, Poth M, Rogers WM, Bernton EW (1991) Dehydroepiandrosterone antagonises the suppressive effects of dexamethasone on lymphocyte proliferation. Endocrinology 129: 3174–3179

Boelen A, Platvoet-Ter-Schiphorst MC, Wiersinga WM (1993) Association between serum interleukin-6 and serum 3,5,3'-triiodothyronine in nonthyroidal illness. J Clin Endocrinol Metab 77: 1695–1699

Bordet P (1931) Contribution l'tude de l'allergie. C R Soc Biol 107: 622–623

Bretscher PA, Wie G, Menon JN, Bielefeldt-Ohmann H (1992) Establishment of stable, cell-mediated immunit that makes "susceptible" mice resistant to Leishmania major. Science 257: 539–542

Brincourt J (1967) Le calcifrol a-t-il une action liqufiante sur le caseum? Poumon Coeur 23: 841–851

Brown JAK, Stone MM, Sutherland I (1966) BCG vaccination of children against leprosy in Uganda. Br Med J 1: 7–14

Cadranel J, Garabedian M, Milleron B, Guillozo H, Akoun G, Hance AJ (1990) 1,25(OH)2 D3 production by T lymphocytes and alveolar macrophages recovered by lavage from normocalcaemic patients with tuberculosis. J Clin Invest 85: 1588–1593

Carswell EA, Old LJ, Kassel RL, Green S, Fiore W, Williamson B (1975) An endotoxin-induced serum factor that causes necrosis of tumours. Proc Natl Acad Sci USA 72: 3666–3670

Casson PR, Andersen RN, Herrod HG, Stentz FB, Straughn AB, Abraham GE, Buster GE (1993) Oral dehydroepiandrosterone in physiologic doses modulates immune function in postmenopausal women. Am J Obstet Gyneco 169: 1536–1539

Chan J, Xing Y, Magliozzo RS, Bloom BR (1992) Killing of virulent Mycobacterium tuberculosis by reactive nitrogen intermediates produced by activated murine macrophages. J Exp Med 175: 1111–1122

Clerici M, Shearer GM (1993) A Th1 to Th2 switch is a critical step in the etiology of HIV infection. Immunol Today 14: 107–111

Cohen IR, Young DB (1991) Autoimmunity, microbial immunity and the immunological homunculus. Immunol Today 12: 105–110

Cohen J (1994) Vaccines get a new twist. Science 264: 503–505

Cooper AM, Dalton DK, Stewart TA, Griffin JP, Russell DG, Orme IM (1993) Disseminated tuberculosis in interferon-γ gene-disrupted mice. J Exp Med 178: 2243–2247

Daynes RA, Araneo BA, Hennebold J, Enioutina E, Mu HH (1995) steroids as essential regulators of the immune response. J Invest Dermatol 105: 145–195

Daynes RA, Araneo BA, Dowell TA, Huang K, Dudley D (1990) Regulation of murine lymphokine production in vivo. III. The lymphoid tissue microenvironment exerts regulatory influences over T helper cell function. J Exp Med 171: 979–996

Daynes RA, Meikle AW, Araneo BA (1991) Locally active steroid hormones may facilitate compartmentalization of immunity by regulating the types of lymphokines prouduced by helper T cell. Res Immunol 142: 40–45

Daynes RA, Araneo BA, Ershler WB, Maloney C, Li G-Z, Ryu S-Y (1993) Altered regulation of IL-6 proudction with normal aging; possible linkage to the age-associated decline in dehydroepiandrosterone and its sulphated derivative. J Immunol 150: 5219–5230

Drucker D, McLaughlin J (1986) Adrenocortical dysfunction in acute medical illness. Crit Care Med 14: 789–791

Ellner JJ (1991) Regulation of the human cellular immune response to Mycobacterium tuberculosis. The mechanism of selective depression of the response to PPD. Bull Int Union Tuberc Lung Dis 66: 129–132

Etemadi A, Farid R, Stanford JL (1992) Immunotherapy for drug-resistant tuberculosis. Lancet 340: 1360–1361

Ferluga J, Doenhoff MJ, Allison AC (1979) Increased hepatotoxicity of bacterial lipopolysaccharide in mice infected with Schistosoma mansoni. Parasite Immunol 1: 289–294

Filley EA, Rook GA (1991) Effect of mycobacteria on sensitivity to the cytotoxic effects of tumor necrosis factor. Infect Immun 59: 2567–2572

Fine PEM (1993) Immunities in and to tuberculosis: implications for pathogenesis and vaccination. In: McAdam KPWJ, Porter JDH (eds) Tuberculosis: back to the future. Proceedings of the London School of Hygience and Tropical Medicine, 3rd annual public health forum. Wiley, Chichester, pp 54–78

Fine PEM, Ponnighaus JM, Maine N, Clarkson JA, Bliss L (1986) The protective efficacy of BCG against leprosy in Northern Malawi. Lancet ii: 499–502

Fischer A, Konig W (1991) Influence of cytokines and cellular interactions on the gulcocorticoid-induced Ig (E,G, A, M) synthesis of peripheral blood monnuclear cells. Immunology 74: 228–233

Fitzgerald TJ (1992) The Th1/Th2 switch in syphilitic infection: is it detrimental? Infect Immun 60: 3475–3479

Flynn JL, Goldstein MM, Triebold KJ, Koller B, Bloom BR (1992) Major histocompatibility complex class I-restricted T cells are required for resistance to Mycobacterium tuberculosis infection. Proc Natl Acad Sci USA 89: 12013–12017

Flynn JL, Chan J, Triebold KJ, Dalton DK, Stewart TA, Bloom BR (1993) An essential role for interferor gamma in resistance to Mycobacterium tuberculosis infection. J Exp Med 178: 2249–2254

Foley N, Lambert C, McNicol M, Johnson NMI, Rook GAW (1990) An ihibitor of the toxicity of tumour necrosis factor in the serum of patients with sarcoidosis, tuberculosis and Crohn's disease. Clin Exp Immunol 80: 395–399

Freudenberg MA, Galanos C (1991) Tumor necrosis factor-α mediates lethal activity of killed gram-negative and gram-positive bacteria in D-galactosamine-treated mice. Infect Immun 59: 2110–2115

Garbe RT, Barathi J, Barnini S, Zhang Y, Abou-Zeid C, Tang D, Mukerjee R, Young DB (1994) Transformation of mycobacterial species using hygromycin resistance as selectable marker. Microbiology 140: 133–138

Grange JM (1992) The mystery of the mycobacterial persister. Tubercle Lung Dis 73: 249–251

Grzych JM, Pearce E, Cheever A, Caulada ZA, Caspar P, Henry S, Lewis F, Sher A (1991) Egg deposition is the stimulus for the the production of Th2 cytokines in murine schistosomiasis mansoni. J Immuno 146: 1322–1340

Harbuz MS, Rees RG, Eckland D, Jessop DS, Brewerton D, Lightman SL (1992) Paradoxical responses of hypothalamic CRF mRNA and CRF-41 peptick and adenohypophyseal POMC mRNA during chronic infammatory stress. Endocrinology 130: 1394–1400

Harbuz MS, Chalmers J, De Sousa L, Lightman SL (1993) Stress-induced activation of CRF and c-fos mRNAs in the paraenticular necleus are not affected by serotonin depletion. Brain Res 609: 169–173

Harlin CA, Parker CR (1991) Investigation of the effect of interleukin-1β on steroidogenesis in the human fetal adrenal gland. Steroids 56: 72–76

Hashimoto H, Igarashi N, Yachie A, Miyawaki T, Sato T (1994) The relationship between serum levels of interleukin-6 and thyroid hormone in children with acute respiratory infection. J Clin Endocrinol Metab 78: 288–291

Hernandez-Pando R, Rook GAW (1994a) The role of TNFα in T cell-mediated inflammation depends on the Th1/Th2 cytokine balance. Immunology 82: 591–595

Hernandez-Pando R, Rook GAW (1994b) T cell helper types and endocrines in the regulation of tissue-damaging mechanisms in tuberculosis. Immunobiology (in press)

Hirsch CS, Ellner JJ, Russell DG, Rich EA (1994) Complement receptor mediated uptake and tumour necrosis factor α-mediated growth inhibition of Mycobacterium tuberculosis by human alveolar macrophages. J Immuno 152: 743–753

Howarth NM, Purohi A, Reed MJ, Potter BVL (1994) Estrone sulphamates: potent inhibitors of estrone sulphatase with therapeutic potential. J Med Chem 37: 219–221

Jäättelä M, Ilvesmäki V, Voutilainen R, Stenman UH, Saksela E (1991) Tumor necrosis factor as a potent inhibitor of adrenocorticotropin-induced cortisol production and sterioidogenci P450 enzyme gene expression in cultured human fetal adrenal cells. Endocrinology 128: 623–629

Kardjito T, Beck JS, Grange JM, Stanford JL (1986) A comparison of the responsiveness to four new tuberculins among Indonesian patients with pulmonary tuberculosis and healthy subjects. Eur J Respir Dis 69: 142–145

Kindler V, Sappino AP, Grau GE, Piguet PF, Vassalli P (1989) The inducing role of tumor necrosis factor in the development of bactericidal granulomas during BCG infection. Cell 56: 731–740

Koch R (1891) Fortsetzung über ein Heilmittel gegen Tuberculose. Dtsch Med Wochenschr 17: 101–102

Koga T, Wand-Wurttenberger A, DeBruyn J, Munk ME, Schoel B, Kaufmann SH (1989) T cells against a bacterial heat shock protein recognise stressed macrophages. Science 245: 1112–1115

Lagrange PH, Mackaness GB (1975) A stable form of delayed hypersensitivity. J Exp Med 141: 82–96

Lurie MB (1964) Resistance to tuberculosis; experimental studies in native and acquired defensive mechanisms. Harvard University Press, Cambridge, Massachussetts

Macrae DE (1947) Calciferol treatment of lupus vulgaris. Br Med J 59: 333–338

Maeda J, Ueki N, Ohkawa T, Iwahashi N, Nakano T, Higashino K (1993) Local production and localization of transforming growth factor-β in tuberculous pleurisy. Clin Exp Immunol 92: 32–38

McDonough KA, Kress Y, Bloom BR (1993) Pathogenesis of tuberculosis: interaction of Mycobacterium tuberculosis with macrophages. Infect Immun 61: 2763–2773

Meikle AW, Dorchuck RW, Araneo BA, Stingham JD, Evans TG, Spruance SL, Daynes RA (1992) The presence of a dehydroepiandrosterone-specific receptor-binding complex in murine T cells. J Steroid Biochem Mol Biol 42: 293–304

Meyaard L, Otto SA, de Jong R, Miedema F (1993) Preferential outgrowth of TH2 cells after HIV infection. Proceedings of the IXth international conference on AIDS. Berlin, 7–11 June, abstract number WS-A16-3: 104

Moreira AI, Sampaio EP, Zmuidzinas A, Frindt P, Smith KA, Kaplan G (1993) Thalidomide exerts ist inhibitory action on tumor necrosis factor-α by enhancing mRNA degradation. J Exp Med 177: 1675–1680

Moreland LW, Koopman WJ (1991) Infection as a cause of arthritis. Curr Opin Rheumato 3: 639–649

Moreno C, Taverne J, Mehlert A, Bate CA, Brealey RJ, Meager A, Rook GAW, Playfair JHL (1989) Lipoarabinomannan from Mycobacterium tuberculosis induces the production of tumour necrosis factor from human and murine macrophages. Clin Exp Immunol 76: 240–245

Myrvik QN, Leake ES, Wright MJ (1984) Disruption of phagosomal membrances of normal alveolar macrophages by the H37Rv strain of Mycobacterium tuberculosis. A correlate of virulence. Am Rev Respir Dis 129: 322–328

Nagao S, Tanaka A (1985) Necrotic inflammatory reaction induced by muramyl dipeptide in guinea-pigs sensititzed by tubercle bacilli. J Exp Med 162: 401–412

Onyebujoh P, Rook GAW (1991) Letter. Lancet 338: 1534

Orme I, Flynn JL, Bloom BR (1993) The role of CD8+ T cells in immunity to tuberculosis. Trends Microbiol 1: 77–78

Ottenhoff TH, Ab BK, van-Embden JD, Thole JE, Kiessling R (1988) The recombinant 65kDa heat shock protein Mycobacterium bovis Bacillus Calmette-Guerin/M. tuberculosi is a target molecule for CD4+ cytotoxic T lymphocytes that lyse human moncytes. J Exp Med 168: 1947–1952

Palmer CE, Long MW (1966) Effects of infection with atypical mycobacteria on BCG and tuberculosis. Am Rev Respir Dis 94: 553–568

Post FA, Soule SG, Willcox PA, Levitt NA (1994) The spectrum of endocrine dysfunction in active pulmonary tuberculosis. Clin Endocrinol (Oxf) 40: 367–371

Rademacher T, Parekh RB, Dwek RA, Isenberg D, Rook GAW, Axford JS, Roitt I (1988) The role of IgG glycoforms in the pathogenesis of rheumatoid arthritis. Springer, Semin Immunophathol 10: 231–249

Reincke M, Allolio B, Petzke F, Heppner C, Mbulamberi D, Vollmer D, Winkelmann W, Chrousos GP (1993) Thyroid dysfunction in African trypanosomiasis: a possible role for inflammaotry cytokine. Clin Endocrinol (Oxf) 39: 455–461

Rigby WF, Denome S, Fanger MW (1987) Regulation of lymphokine production and human T lymphocyte activation by 1,25-dihydroxyvitamin D3. Specific inhibition at the level of messenger RNA. J Clin Invest 79: 1659–1664

Rook GA, Steele J, Fraher L, Barker S, Karmali R, O'Riordan J, Stanford J (1986) Vitamin D3, gamma interferon, and control of proliferation of Mycobacterium tuberculosis by human monocytes. Immunology 57: 159–163

Rook GAW (1988) The role of vitamin D in tuberculosi (editorial). Am Rev Respir Dis 138: 768–770

Rook GAW, Al Attiyah R (1991) Cytokines and the Koch phenomenon. Tubercle 72: 13–20

Rook GAW, Taverne J, Leveton C, Steele J (1987a) The role of γ-interferon, vitamin D3 metabolites and tumour necrosis factor in the pathogenesis of tuberculosis. Immunology 62: 229–234

Rook GAW, Taverne J, Steele J, Altes C, Stanford JL (1987b) Interferon-γ, cholecalciferol metabolites, and the regulation of antimycobacterial and immunopathological mechanisms in human and murine macrophages. Bull Int Union Against Tuberc 62: 41–43

Rook GAW, Lydyard PM, Stanford JL (1993a) A reappraisal of the evidence that rheumatoid arthritis, and several other idiopathic diseases, are slow bacterial infections. Ann Rheum Dis 52 [Suppl]: S30–S38

Rook GAW, Onyebujoh P, Stanford JL (1993b) TH1→TH2 switch and loss of CD4 cells is chronic infections; an immuno-endocrinological hypothesis not exclusive to HIV. Immunol Today 14: 568–569

Rook GAW, Onyebujoh P, Wilkins E, Ho Minh Ly, Al Attiyah R, Bahr GM, Corrah T, Hernandez H, Stanford JL (1994) A longitudinal study of % agalactosys IgG in tuberculosis patients receiving chemotherapy, with or without immunotherapy. Immunology 81: 149–154

Rose FA (1982) The mammalian sulphatases and placental sulphatase deficiency in man. J Inherited Metab Dis 5: 145

Rothstein J, Schreiber H (1988) Synergy between tumor necrosis factor and bacterial products cuases hemorrhagic necrosis and lethal shock in normal mice. Proc Natl Acad Sci USA 85: 607–611

Salgame PR, Abrams JS, Clayberger C, Goldstein H, Modlin RL, Convit J, Bloom BR (1991) Differing lymphokine profiles of functional subsets of human CD4 and CD8 T cell clones. Science 254: 279–282

Salk J, bretscher PA, Salk PL, Clerici M, Shearer GM (1993) A strategy for prophylactic vaccination against HIV. Science 260: 1270–1272

Scoot GM, Murphy PG, Gemidjioglu ME (1990) Predicting deterioration of treated tuberculosis by corticosteroid reserve and C-reactive protein. J Infect 21: 61–69

Semple CG, Gray CE, Beastall GH (1987) Male hypogonadism – a non-specific consequence of illness. J Med 64: 601–607

Shands JW, Senterfitt VC (1972) Endotoxin-induced hepatic damage in BCG-infected mice. Am J Pathol 67: 23–40

Shepard CC (1958) A comparison of the growth of selected mycobacteria in HeLa, monkey kidney, and human amnion cells in tissue culture. J Exp Med 107: 237–246

Shoenfeld Y, Isenberg DA (1988) Myccobacteria and autoimmunity. Immunol Today 9: 178–182

Shwartzman G (1937) Phenomenon of local tissue reactivity and its immunological, pathological, and clinical significance. Hoeber, New York, p 461

Siegele DA, Kolter R (1992) Life after log. J Bacteriol 174: 345–348

Sieling PA, Abrams JS, Yamamura M, Salgame P, Bloom BR, Rea TH, Modlin RL (1993) Immunosuppressive roles for IL-10 and Il-4 in human infection. In vitro modulation of T cell responses in leprosy. J Immunol 150: 5501–5510

Silva CI, Faccioli LH (1988) Tumor necrosis factor (cachectin) mediates induction of cachexia by cord factor from mycobacteria. Infect Immun 56: 3067–3071

Stanford JL, Grange JM (1994) The promise of immunotherapy for tuberculosis. Respir Med 88: 3–7

Stanford JL, Onyebujoh PC, Rook GAW, Grange JM, Pozniak A (1993) Old plague, new plague and a treatment for both? AIDS 7: 1275–1277

Stankovic AK, Dion LD, Parker CR (1994a) Effects of TGFβ on human fetal adrenal steroid production. Mol Cell Endocrinol 99: 145–151

Stankovic AK, Falany CN, Grizzle WE, Parker CR (!994b) Effects of tumour necrosis factor α on growth and the steroid biosynthetic pathway of human foetal adrenal cells. Proceedings of the 41st annual meeting of the Society for Gynecologic Investigation, abstract, p 236

Stuehr DJ, HJ, Kwon NS, Weise MF, Nathan CF (1991) Purification and characterization of the cytokine-induced macrophage nitric oxide synthase: an FAD- and FMN-containing flavoprotein. Proc Natl Acad Sci USA 88: 7773–7777

Surcel HM, Troye-Blomberg M, Paulie S, Andersson G, Moreno C, pasval G, Ivanyi J (1994) Th1/Th2 profiles in tuberculosis based on proliferation and cytokine response of peripheral blood lymphocytes to mycobacterial antigens. Immunology 81: 171–176

Suzuki T, Suzuki N, Daynes RA, Engleman EG (1991) Dehydroepiandrosterone enhances IL2 production and cytotoxic effector function of human T cells. Clin Immunol Immunopatho 61: 202–211

Takashima T, Ueta C, Tsuyuguchi I, Kishimoto S (1990) Production of tumor necrosis factor alpha by monocytes from patients wilth pulmonary tuberculosis. Infect Immun 58: 3286–3292

Valone SE, Rich EA, Wallis RS, Ellner JJ (1988) Expression of Tumour necrosis factor in vitro by human monnuclear phagocytes timulated with whole Mycobacterium bovis BCG and mycobacterial antigens. Infect Immun 56: 3313–3315

van Diggeln OP, Konstantinidoe AE, Bousema NT, Boer M, Bakx T, Jöbsis AC (1989) A fluorimetric assay of steroid sulphatase in leukocytes: evidence for two genetically different enzymes with arysulphatase C activity. J Clin Inherited Metab Dis 12: 273–280

van-der-Poll T, Romiji JA, Wiersinga WM, Sauerwein HP (1990) Tumor necrosis factor: a putative mediator of the sick euthyroid syndrome in man. J Clin Endocrinol Mestab 71: 1567–1572

van-Eden W, Thole JER, van-der-Zee R, Noordzij A, van-Embden JDA, Hensen EJ, Cohen IR (1988) Cloning of the mycobacterial epitope recognised by T lymphocytes in adjuvant arthritis. Nature 331: 171–173

Wilson GS, Schwabacher H, Maier I (1940) The effect of the desensitisation of tuberculous guinea-pigs. J Pathol Bacteriol 50: 89–109

Winsniewski TL, Hilton CW, Morse EV, Svec F (1993) The relationship of serum DHEA-S and cortisol levels to measures of immune function in human immunodefinciency virus-related illness. Am J Med Sci 305: 79–83

Yong AJ, Grange JM, Tee RD, Beck JS, Botamley GH, Kemeny DM, Kardjito T (!989) Total and anti-mycobacterial IgE levels in serum from patients with tuberculosis and leprosy. Tubercle 70: 273–279

Zwilling BS (1992) Stress affects disease outcomes. ASM News 58: 23–25

BCG Vaccination in the Control of Tuberculosis

R.E. Huebner

1 The History of BCG

The history of the bacillus of Calmette and Guérin (BCG) begins in 1908 when Calmette and Guérin began their work with *Lait Nocard*, a virulent bovine strain of *Mycobacterium tuberculosis* isolated by Nocard from a cow that had tuberculous mastitis. Calmette and Guérin subcultured the organism every 3 weeks on a glycerinated beef-bile-potato medium. In 1921, after 13 years and 230 subcultures, BCG had lost its virulence for animals (Grange et al. 1983; Gardner 1932; Crispen 1989). The first child immunized with BCG in July 1921 was a newborn whose grandmother had tuberculosis. Since then more than 1 billion children in more than 182 countries throughout the world have been vaccinated with BCG.

Division of Tuberculosis Elimination, Centers for Disease Control and Prevention, 1600 Clifton Road, Mailstop E10, Atlanta, 30333, USA

The original strain of BCG, maintained at the Pasteur Institute in Lille, France, produced hundreds of "daughter" strains and is the progenitor of the most commonly used vaccines, the Copenhagen strain, originating in 1931 as the 423rd transfer; the Tokyo strain, sent from France as a seed culture in 1925; and the Glaxo strain, derived from the 1077th transfer of the Copenhagen strain (OSBORN 1983).

Early BCG strains were maintained as freshly prepared cultures through serial passage on potato-bile or potato-Sauton media. Such repeated subculturing resulted in BCG strains that differed markedly from each other. Heterogeneous in vitro characteristics such as colony morphology (spreading vs nonspreading), viability on culture medium (GHEORGHIU and LAGRANGE 1983), biochemical composition and activity (MINNIKIN et al. 1984; ABOU-ZEID et al. 1986; INTERNATIONAL UNION AGAINST TUBERCULOSIS 1978), drug resistance (HESSELBERG 1972), immunogenicity in animals and humans (GHEORGHIU and LAGRANGE 1983; GRANGE and GIBSON 1986; NYBOE and BUNCH-CHRISTENSEN 1966) and virulence in animals (JESPERSEN 1971) have been found among currently available vaccines. All BCG strains are resistant to pyrazinamide. In 1950, the WORLD HEALTH ORGANIZATION (WHO), in an effort to limit the genetic variability among strains of BCG, issued the first in a series of recommendations for BCG production. Since 1960 the availability of lyophilized vaccine has allowed laboratories worldwide to produce seed lots of bacteria which are used to inoculate working vaccine cultures. The homogeneity of the vaccine strain can be maintained by the use of organisms that are no more than four generations from the primary seed lot (WHO TECHNICAL REPORT SERIES 1977; WHO EXPERT COMMITTEE ON BIOLOGICAL STANDARDIZATION 1985). In 1982 the responsibility for the international quality control of BCG vaccines was given to the Statens Seruminstitut in Copenhagen, Denmark. The Statens Seruminstitut provides training in vaccine production, distributes reference and seed lots, and oversees the standardization of BCG strains. Three parent strains (Glaxo, Tokyo, and Pasteur) now account for more than 90% of the vaccines used worldwide (MILSTIEN and GIBSON 1990).

2 Mechanism of Immunity

While no in vitro test has been found to correlate with protection in humans, the dose of BCG and the proportion of viable bacilli to nonviable bacilli in the preparation are thought to influence vaccine efficacy (WHO TECHNICAL REPORT SERIES 1977). An estimated 19 683 vaccine test systems, using different animal models and varying routes of vaccination and challenge, are available for evaluating BCG and the events leading to protective immunity. Although rabbits immunized with BCG are protected against challenge with virulent *M. bovis*, the passive transfer of sera from such animals does not protect *M. bovis*-naive rabbits (Smith 1985).The role of cell-mediated immunity in protection against tuberculosis was demonstrated by experiments in which decreased numbers of bacilli were recovered from the lungs and spleens of *M. tuberculosis*-challenged mice given T lymphocytes from BCG-vaccinated animals (Lefford 1977).

BCG vaccine appears to be able to elicit protective cell-mediated immune responses in the absence of progressive disease. What is the specific mechanism of protective immunity conferred by BCG? Levy et al. (1961) demonstrated that BCG vaccination of mice did not prevent infection or the establishment of primary foci. Smith et al. (1970) found that the number of organisms recovered from the lung, spleen, and lymph nodes of vaccinated guinea pigs was not significantly different from those isolated from unvaccinated animals until 14 days post-challenge. Vaccinated animals showed fewer organisms in the spleen and no dissemination of bacilli to lobes of the lungs not infected at the time of challenge. Thus, vaccination with BCG does not protect against infection, but rather against uncontrolled replication and dissemination of *M. tuberculosis* from the primary foci to other parts of the lung and body.

3 Vaccine Administration

BCG vaccine is most commonly administered by the intradermal method with 0.1 ml of the vaccine delivered into the upper layers of the skin through a 27-gauge syringe. Children younger than 30 days old should receive one half the usual dose. If the indications for vaccination persist, they should receive a full dose of the vaccine after they are one year of age if they have a negative skin test reaction to 5 tuberculin units of PPD tuberculin.

Percutaneous scarification or multiple puncture can also be used for administration of BCG. This technique involves placing several drops of the vaccine on the arm and then introducing the vaccine into the skin through multiple small needles on the puncture device. Though simpler and faster than the intradermal method, this procedure requires a stronger concentration of vaccine. Furthermore, because of the nature of the technique, it is impossible to determine the actual dose of the vaccine administered (Rosenthal 1980).

Normal reactions to the vaccine are characterized by a host immune response and eventual scar formation. With the intradermal method, an indurated papule about 5 mm in diameter forms within 2–3 weeks of BCG administration. A pustule develops by 6–8 weeks which ulcerates and heals in 3 months, leaving a scar at the vaccination site (Myint et al. 1985). With the multiple puncture technique, 2–3 mm papules develop within 2 weeks and begin to subside by 6 weeks after vaccination. After 3–4 months the lesions disappear, leaving only discoloration of the skin or slight pitting (Rosenthal 1980).

4 Sensitivity to Tuberculin After Vaccination

Tuberculin reactivity develops after vaccination with BCG. Sensitivity appears as early as 2 weeks after vaccination; maximal responses develop after 6–12 weeks (STEWART 1968). The size of the reaction depends on a number of factors related both to the vaccine and the host. These factors include the number of viable bacilli in the vaccine (EDWARDS and GELTING 1950; BUNCH-CHRISTENSEN 1977; ASHLEY and SIEBENMAN 1967; LEHMANN et al. 1979), familial genetics (PALMER and MEYER 1951), and the age (MANERIKAR et al. 1976; DAWODU 1985; KATHIPARI et al. 1982) and the nutritional status at the time of vaccination (HEYWORTH 1977).

Postvaccination tuberculin sensitivity in BCG-vaccinated children ranges in size from 3 mm to 19 mm; this reactivity wanes with time and is unlikely to persist beyond 10 years post-vaccination (HORWITZ and BUNCH-CHRISTENSEN 1972; KARALLIEDDE et al. 1987; COMSTOCK et al. 1971; ROSENTHAL 1980). Repeated tuberculin skin testing may prolong sensitivity in BCG-vaccinated persons (GULD et al. 1968; YOUNG and MIRDAD 1992). Tuberculin reactivity that has waned may also be increased or boosted by a subsequent tuberculin skin test given 1 week to 1 year after the initial test. After repeated skin testing an increase of 6 mm or greater induration was noted in 13% of children attending day care centers in South Africa (FRIEDLAND 1990). Finally, revaccination with BCG also can increase the size of tuberculin reactions (SHAABAN et al. 1990).

Although tuberculin sensitivity occurs after BCG vaccination, the presence or size of postvaccination tuberculin skin test reactions does not reliably predict the degree of protection against active tuberculosis (COMSTOCK 1988). Some studies that have shown high tuberculin conversion rates demonstrate little protective efficacy, while other studies with low conversion rates have found good protection after vaccination (COMSTOCK 1988; HART et al. 1967). Nevertheless, in most developed countries where BCG is used, children are revaccinated if their tuberculin skin test reactions do not convert to positive within 2–3 months of vaccination.

After BCG vaccination it is difficult to distinguish between a tuberculin skin test reaction caused by virulent *M. tuberculosis* infection or by vaccination itself. Several factors increase the probability that a reaction was caused by virulent mycobacterial infection: large reaction size, a history of close contact with a person with active tuberculosis, residence in an area with a high prevalence of tuberculosis, and a long time interval since vaccination (SNIDER 1985). However, tuberculosis should always be considered in the differential diagnosis for symptomatic tuberculin skin test-positive persons who have been vaccinated with BCG.

5 Vaccine Side Effects

The safety of early BCG vaccines was questioned in 1930 when active tuberculosis developed in 72 of 251 children given oral vaccination during a mass vaccination program in Lubeck, Germany. All 72 children died (Wilson 1967). Subsequent investigations found that a culture of virulent *M. tuberculosis* had been mistaken for the BCG strain, which had been kept in the same incubator. The directors and others in the laboratory were subsequently tried and convicted of negligence and malpractice.

Today BCG is considered to be one of the safer vaccines available. The most frequent nonfatal side effects after BCG vaccination are regional suppurative adenitis and osteitis.

5.1 Regional Lymphadenitis

Regional lymphadenitis most commonly occurs within the first 5 months after vaccination (Lotte et al. 1984). In 1983, a prospective survey of 2 million infants vaccinated between 1979 and 1981 in six European countries (Denmark, German Democratic Republic, Hungary, Romania, and four areas within Croatia and the Federal Republic of Germany) found the risk of lymphadenitis to be 0.025 per 1000 (Lotte et al. 1988).

Several factors contribute to the occurrence of regional lymphadenitis including vaccine strain, the total number of viable and nonviable bacilli in the vaccine preparation, and the dose of BCG given. Regional lymphadenitis is also more common in young children; the risk of lymphadenitis is five to ten times greater for newborns than for preschool and school-age children (Lotte et al. 1988). Outbreaks of lymphadenitis associated with the use of the Pasteur vaccine in Africa and the Caribbean prompted the 1988 revised recommendations lowering the dose of vaccine given to newborns and infants younger than one year old. Poorly administered intradermal vaccinations also were implicated in one outbreak in Zimbabwe (Ray et al. 1988).

Opinions on the treatment of BCG adenitis range from administering no treatment, to providing antibiotic chemotherapy, to surgically excising the nodes. Data on the efficacy of the different treatment options are not available from statistically proven and controlled studies. The WHO suggests that drainage and direct instillation of an antituberculosis drug into the lesion be considered for adherent or fistulated lymph glands. Lesions that are not adherent will heal spontaneously without treatment. Systemic treatment with antituberculosis drugs is ineffective (World Health Organization 1986).

5.2 Osteitis

Osteitis can also develop after vaccination with BCG. The risk of developing osteitis after BCG vaccination varies from country to country. Between 1948 and 1974, 291 cases of disseminated BCG involving lesions of the musculoskeletal system were reported from 13 countries (LOTTE et al. 1988). Japan showed the lowest risk (0.1 per 100 000); Sweden and Finland showed the highest risks (325 and 434 per 100 000, respectively). In Sweden the incidence of BCG osteitis increased significantly from 1 per 100 000 to 2.5 per 100 000 in the years 1950–1969 to an annual rate of 25 per 100 000 to 33 per 100 000 between 1972 and 1975 (BOTTIGER et al. 1982). Osteitis developed within 4 months to 144 months of vaccination; most cases occurred within 7–24 months. Osteitis was most common in the epiphysis of the long bones, particularly in the legs. Similar results were seen in Finland (PELTOLA et al. 1984). The increases in BCG osteitis in both countries remain unexplained. Cases appeared to coincide with a change in vaccine production. In 1971, production of the Gothenburg strain, used in both countries, was transferred from Sweden to the Statens Seruminstitut in Copenhagen. Other countries also have experienced increases in BCG osteitis after changes in either the vaccine strain or the method of production (MILSTIEN and GIBSON 1990). The compulsory notification of BCG reactions to the Swedish Adverse Drug Reaction Committee, mandated in 1972, and the dissemination of information concerning the untoward reactions also likely prompted the expanded surveillance and reporting of cases of osteitis in both countries. As a result of these findings, Sweden discontinued the routine vaccination of neonates in 1975. In Sweden, selective BCG vaccination is now practiced for most at-risk groups, such as foreign-born children or those with a family history of tuberculosis. Finland continued routine immunization using the Glaxo strain of vaccine.

Antituberculosis therapy, in some cases in conjunction with surgical curettage and debridement, usually results in healing of the skeletal lesions.

5.3 Disseminated BCG Disease

The most serious complication of BCG vaccination is disseminated BCG disease. Although rare, BCGitis is usually fatal. Hematogenous dissemination of the bacilli after vaccination has been documented. In most persons, the spread of the organisms is relatively benign and does not cause overt disease; however, at least 35 fatalities were reported between 1948 and 1972 (LOTTE et al. 1988). Immunodeficiencies, such as severe combined immunodeficiency syndrome or chronic granulomatous disease, are risk factors for the development of BCGitis in children. Disease rarely develops in the presence of an intact immune system; however, two cases of culture-confirmed BCG meningitis in immunocompetent children immunized 18 months to 5 years previously have been reported (TARDIEU et al. 1988). Although most cases of BCGitis occur within the first 6 months of vaccination, longer latent periods have been noted. Culture-confirmed BCG disease with widespread lymph node, liver, bone, and lung involvement developed in an 18 year old boy 6 years after vaccination (MacKay et al. 1980). The patient was subsequently diagnosed with several immunological defects, including abnormal macrophage and T lymphocyte functions. The patient died despite prolonged antimycobacterial treatment.

Treatment for disseminated BCG disease is similar to the treatment for active tuberculosis with one exception; pyrazinamide is not used to treat BCGitis because all BCG strains are resistant to this antibiotic. Even when treated with potentially effective chemotherapeutic agents, many cases of disseminated BCG are fatal, particularly if the patient has underlying immunosuppression.

5.4 BCG and Human Immunodeficiency Virus Infection

Infection with HIV is known to be the greatest risk factor for the development of active tuberculosis. In developing countries, tuberculosis is one of the most frequent opportunistic infections in HIV-infected persons (Colebunders et al. 1987). WHO has recommended BCG vaccination as part of the Expanded Programme on Immunization; however, there now is concern that many children in developing countries may be infected with HIV and, as a result, may be at increased risk of disseminated BCG disease after vaccination. Weltman and Rose (1993) reviewed the international literature published between 1982 and 1992 and were unable to find sufficient evidence to prove or disprove an increased risk of complications after BCG vaccination of HIV-infected persons. To date, there have been no controlled studies of the safety of BCG vaccination in either children or adults infected with HIV. Two prospective studies of infants at risk for perinatal HIV infection found no increase in untoward reactions to BCG vaccination. Lallemont-La Coeur et al. (1991) followed 64 children who were born to HIV-infected mothers and 130 control children born to HIV-uninfected mothers who were vaccinated during the first month of life. After 40 months of observation, no significant differences in the risk of lymphadenitis were seen between the HIV-infected and uninfected children (24% vs 18%). All the cases resolved spontaneously and there were no cases of disseminated BCG.

Another prospective study in Rwanda found six cases of suppuration at the injection site among 377 children vaccinated during the first week of life; one child was HIV infected, one was born to an HIV-infected mother but was seronegative at 15 months of age; and four were born to HIV-uninfected mothers. No other untoward reactions were reported (Centers for Disease Control 1991).

In contrast to the prospective studies, case reports have documented dissemination of BCG and subsequent disease in HIV-infected persons. *M. bovis* BCG was cultured from the lymph node and cerebrospinal fluid of an HIV-infected child presenting with fever, adenitis, and diarrhea 4 months after vaccination (NINANE et al. 1988). The symptoms improved after antituberculosis treatment.

M. bovis BCG was isolated from blood, gastric and tracheal aspirates, and the lung of a Native American child vaccinated at 3 months of age (HOUDE and DERY 1988). The vaccination site was crusted and oozing and the child had symptoms of AIDS.

Complications of BCG vaccination also have been reported in adults who had either prior or subsequent HIV infection. Ulceration at the injection site and regional lymphadenopathy developed 4 months after vaccination of one AIDS patient; *M. bovis*, BCG strain, was identified in lesion and blood samples from this patient (CENTERS FOR DISEASE CONTROL 1985). BCG adenitis and disseminated disease can occur from months to years after vaccination (LUMB and SHAW 1992; SMITH et al. 1992; BLONDON et al. 1991;) cases were reported to have occurred in two persons 30 years after they were vaccinated with BCG (REYNES et al. 1989; ARMBRUSTER et al. 1990).

6 Vaccine Efficacy

6.1 Efficacy Studies

6.1.1 Major Controlled Trials

A total of 17 controlled trials of the efficacy of BCG vaccine in humans have been carried out (Table 1). The eight major trials are described in detail.

The first large field trial of BCG efficacy began in 1935 among eight North American Indian tribes (ARONSON et al. 1958). The efficacy of a vaccine prepared by the Henry Phipps Institute of Philadelphia was evaluated in tuberculin-negative children and adolescents age 0–20 years. The duration of follow-up was 9–11 years for the development of active tuberculosis; tuberculosis mortality was assessed after 20 years. Roentgenographic evidence of pulmonary tuberculosis was found in 238 of the 1457 unvaccinated persons and in 64 of the 1551 persons given BCG vaccination; the protective efficacy against active tuberculosis was 75%. There was an 82% reduction in deaths resulting from tuberculosis.

ROSENTHAL and colleagues (1961b) also found that BCG conferred a high degree of protection in high-risk infants in Chicago. A total of 3381 children born at Cook County Hospital between 1937 and 1948 were randomly assigned to receive either the Tice strain of BCG or to remain unvaccinated. The length of follow-up ranged from 12 to 23 years. Tuberculosis developed in 59 of the 1665 unvaccinated persons while active disease occurred in only 16 of 1716 vaccinated individuals (efficacy = 75%).

The Copenhagen strain of vaccine was evaluated in a large controlled trial conducted by the Medical Research Council of Great Britain (MRC) (HART and

Table 1. Controlled trials of BCG efficacy against tuberculosis

Population	*n*	Year	Years of follow-up	Reduction in cases	Reference
Infants	237	1927	3–48	83 (deaths)	Aronson and Dannenberg (1935)
Infant contacts	1092	1933	5–11	8 (deaths)	Levine and Sackett (1946)
Newborns	609	1933	15	81	Ferguson and Simes (1949)
Newborns	41 301	1935	1–11	11 (deaths)	Sergent et al. (1960)
Indian children	2996	1935	9–11	75	Aronson et al. (1958)
Newborns	3381	1937	12–23	75	Rosenthal et al. (1961a)
Indian newborns	262	1938	6–8	59	Aronson (1948)
Newborn contacts	451	1941	19	75	Rosenthal et al. (1961b)
Children	4839	1947	20	–57	Comstock and Webster (1969)
Children	1025	1947	13	–40	Bettag et al. (1964)
Children	77 972	1949	18–20	29	Comstock et al. (1974)
Adults	13 914	1950	20	33	Comstock et al. (1976)
Villagers	21 000	1950	16–21	20	Frimodt-Moller et al. (1973)
Adolescents	27 400	1950	18–20	77	Hart and Sutherland (1977)
Miners	16 314	1965	3	38	Coetzee and Berjak (1968)
Villagers	2174	1965	3	67	Vandiviere et al. (1973)
Villagers	260 000	1968	15	–2	Tripathy (1987)

Sutherland 1977). Between 1950 and 1952, a total of 14 100 tuberculin-negative school children between the ages of 14 and 16 were randomized to receive BCG; 13 300 children remained unvaccinated. The protective efficacy of the vaccine was 84% during the first 5 years of the study. By 15 years after vaccination, 243 cases of tuberculosis had developed in the unvaccinated control group, compared with only 56 cases in those who had been given BCG (efficacy = 77%). The prevalence and incidence of tuberculosis in Great Britain decreased significantly during the study period. The study was discontinued after 19 years of follow-up because of the small number of cases in the unvaccinated control group.

A study in a rural population in southern India found a 60% reduction in the incidence of tuberculosis in the vaccinated group as compared with the unvaccinated group 7 years after vaccination (Frimodt-Moller et al. 1973). By 20 years after vaccination, the vaccine efficacy had declined to 20%.

The United States Public Health Service sponsored a series of controlled trials of BCG efficacy, the first of which began in 1947. A total of 2498 tuberculin-negative school children in Muscogee County, Georgia, were vaccinated with the Tice strain of vaccine; 2341 tuberculin-negative children were left unvaccinated (Comstock and Webster 1969). Allocation to vaccine group was done by even or odd year of birth. During the 20 year study period, three cases of tuberculosis occurred in unvaccinated persons and five cases occurred in those given BCG; no protective efficacy was detected in this study.

The incidence of tuberculosis in school children in Muscogee County was not as high as originally expected; therefore, a second study was begun in adults in the county as well as in neighboring Russell County, Alabama (Comstock et al. 1976). The 20 year evaluation found equal numbers of cases in the vaccinated and control groups. Vaccination with BCG offered no protection against tuberculosis in this study.

Finally, a 29% reduction in tuberculosis incidence was observed in another United States Public Health Service-sponsored study of Puerto Rican children age 1–18 years who were given a vaccine prepared by the New York State Department of Health (Birkhaug strain) (Comstock et al. 1974).

The variable efficacy observed in the published trials of BCG efficacy and the potential differences in potency of the vaccines evaluated in previous studies resulted in the largest controlled trial ever designed to examine the efficacy of BCG vaccine. The trial, cosponsored by the Indian Council of Medical Research, WHO, and the United States Public Health Service, took place in the Chingleput district near Madras, India (Tripathy 1987). Between 1968 and 1971, 265 172 villagers older than 1 month of age were randomized to receive either a low or high dose of the French or Danish vaccine or dextran placebo. Participants were included in the study regardless of their initial tuberculin sensitivity or their radiographic findings at the time of study recruitment. The French and Danish vaccines were selected for the study as they had demonstrated the best test activity in animal models and were available as freeze-dried preparations. Follow-up data were analyzed according to the degree of initial skin test sensitivity. During the 15 year follow-up period, 533 cases of pulmonary tuberculosis developed in those persons with less than 7 mm induration to tuberculin at enrollment; the number of cases was divided evenly among those receiving the placebo or the low or high doses of vaccine. Vaccination did provide some protection (17%) against pulmonary tuberculosis in children less than 14 years of age; adult vaccinees had more tuberculosis than the unvaccinated controls.

6.1.2 Case Control and Cohort Studies

Rather than resolving the issue of the efficacy of BCG, the Chingleput study brought forward even more questions about the usefulness of vaccination. The lack of protective efficacy observed in the study forced a reevaluation of the WHO policy on the use of BCG in developing countries. Other than the two southern India trials, none of the controlled studies were conducted in Africa or Asia, areas of the world with the highest incidence and prevalence of tuberculosis and where BCG vaccination might be of greatest value. Another randomized, placebo-controlled trial of the efficacy of BCG would be expensive, logistically difficult, and would require many years of follow-up to reach a statistically valid endpoint. Therefore, the WHO sponsored several case control studies of BCG vaccination to assess the impact of BCG on tuberculosis control in developing countries. The studies, conducted in Brazil, Burma, Sri Lanka and Argentina, concentrated on childhood tuberculosis and used a variety of control populations, such as patients hospitalized for diseases other than tuberculosis and neighbors or siblings of tuberculosis patients (Smith 1987). As in the previous controlled prospective trials, the results varied widely: the effectiveness of the vaccination programs ranged from 2% to 84%. Protection conferred by vaccination with BCG was highest against disseminated forms of tuberculosis, such as meningitis. In recent years, numerous other case control investigations have found similarly variable results (SHAPIRO et al. 1985; YOUNG and HERSHFIELD 1986; HOUSTON et al. 1990; BLIN et al. 1986). Again, the majority of studies have focused on vaccine effectiveness in children and have found protection ranging from 16% to 80%. A small number of cohort studies of children who were contacts of cases of newly diagnosed, culture-confirmed tuberculosis also have been funded through WHO. In these studies, the incidence of tuberculosis decreased by 53% to 74% in children who were given BCG prior to exposure to tuberculosis compared to those who were not vaccinated with BCG (PADUNGCHAN et al. 1986; TIDJANI et al. 1986).

6.2 Hypotheses to Explain the Variability in Efficacy

Hypotheses have been put forward to explain the variability in results of field trials with BCG. These hypotheses have been reviewed by FINE (1988). One explanation is that prior infection with nontuberculous mycobacteria confers protection against active tuberculosis. Natural infection with these mycobacteria could decrease the susceptibility of the unvaccinated control group, thereby diluting the protective effects in the BCG vaccinated groups. Studies in both animals and humans have demonstrated that infection with environmental mycobacteria can increase immunity to tuberculosis (PALMER and LONG 1966). Trials showing the lowest BCG efficacy were in areas, such as the southeastern United States, where nontuberculous mycobacteria are more prevalent. However, data from studies in Puerto Rico showed no difference in protection after vaccination of persons who reacted only to a large dose (10 tuberculin units) of tuberculin, indicative of infection with nontuberculous mycobacteria (COMSTOCK et al. 1974). In addition, two trials in the U.S. which excluded reactors to the large dose of tuberculin showed the worst efficacy estimates of all of the trials.

Some have suggested that previous infection with environmental mycobacteria may either enhance or oppose protective immunity. ROOK et al. (1981) postulate that two types of cell-mediated immune responses may occur as a result of mycobacterial infection. The "Koch phenomenon" develops in animals 4–6 weeks after infection with *M. tuberculosis*, causing necrosis at the site of tuberculin skin testing. The second cell-mediated immune response, described by Mackaness as a listeria-type response, occurs within days of infection and correlates with the appearance of macrophage-activating T lymphocytes. No necrosis is seen after skin testing. Different species of mycobacteria have been demonstrated in animal models to produce varying levels of these two immune responses. For example, prior infection with *M. vaccae* or any of the nontuberculous mycobacteria which produce the listeria-type response may potentiate the protective effect of subsequent vaccination with BCG. In contrast, infection with *M. scrofulaceum* or other mycobacteria that induce the Koch phenomenon, may inhibit the development of protective immunity following

BCG, thereby increasing susceptibility to disease from *M. tuberculosis*. It is unlikely, however, that exposure to nontuberculous mycobacteria accounts for all of the observed variability in the efficacy of BCG.

Differences in the potency or immunogenicity of individual vaccine strains also may contribute to the variability in the efficacy of BCG. Results in animal models do not predict the efficacy of BCG vaccines in humans; therefore, protection can only be assessed in field trials in humans. Microbiological differences between the vaccine strains do exist; however, studies using the same vaccines in different settings have shown both similar and varying efficacies. Studies using different vaccines in the same setting have also shown inconsistent results (Fine 1988).

The efficacy of BCG is believed by some to be influenced by differences in the risk of infection with *M. tuberculosis* of the study populations, differences in the virulence of *M. tuberculosis* isolates, or differences in the pathogenesis of disease. Some *M. tuberculosis* isolates from the Chingleput area are of low virulence for guinea pigs; BCG may not be able to protect against these "South Indian variants." Although protection may be influenced by the virulence of the challenge strain of *M. tuberculosis*, there are no laboratory data which suggest that BCG provides poor protection against low-virulence South Indian variants (Hank et al. 1981).

Host genetics also may play a role in the efficacy of BCG. The immune response to BCG in mice is regulated by a gene on chromosome 1; this gene also has been found on chromosome 2 in humans (Blackwell 1988). However, no statistically significant differences in the efficacy of BCG have been found between racial groups in field trials which suggest that genetic variability influences protection following vaccination.

Chronic moderate dietary protein deficiency is known to impair the ability of BCG vaccine to protect guinea pigs against challenge with virulent *M. tuberculosis* (McMurray et al. 1985). However, data from human trials have failed to support nutrition as a factor in the variability in BCG efficacy.

In animal models BCG protects against dissemination of organisms from the primary site of infection. The efficacy of BCG may depend on whether the mechanism of pathogenesis is endogenous reactivation or exogenous reinfection. In the latter, the progression to disease would depend on the primary infection. Thus, in areas where the probability of reinfection is high, such as in southern India, the efficacy of BCG may be lower than in areas where reactivation disease predominates. This hypothesis is supported by data from the Chingleput trial, which showed a high risk of infection with *M. tuberculosis* in the study area, but a low incidence of tuberculosis in uninfected persons, a high incidence of tuberculosis in the infected population, and a long interval between tuberculin conversion and evidence of active disease (Ten Dam and Pio 1982). In contrast, data from the earlier trial in Puerto Rico contradict this hypothesis. At the time of study recruitment, the infection rate in Puerto Rican children 1–3 years of age was approximately 2.5% per year. The study area also had a high tuberculosis rate. In the study population, uninfected persons had a relatively low tuberculosis rate, while persons infected with *M. tuberculosis* had a relatively high rate. During the 20 year follow-up, the rate of infection decreased, yet the efficacy of BCG remained constant (Comstock and Edwards 1972).

Finally, the different study methodologies used in the controlled studies have been blamed for the variable study results. Clemens and colleagues (1983) reviewed the methodological and statistical approaches used in the eight major controlled trials of BCG efficacy. The published reports of the trials were compared on the basis of: (1) the unbiased allocation of study participants to either the vaccine or control group, as such random allocation would ensure equal susceptibility to tuberculosis in both groups; (2) the equal follow-up and surveillance of the two study groups for symptoms of tuberculosis and; (3) the unbiased detection of tuberculosis in vaccinated and unvaccinated persons. The equal frequency of diagnostic tests in the vaccinated and control groups and the interpretation of clinical information in a manner blinded to vaccination status would minimize "detection bias." Using these criteria, the investigators concluded that the major source of bias in the trials was in the detection and diagnosis of cases of tuberculosis. Reliance on chest radiographs of only symptomatic persons could result in a biased estimate of BCG efficacy as the likelihood of obtaining a chest film would depend on the motivation of the patient to seek medical care and the physician's clinical suspicion of tuberculosis. Systematic chest roentgenographic surveys of all study participants with blinded reading of the chest films was done in only three of the trials, namely the trial in North American Indians, the trial in children in Chicago, and the trial in England. The efficacy of BCG vaccine in these trials was 75%–80%.

Clemens et al. (1983) also estimated the statistical precision of each of the trials by calculating the 95% confidence intervals for each study efficacy value. Trials with narrower confidence intervals were considered to have a more precise estimate of vaccine efficacy than those trials with a wider range of efficacy values. Studies with the greatest statistical precision reported the higher vaccine efficacies; trials demonstrating little or no protection after BCG vaccination had wider confidence intervals.

From these qualitative analyses, the authors concluded that BCG is protective against tuberculosis and that the disparate results found in the controlled trials may have resulted, at least in part, from methodological differences in the studies.

6.3 Meta-analyses of BCG Efficacy

Rodrigues et al. (1993) performed a meta-analysis on the efficacy of BCG in preventing tuberculosis. In meta-analysis, data from clinical trials and case control studies are combined to increase the statistical power for comparing endpoints, such as vaccine efficacy, risk of disease after occupational exposures and behavioral characteristics and disease. Rodrigues et al. (1993) found that the published reports of ten randomized controlled trials and eight case control studies met the inclusion criteria for analysis, namely, that they showed evidence of control for bias and confounding and presented site-specific information. The protective effects of the vaccine for pulmonary tuberculosis were highly variable, ranging from –88% to 79% in the randomized trials and from 16% to 76% in the case control studies. Tests of heterogeneity were statistically significant; therefore, no summary measures of protective effect for pulmonary tuberculosis were calculated. The protective effects against miliary and meningeal tuberculosis showed less variability (46% to 100% in the randomized trials and 58% to 100% in the case control studies) and were statistically homogeneous. The summary protective effect against miliary and meningeal tuberculosis was 86% (95% confidence interval, 65%–95%) in the randomized trials and 75% (95% confidence interval, 61%–84%) in the case control studies. The authors speculated that the differences in protective effects seen for pulmonary as compared with miliary and meningeal tuberculosis may result from the differences in age of patients with the different forms of disease. Miliary and meningeal tuberculosis occur most commonly in younger age groups. Exposure to and potential protection from infection with nontuberculous mycobacteria would be less in younger children. Most children are vaccinated with BCG at birth; therefore, younger children would also show less waning of immunity and greater protection after vaccination. Finally, different immunological mechanisms may be responsible for protection against the different forms of disease. Vaccination with BCG may be more effective against forms of disease, such as miliary and meningeal tuberculosis, which result from the hematogenous spread of the organism. Pulmonary tuberculosis may result from reactivation of latent *M. tuberculosis* infection, reinfection, or progression of primary disease; these differing mechanisms of pathogenesis may be less influenced by vaccination with BCG.

The uncertain efficacy of BCG vaccine has made policy decisions concerning the immunization of groups at increased risk for tuberculous infection, such as health care workers, problematic. In 1992, the Centers for Disease Control and Prevention funded investigators at the Harvard School of Public Health to review the available published literature and data bases and perform a quantitative meta-analysis on the efficacy of BCG in preventing tuberculosis. Colditz and colleagues (1994) included individual studies in the analysis if BCG vaccine was used, if there was a vaccinated and concurrent control group, and if vaccine efficacy was measured by the detection of at least one tuberculosis case during the study. A numerical ranking score was given to each study using the following criteria: (1) the inclusion in the report of a description of the vaccine strain, dose and route of administration; (2) reporting of the number of participants lost to follow-up during the study; (3) demonstration of equal surveillance for tuberculosis in the vaccinated and control groups; and (4) the quality of the diagnostic confirmation of tuberculosis disease. The studies were added to the analysis in order of their relative score. Using data from the prospective clinical trials, the protective effect of BCG decreased as studies with lower ranking scores were included in the analysis. The seven clinical trials that used random allocation of participants to receive or not to receive the vaccine gave a protective effect of 63%, with a 95% confidence interval of 26%–82%. Inclusion of the six studies that used less stringent alternate or systematic allocation schemes resulted in a protective effect of 51% (95% confidence interval 30%–66%). Data from ten case control studies of BCG efficacy showed similar protection after vaccination; the number of tuberculosis cases was reduced by 50% after BCG vaccination.

Subgroup analyses were also done to examine the influence of other factors on BCG efficacy. Age at vaccination was found to be weakly associated with protection; vaccine efficacy decreased with increasing age. The protective effect was 85% at birth and decreased to 52% for those vaccinated at 20 years of age. A more stringent analysis of the effects of age on vaccine efficacy was not possible due to the paucity of available data on BCG vaccination in adults.

The geographical location of the study and the incidence of active tuberculosis in the study control group showed the strongest associations with BCG efficacy. Lower protective effects were seen in studies performed closer to the equator. The investigators postulate that geography may be a surrogate for factors such as the prevalence of Nontuberculous mycobacteria infections, climate and dietary factors, or the genetic susceptibility of individual populations.

Vaccination also appeared more protective in populations with a higher incidence of active tuberculosis. This association has been seen in other trials; however, in the earlier trials, the association resulted from the intrinsic correlation between the ratio of the rates of tuberculosis in the vaccinated and unvaccinated groups (incidence) (SUTHERLAND 1972). Thus, the significant effect of the background incidence of tuberculosis on BCG efficacy is controversial and will require additional analyses to confirm.

Many different BCG strains have been used in both clinical trials and case control studies of vaccine efficacy. A rigorous analysis of the effects of individual strains on BCG efficacy was not possible; however, a significant protective effect of vaccination was seen across many study designs, populations and strains. COLDITZ and colleagues concluded that many different strains were able to confer some protection. The use of nonlyophilized vaccines in earlier prospective studies and the changes that have occurred in vaccine strains over time raise questions as to the comparability of the vaccine strains evaluated for efficacy and those available today. The issue of strain differences remains controversial despite the current analysis.

One group often suggested to receive BCG vaccine is health care workers. A total of eight studies of the efficacy of BCG vaccination in health care workers were reviewed for the meta-analysis. Although the published studies did suggest a protective effect, methodologic weaknesses in the studies prevented their inclusion in the formal meta-analysis.

It was also not possible to evaluate the duration of protection after BCG vaccination. Long-term surveillance of vaccinees is hampered by increasing losses to follow-up over time and the decreases in tuberculosis incidence in most countries over the last several decades.

The compilation of the global data on BCG using meta-analysis has provided some insight into the true efficacy of the vaccine. However, applying the results of the analysis to public health practice should be done with caution. Estimates of vaccine efficacy in populations may not directly translate to equal protective effects in individuals. While BCG vaccine may protect 50% of a group of vaccinated persons from developing active tuberculosis, vaccination may not reduce an individual's risk of developing tuberculosis by 50%. The meta-analysis has raised many interesting questions about the role of environmental and genetic factors in vaccine efficacy. Current technologies may provide the tools to unravel the mysteries of protective immunity to *M. tuberculosis* and the role of BCG vaccination.

7 Use of BCG in the Control of Tuberculosis

Given the high incidence of tuberculosis in many developing countries and the inability to control the spread of infection through chemoprophylaxis and the prompt diagnosis and effective treatment of infectious cases, WHO continues to recommend the use of BCG vaccine on a worldwide basis. BCG vaccine is given as part of the WHO Expanded Programme on Immunization; the vaccine is administered to infants as soon as possible after birth (WHO Study Group 1980).

In the United States, the general population is at low risk for acquiring tuberculous infection. Thus, the control of tuberculosis is predicated on periodic tuberculin skin testing of high-risk children and adults and with administration of isoniazid preventive therapy to selected persons with positive tuberculin tests. Widespread BCG immunization would undermine this strategy by rendering the tuberculin test of BCG recipients difficult if not impossible to interpret, thus confounding the use of preventive therapy. The Immunization Practices Advisory Committee and the Advisory Committee for the Elimination of Tuberculosis have recommended that BCG vaccination be considered only for children with negative tuberculin skin tests who cannot be placed on isoniazid preventive therapy but who have continuous exposure to persons with active tuberculosis, particularly isoniazid and rifampin-resistant disease (Centers for Disease Control 1988). BCG vaccination is not recommended for health care workers; however, in light of recent nosocomial outbreaks of drug-resistant tuberculosis, these recommendations are currently under review (CDC, personal communication).

The use of BCG vaccine in persons infected with HIV remains controversial. There are no data from controlled clinical trials on the efficacy of BCG in either HIV-infected adults or children. However, given the lack of data suggesting a harmful effect of BCG vaccination on HIV-infected children and the high risk this population faces of being infected with *M. tuberculosis* and developing disease, WHO recommends BCG vaccination for asymptomatic HIV-infected children who are at increased risk of tuberculous infection. The WHO does not recommend BCG vaccination for children or adults with symptomatic HIV infection or for persons known or suspected to be infected with HIV, who have minimal risk of infection with *M. tuberculosis* (World Health Organization and the International Union against Tuberculosis 1989). The latter recommendation would apply to most populations in the United States for whom BCG might be considered (Centers for Disease Control 1988).

References

Abou-Zeid C, Smith I, Grange J, Steele J, Rook G (1986) Subdivision of daughter strains of Bacille Calmette-Guérin (BCG) according to secreted protein patterns. J Gen Microbiol 132: 3047–3053

Armbruster C, Junker W, Vetter N, Jaksch G (1990) Disseminated Bacille Calmette- Guérin infection in an AIDS patient 30 years after BCG vaccination. J Infect Dis 162: 1216

Aronson JD (1948) Protective vaccination against tuberculosis with social references to BCG vaccination. Am Rev Tuberc 58: 275–281

Aronson JD, Dannenberg AM (1935) Effect of vaccination with BCG on tuberculosis in infancy and in childhood. Am J Dis Child 50: 1117–1130

Aronson JD, Aronson CF, Taylor HC (1958) A twenty-year appraisal of BCG vaccination in the control of tuberculosis. Arch Intern Med 101: 881–893

Ashley MJ, Siebenman CO (1967) Tuberculin skin sensitivity following BCG vaccination with vaccines of high and low viable counts Can Med Assoc J 97: 1335–1338

Bettag OL, Kaluzny AA, Morse D, Radner DB (1964) BCG study at a state school for mentally retarded. Dis Chest 45: 503–507

Blackwell JM (1988) Bacterial infections. In: Wakelin D, Blackwell JM (eds) Genetics of resistance to bacterial and parasitic infections. Taylor and Francis, London

Blin P, Delolme HG, Heyraud JD, Charpak Y, Sentilhes L (1986) Evaluation of the protective effect of BCG vaccination by a case-control study in Yaounde, Cameroon. Tubercle 67: 283–288

Blondon H, Guez T, Paul G, Truffot-Pernot C, Sicard D (1991) Adenite a BCG 6 ans après la vaccination au cours d'un SIDA. Presse Med 20: 1091

Botttiger M, Romanus V, de Verdier C, Boman G (1982) Osteitis and other complications caused by generalized BCG-itis. Acta Paediatr Scand 71: 471–478

Bunch-Christensen K (1977) Evaluation of BCG vaccines in children, the effect of strain and dose. J Biol Stand 5: 159–164

Calmette, Guérin (1908) Lait Nocard

Centers for Disease Control (1985) Disseminated *Mycobacterium bovis* infection from BCG vaccination of a patient with acquired immunodeficiency syndrome. MMWR 34: 227–228

Centers for Disease Control (1988) Recommendations of the Immunization Practices Advisory Committee (ACIP). Use of BCG vaccines in the control of tuberculosis: a joint statement by the ACIP and the Advisory Committee for Elimination of Tuberculosis. MMWR 37: 663–664, 669–675

Centers for Disease Control (1991) BCG vaccination and pediatric HIV infection - Rwanda, 1988–1990. MMWR 40: 833–836

Clemens JB, Chuong JJH, Feinstein AR (1983) The BCG controversy: a methodological and statistical reappraisal. JAMA 249: 2361–2369

Coetzee AM, Berjak J (1968) BCG in the prevention of tuberculosis in an adult population. Proc Mine Med Off Assoc 48: 41–53

Colditz G, Brewer T, Berkey C, Wilson M, Burdick E, Fineberg H, Mosteller F (1994) Efficacy of BCG vaccine in the prevention of tuberculosis. JAMA 271: 698–702

Colebunders R, Mann JM, Francis H, Bila K, Izaley L, Kakonde N, Kabasele K, Ifoto L, Nzilambi N, Quinn TC, Van der Groen G, Curran J, Vercauteren G, Piot P (1987) Evaluation of a clinical case-definition of acquired immunodeficiency syndrome in Africa. Lancet 1: 492–494

Comstock GW (1988) Identification of an effective vaccine against tuberculosis. Am Rev Respir Dis 138: 479–480

Comstock GW, Edward PQ (1972) An American view of BCG vaccination, illustrated by results of a controlled trial in Puerto Rico. Scand J Respir Dis 53: 207–217
Comstock GW, Webster RG (1969) Tuberculosis studies in Muscogee County, Georgia. VII. A twenty-year evaluation of BCG vaccination in a school population. Am Rev Respir Dis 100: 839–845
Comstock, GW, Edwards LB, Nabangxang H (1971) Tuberculin sensitivity eight to fifteen years after BCG vaccination. Am Rev Respir Dis 103: 572–575
Comstock GW, Livesay VT, Woolpert SF (1974) Evaluation of BCG vaccination among Peurto Rican children. Am J Public Health 64: 283–291
Comstock GW, Woolpert SF, Livesay VT (1976) Tuberculosis studies in Muscogee County, Georgia. Twenty-year evaluation of a community trial of BCG vaccination. Public Health Rep 91: 276–280
Crispen R (1989) History of BCG and its substrains. In: BCG in superficial bladder cancer. EORTC Genitour Group Monogr 6: 35–50
Dawodu AH (1985) Tuberculin Conversion following BCG vaccination in preterm infants. Acta Paediatr Scand 74: 564–567
Edwards LB, Gelting AS (1950) BCG-vaccine studies. Bull World Health Organ 3: 1–24
Ferguson RS, Simes AB (1949) BCG vaccination of Indian infants in Saskatchewan. Tubercle 30: 5–11
Fine PEM (1988) BCG vaccination against tuberculosis and leprosy. Br Med Bull 44: 691–703
Friedland IR (1990) The booster effect with repeat tuberculin testing in children and its relationship to BCG vaccination. S Afr Med J 77: 387–389
Frimodt-Moller J, Acharyulu GS, Kesava Pillai K (1973) Observations on the protective effect of BCG vaccination in a South Indian rural population: fourth report. Bull Int Union Tuberc 48: 40–49
Gardner LV (1932) The history of the R1 strain of tubercle bacillus. Am Rev Tuberc 25: 577–590
Gheorghiu M, LaGrange PH (1983) Viability, heat stability and immunogenicity of four BCG vaccines prepared from four different BCG strains. Ann Immunol 134C: 125–147
Grange JM, Gibson JA (1986) Strain to strain variation in the immunogenicity of BCG. Dev Biol Stand 58: 37–41
Grange JM, Gibson J, Osborn TW (1983) What is BCG? Tubercle 64: 129–139
Guld J, Waaler H, Sundaresan TK, Kaufmann PC, Ten Dam HG (1968) The duration of BCG-induced tuberculin sensitivity in children and its irrelevance for revaccination. Bull World Health Organ 39: 829–836
Hank JA, Chan JK, Edwards ML, Muller D, Smith DW (1981) Influence of the virulence of *Mycobacterium tuberculosis* on protection induced by Bacille-Calmette-Guérin in guinea pigs. J Infect Dis 143: 734–738
Hart PD, Sutherland I (1977) BCG and vole bacillus vaccines in the prevention of tuberculosis in adolescence and early adult life. Med J 2: 293–295
Hart PD, Sutherland I, Thomas J (1967) The immunity conferred by effective BCG and vole bacillus vaccines in relation to individual variations in tuberculin sensitivity and to technical variations in the vaccines. Tubercle 48: 201–210
Hesselberg I (1972) Drug resistance in the Swedish/Norwegian BCG strain. Bull World Health Organ 46: 503–507
Heyworth B (1977) Delayed hypersensitivity to PPD-S following BCG vaccination in African children - an 18-month field study. Trans R Soc Trop Med Hyg 71: 251–253
Horwitz O, Bunch-Christensen K (1972) Correlation between tuberculin sensitivity after 2 months and 5 years among BCG vaccinated subjects. Bull World Health Organ 47: 49–58
Houde C, Dery P (1988) *Mycobacterium bovis* sepsis in an infant with human immunodeficiency virus infection. Pediatr Infect Dis J 7: 810–812
Houston S, Fanning A, Soskolne CL (1990) The effectiveness of bacillus Calmette-Guérin (BCG) vaccination against tuberculosis: a case- control study in Treaty Indians, Alberta, Canada. AM J Epidemiol 340–348
International Union against Tuberculosis (1978) Phenotypes of BCG-vaccines seed lot strains: results of an International Cooperative Study. Tubercle 59: 139–142
Jespersen A (1971) The potency of BCG vaccines determined on animals. Ph D thesis, University of Copenhagen
Karalliedde S, Katugaha LP, Uragoda CG (1987) Tuberculin Response of Sri Lankan children after BCG vaccination at birth. Tubercle 68: 33–38
Kathipari K, Seth V, Sinclair S, Arora NK, Kukreja N (1982) Cell mediated immune response after BCG as a determinant of optimum age of vaccination. Indian J Med Res 76: 508–511
Lallemant-Le Coeur S, Lallemant M, Cheynier D, Nzingoula S, Drucker J, Larouze B (1991) Bacillus Calmette-Guérin immunization in infants born to HIV-1-seropositive mothers. AIDS 5: 195–199

Lefford MJ (1977) Induction and expression of immunity after BCG immunization. Infect Immun 18: 646–653

Lehmann HG, Engelhardt H, Freudenstein H, Hennessen W, Widmark R (1979) BCG vaccination of neonates, infants, schoolchildren and adolescents, part I: dose finding studies with BCG strain 1331 Copenhagen. Dev Biol Stand 43: 127–132

Levine MI, Sackett MF (1946) Results of BCG immunization in New York City. Am Rev Tuberc 53: 517–532

Levy FM, Conge GA, Pasquier JF, Mauss H, Dubos R, Schaedler R (1961) The effect of BCG vaccination on the fate of virulent tubercle bacilli in mice. Am Rev Respir Dis 84: 28–36

Lotte A, Wasz-Hockert O, Poisson N, Dumitrescu N, Verron M, Couvet E (1984) BCG complications. Adv Tuberc Res 21: 107–193

Lotte A, Wasz-Hockert O, Poisson N, Engbaek H, Landmann H, Quast U, Andrasofszky B, Lugosi L, Vadasz Mihailescu P, Pal D, Sudic D (1988) Second IUATLD study on complications induced by intradermal BCG-vaccination. Bull Int Union Tuberc Lung Dis 63: 47–59

Lumb, R, Shaw D (1992) *Mycobacterium bovis* (BCG) vaccination. Progressive disease in a patient symptomatically infected with the human immunodeficiency virus. Med J Aust 156: 286–287

MacKaness GB (1968) The immunology of antituberculous immunity. Am Rev Resp Dis 97: 337

MacKay A, Alcorn MJ, Macleod IM, Stack BHR, Macleod T, Laidlaw M, Millar JS, White RG (1980) Fatal disseminated BCG infection in an 18-year-old boy. Lancet 2: 1332–1334

Manerikar SS, Malaviya AN, Singh MB, Rajgopalan P, Kumar R (1976) Immune status and BCG vaccination in newborns with intrauterine growth retardation. Clin Exp Immunol 26: 173–175

McMurray DN, Carlomagno MA, Mintzer CL, Tetzlaff CL (1985) *Mycobacterium bovis* BCG vaccine fails to protect protein-deficient guinea pigs against respiratory challenge with virulent *Mycobacterium tuberculosis*. Infect Immun 50: 555-559

Milstien JB, Gibson JJ (1990) Quality control of BCG vaccine by WHO: a review of factors that may influence vaccine effectiveness and safety. Bull World Health Organ 68: 93–108

Minnikin DE, Parlett JH, Magnusson M, Ridell M, Lind A (1984) Mycolic acid patterns of representatives of *Mycobacterium bovis* BCG. J Gen Microbiol 130: 2733–2736

Myint TT, Yin Y, Yi MM, Aye HH (1985) BCG test reaction in previously BCG vaccinated children. Ann Trop Paediatr 5: 29–31

Ninane J, Grymonprez A, Burtonboy G, Francois A, Cornu G (1988) Disseminated BCG in HIV infection. Arch Dis Child 63: 1268–1269

Nyboe J, Bunch-Christensen K (1966) Assay in man of different BCG products. Bull World Health Organ 35: 645–650

Osborn TW (1983) Changes in BCG strains. Tubercle 64: 1–13

Padungchan S, Konjanart S, Kasiratta S (1986) The effectiveness of BCG vaccination of the newborn against childhood tuberculosis in Bangkok. Bull World Health Organ 64: 247–258

Palmer CE, Long MW (1966) Effect of infection with atypical mycobacteria on BCG vaccination and tuberculosis. Am Rev Respir Dis 94: 533-568

Palmer C, Meyer SN (1951) Research contributions of BCG vaccination programs. I. Tuberculin allergy as a family trait. Public Health Rep 66: 259–276

Peltola H, Salmi I, Vahvanen V, Ahlqvist J (1984) BCG vaccination as a cause of osteomyelitis and subcutaneous abscess. Arch Dis Child 59: 157–161

Ray CS, Pringle D, Legg W, Mbengeranwa OL (1988) Lymphadenitis associated with BCG vaccination: a report of an outbreak in Harare, Zimbabwe. Cent Afr J Med 34: 281–286

Reynes J, Perez C, Lamaury I, Janbon F, Bertrand A (1989) Bacille Calmette- Guérin adenitis 30 years after immunization in a patient with AIDS. J Infect Dis 160: 727

Rodrigues L, Diwan V, Wheeler J (1993) Protective effect of BCG against tuberculous meningitis and miliary tuberculosis: A meta-analysis. Int J Epidemiol 22: 1154–1158

Rook GAW, Bahr GM, Stanford JL (1981) The effect of two distinct forms of cell-mediated response to mycobacteria on the protective efficacy of BCG. Tubercle 62: 63–68

Rosenthal SR (1980) Routes and methods of administration. In: Rosenthal SR (ed) BCG vaccine: tuberculosis - cancer. PSG, Littleton, chap 11,12

Rosenthal SR, Loewinsohn E, Graham M, Liveright D, Thorne M, Johnson V (1961a) BCG vaccination in tuberculous households. Am Rev Respir Dis 84: 690–704

Rosenthal SR, Loewinsohn E, Graham M, Liveright D, Thorne M, Johnson V (1961b) BCG vaccination against tuberculosis in Chicago. A twenty-year study statistically analyzed. Pediatrics 28: 622–641

Sergent E, Catanei A, Ducros-Rougebief H (1960) Premunition anti-tuberculeuse par le BCG. Campagne controlee poursuivie a Alger depuis 1935. Troisieme note. Arch Inst Pasteur d'algerile 38: 131–137

Shaaban MA, Ati Abdul M, Bahr GM, Stanford JL, Lockwood DNJ, McManus IC (1990) Revaccination with BCG: its effects on skin tests in Kuwaiti senior school children. Eur Respir J 3: 187–191
Shapiro C, Cook N, Evans D, Willett W, Fajardo I, Koch-Weser D, Bergonzoli G, Bolanos O, Guerroero R, Hennekens C (1985) A case-control study of BCG and childhood tuberculosis in Cali, Colombia. Int J Epidemiol 14: 441–446
Smith DW, (1985) Protective effect of BCG in experimental tuberculosis. Adv Tuberc Res 22: 66–73
Smith DW, McMurray DN, Wiegeshaus EH, Grover AA, Harding GE (1970) Host-parasite relationships in experimental tuberculosis. IV. Early events in the course of infection in vaccinated and nonvaccinated guinea pigs. Am Rev Respir Dis 102: 937–949
Smith E, Thybo S, Bennedsen J (1992) Infection with *Mycobacterium bovis* in a patient with AIDS: a late complication of BCG vaccination. Scand J Infect Dis 24: 109–110
Smith PG (1987) Case-control studies of the efficacy of BCG against tuberculosis. Bull Int Union Tuberc 62: 73–79
Snider DE (1985) Bacille Calmette-Guérin vaccinations and tuberculin skin tests. JAMA 253: 3438–3439
Stewart CJ (1968) Skin sensitivity to human, avian, and BCG PPDs after BCG vaccination. Tubercle 49: 84–91
Sutherland I (1972) Letter. Tubercle 53: 150–151
Tardieu M, Carriere JP, Truffot-Pernot C, Dupic Y (1988) Tuberculous meningitis due to BCG in two previously healthy children. Lancet 1: 440–441
Ten Dam HG, Pio A (1982) Pathogenesis of tuberculosis and effectiveness of vaccination. Tubercle 63: 225–233
Tidjani O, Amedome A, Ten Dam HG (1986) The protective effect of BCG vaccination of the newborn against childhood tuberculosis in an African community. Tubercle 67: 269–281
Tripathy SP (1987) Fifteen-year follow-up of the Indian BCG prevention trial. Bull Int Union Tuberc 62: 69–72
Vandiviere HM, Dworski M, Melvin I, Watson K, Begley J (1973) Efficacy of bacillus Calmette-Guérin and isoniazid-resistant bacillus Calmette-Guérin with and without isoniazid chemoprophylaxis from day of vaccination. Am Rev Respir Dis 108: 301–313
Weltman A, Rose D (1993) The safety of Bacille Calmette-Guérin vaccination in HIV infection and AIDS. AIDS 7: 149–157
World Health Organization (1977) WHO-sponsored international quality control of BCG vaccine. WHO Technical report series 8
World Health Organization (1986) BCG vaccination of the newborn. Rationale and guidelines for country programmes. WHO/TB/86.147
World Health Organization Expert Committee on Biological Standardization (1985) Requirements for dried BCG vaccine. Requirements for biological substances, no 11
World Health Organization/International Union against Tuberculosis (1989) Tuberculosis and AIDS. Statement on AIDS and tuberculosis. Bull Int Union Tuberc Lung Dis 64: 8–11
World Health Organization Study Group (1980) BCG vaccination policies. WHO technical report series 652
Wilson GS (1967) The hazards of immunization. Athlone, London
Young TK, Hershfield ES (1986) A case-control study to evaluate the effectiveness of mass neonatal BCG vaccination among Canadian Indians. Am J Public Health 76: 783–786
Young TK, Mirdad S (1992) Determinants of tuberculin sensitivity in a child population covered by mass BCG vaccination. Tubercle Lung Dis 73: 94–100

Functional T Cell Subsets in Mycobacterial and Listerial Infections: Lessons from Other Intracellular Pathogens

G. SZALAY[1] and S.H.E. KAUFMANN[1,2]

1 Introduction

Intracellular bacteria like *Mycobacterium tuberculosis, M. bovis, M. leprae, Listeria monocytogenes, Salmonella typhi, S. paratyphi* and *Legionella pneumophila* are human pathogens. They all share the capacity to escape antibody and complement-mediated defense mechanisms afforded by the humoral immune system. The common denominator of intracellular bacteria is their capacity to invade and replicate inside mononuclear phagocytes (MPS). As long as these cells are quiescent, they provide a niche for intracellular bacteria. Once they are activated, they express potent effector functions which lead to the elimination of many bacterial invaders. Upon stimulation by cytokines, MPS express several defense mechanisms including production of defensins, acidification of the phagosome, phagosome-lysosome fusion, depletion of iron, and production of reactive oxygen and nitrogen intermediates (ROI/RNI) (LEHRER et al. 1991; HORWITZ 1988; MOULDER 1985; ANDREW et al. 1985; WEINBERG 1992; HIBBS et al. 1988;

[1]Department of Immunology, University of Ulm, Albert-Einstein-Allee 11, 89081 Ulm, Germany
[2]Max-Planck-Institute for Infection Biology, Jaegerstrasse 10/11, 10117 Berlin, Germany

BABIOR 1984; LIEW and COX 1990). Together these mechanisms are part of the nonspecific host response.

MPS, however, are also involved in the induction of specific immunity. These cells not only express on their surface major histocompatibility complex (MHC) class I molecules, like all nucleated cells, they also express MHC class II molecules on their surface, a property of only few cell types, such as dendritic cells and B cells, which are involved in the induction of immunity. The specific immune response to intracellular bacteria requires T lymphocytes, which do not respond to antigens directly, as do B cells and their antibodies. Rather, MHC molecules present peptides derived from bacterial antigens to T lymphocytes. These cells carry either the α/β T cell receptor (TCR), together with one of two accessory molecules, CD4 or CD8, or the γ/δ TCR; such cells lack both accessory molecules and hence are double negative. The α/β T cells recognize bacterial antigens on host cells after processing and association with MHC molecules. It is generally thought that extracellular proteins, including bacterial antigens, enter the endosomal pathway, are processed in the early and late endosomes, and associate with and are surface expressed by MHC class II molecules (GERMAIN and MARGULIES 1993). CD4 T cell recognize the antigen/MHC class II complex and are thus activated to perform T helper (TH) functions. The TH cells produce cytokines thereby activating effector cells. CD4 T cells can be further subdivided into TH-1 and TH-2 cells according to the cytokines they produce. TH-1 cells secrete interleukin (IL) 2, interferon γ (IFNγ) and tumor necrosis factorβ (TNFβ). TH-2 cells produce IL-4, IL–5, IL-6 and IL-10. Both subsets produce IL-3 and granulocyte/macrophage colony–stimulating factor (MOSMANN and COFFMAN 1989; SHER and COFFMAN 1992). Thus, by producing a variety of cytokines, CD4 T cells influence the course of bacterial infection indirectly.

In contrast, CD8 T cells lyse infected host cells and hence control infection directly. Until recently, the cytolytic functions of CD8 T cells have been demonstrated only under in vitro conditions. Very recently, however, target cell lysis has been shown to also occur in vivo (KÄGI et al. 1994a). Antigenic peptides associate with MHC class I molecules in the cytosolic compartment and surface expression of this complex is a prerequiste for CD8 T cell recognition. The central role of this mechanism in host defense against viral infections is unequivocal, whereas its contribution to antibacterial defense is less well understood. The subcellular localization of bacteria is thought to be important for effective T cell mediated immunity. Accordingly, peptides from bacteria which are capable of hiding in the cytosol of a host cell and thus grow in this compartment, readily associate with MHC class I molecules and are presented to CD8 T cells. In contrast, bacteria which remain in the endosome are generally considered to be excluded from the MHC class I processing pathway.

L. monocytogenes infection in mice is the most extensively studied model of T cell mediated immunity against intracellular bacteria (MACKENESS 1962; HAHN and KAUFMANN 1981; KAUFMANN 1993). Protective immunity against listerial infection largely depends on MHC class I restricted CD8 T cells (KAUFMANN 1988). This CD8 T cell dependence most likely does not hold true for infection with

M. bovis BCG and a role for CD8 T cells is just recently being appreciated for protection against *M. tuberculosis* (FLYNN et al. 1992). In the following we will describe the properties of listerial infection in the murine system with an emphasis on CD8 T cells and on cytolytic T cell mechanisms, and we will compare this model with mycobacterial infections.

2 Similarities and Differences Between *Mycobacterium bovis* BCG and *Mycobacterium tuberculosis* Infections

M. tuberculosis is by far the most frequent etiologic agent of tuberculosis. The attenuated strain *M. bovis* BCG serves as a vaccine; however, its efficacy has been found to differ markedly (COLDITZ et al. 1994). To explain the failure of *M. bovis* BCG vaccination, differential immune responses induced by *M. bovis* BCG and required for efficient defense against tuberculosis have been proposed. A shared feature of *M. bovis* BCG and *M. tuberculosis* is their intracellular habitat: MPS serve as the major, if not exclusive, habitat and in this niche the mycobacteria survive, replicate and establish stable infection. During *M. bovis* BCG infection, however, most organisms are destroyed by activated MPS and only few bacteria persist in the phagosome, causing chronic infection without clinical disease. The T lymphocytes responsible for MP activation are CD4 T cells. The predominant role of CD4 T cells in infection with *M. bovis* BCG has been demonstrated in vivo and in vitro. Adoptive T cell transfers and monoclonal antibody depletion studies have revealed the relevance of CD4 T cells in vivo (ORME and COLLINS 1983; ORME 1987; PEDRAZZINI et al. 1987). *M. bovis* BCG infection induces antigen-specific CD4 T cells, which secret IFNγ in vitro. This cytokine plays a decisive role in the activation of antimycobacterial MP activities (PEDRAZZINI and LOUIS 1986; EMMERICH et al. 1986; MAGEE and WING 1988; HUYGEN et al. 1992; KAWAMURA et al. 1992). IFNγ production by CD4 T cells is supported by IFNγ from CD8 T cells, but it is likely that this T cell subset is only a minor source of IFNγ (DELIBERO et al. 1988). A second, but probably inferior, function of CD4 T cells is their cytolytic activity (DELIBERO et al. 1988; OTTENHOFF et al. 1988). Studies with β_2 microglobulin (β_2m) and H2-IAß gene deficient mice, which lack functionally active CD8 T cells or CD4 T cells, respectively, confirm that CD4 T cells are essential and sufficient for control of *M. bovis* BCG infection during the first few months (LADEL et al. 1995). However, at high inocula, CD8 T cells also contribute to acquired resistance. So, control of *M. bovis* BCG infection largely depends on CD4 T cells, which, by producing cytokines, activate antimycobacterial MP functions.

In contrast, activation of MPS alone is insufficient for defense against *M. tuberculosis*. As in the case of *M. bovis* BCG infection, MP activation is a function of CD4 T cells and only to a lesser extent of CD8 T cells (KAUFMANN and

FLESCH 1986; BOOM et al. 1987; DELIBERO et al. 1988; ORME et al. 1992). However, *M. tuberculosis* has the capacity to deactivate macrophages by various mechanisms (reviewed by BRITTON et al. 1994). These bacteria inhibit phagosome-lysosome fusion by production of ammonia and sulfatides. They bypass the oxidative burst by abusing the complement receptors CR1 and CR3 for uptake and by adhering to fibronectin and vitronectin receptors. Cell wall components of mycobacteria modulate cytokine effects and also interfere with ROI synthesis or other effector functions of MPS. Although evasion from attack by RNI is incompletely understood, evidence exists that this may also occur (for overview see CHAN and KAUFMANN 1994).

Thus, for optimal protection against *M. tuberculosis* infection, it is necessary to destroy the infected MPS, so that *M. tuberculosis* organisms can be engulfed by more efficacious MPS. The cells which destroy the deactivated MPS are probably cytolytic T lymphocytes (CTLS) (KAUFMANN 1988). Cytolytic function is predominantly expressed by CD8 T cells, and cytolytic activities of mycobacteria-specific CD8 T cells in vitro have been reported (D ELIBERO et al. 1988; OTTENHOFF et al. 1988). The importance of CD8 T cells in vivo has been studied by infection of $ß_2$m gene deficient mice infected with *M. tuberculosis* (FLYNN et al. 1992). These data strongly suggest that CD8 T cells represent the relevant T cell subpopulation in protection against *M. tuberculosis* infection. At the same time, the studies with these mutant mice confirm previous data from depletion studies and adoptive transfers, which together point to protective T cell mechanisms in addition to IFNγ production by CD4 T cells (MÜLLER et al. 1987; ORME 1987; PEDRAZZINI et al. 1987).

Therefore, failure of *M. bovis* BCG vaccination may be explained, at least partially by insufficient activation of CD8 T cells. The proportion of CD8 T cells in *M. tuberculosis* infection may be underrepresented and elevation of the CD8 T cell response may improve control. Further clues to the role of CD8 T cells in protection against tuberculosis can be obtained from experimental infection studies with *L. monocytogenes*, to be described in the following.

3 *Listeria monocytogenes* Infection

Similar to mycobacteria, *L. monocytogenes* organisms infect MPS. In contrast to mycobacteria, *L. monocytogenes* readily gains access to the host cytosol, where listerial proteins are processed and the resulting listerial peptides associate with MHC class I molecules. Subsequently this complex is transported to the surface and then recognized by CD8 T cells. In contrast to *M. tuberculosis*, non-phagocytic hepatocytes are additional host cells for *L. monocytogenes* (ROSEN et al. 1989). *L. monocytogenes* is able to grow in both hepatocytes and quiescent MPS in a more or less unrestricted fashion. This has consequences for the T cell response, because hepatocytes are MHC class II negative. Therefore T cells, which are able to recognize antigen-MHC class I complexes, are required. Destruction of hepatocytes and uptake by activated macrophages is necessary for defense. As a corollary, in listerial infections CD8 T cells are of central importance. Activation of MPS by IFNγ interferes with listerial egression into the cytosol and the pathogen is then unable to resist the defense mechanisms expressed in the phagosome-lysosome (Portnoy et al. 1989). Therefore, the course of *L. monocytogenes* infection is acute and immunity causes sterile eradication of bacteria. Yet, *L. monocytogenes* has developed several factors which facilitate intracellular survival and contribute to virulence. These factors are also thought to serve as important T cell antigens.

4 Virulence Factors of *Listeria monocytogenes*

Listeriosis is a food-borne disease and therefore the gut is the natural port of entry for *L. monocytogenes*. The bacteria invade epithelial cells and are translocated through the gut and then on to the blood stream and their target organs, liver and spleen. There they grow in the intracellular milieu and cause systemic infection. For effective infection of epithelial cells, *L. monocytogenes* has developed at least two independent factors which promote bacterial uptake into epithelial cells of the gut. The responsible genes are referred to as iap and inlA and the corresponding proteins p60 and internalin (Kuhn and Goebel 1989; Köhler et al. 1991; Gaillard et al. 1991; Kocks et al. 1992). Mutations of the genes reduce invasion into epithelial cells and fibroblasts (Kuhn and Goebel 1989; Gaillard et al. 1991). The p60 protein and internalin are virulence factors, because the mutant strains have lost their capacity to cause stable infection in mice. For uptake by professional phagocytes in liver and spleen, however, these factors are of minor importance. Therefore, in systemic experimental infections initiated by intraveneous injection of bacteria, these factors only slightly influence virulence. Probably, *M. tuberculosis* invades the host within MPS which are the only known habitat. The possible impact of invasion factors on tuberculosis, therefore, is less clear. However, in vitro, *M. tuberculosis* infects various nonphagocytic cells including epithelial cells and fibroblasts. Recently, a mycobacterial invasion factor has been described, which may be involved in the transcytosis of *M. tuberculosis* from the bronchial space through lung epithelial cells to blood monocytes (Arruda et al. 1993).

In the liver, *L. monocytogenes* organisms are found in both professional phagocytes and in hepatocytes. For evasion from macrophage attack, *L. monocytogenes* expresses additional virulence factors (for overview see Table 1). First, listeriolysin, together with phospholipase C, promotes egression from the phagosome to the cytosol (Katharıou et al 1987; Gaillard et al. 1986, 1987; Leimeister-Wächter et al. 1991; Mengaud et al. 1987). Listeriolysin is activated at the low pH of the phagosome and forms pores in the phagosome

Table 1. Virulence genes and factors of *Listeria monocytogenes*

Gene	Factor	Function	Regulation
iap	p60-Protein	Invasion	Unknown
inlA	Internalin	Invasion	Positive by PRF
hly	Listeriolysin	Lysis of phagosomal membrane	positive by PRF
plcA	PhospholipaseC	Lysis of phagosomal membrane	Positive by PRF
mlp	Metalloprotease	Maturation of lecithinase	Positive by PRF
plcB	Lecithinase	Lysis of double membrane	Positive by PRF
actA	Actin polymerization	Intracellular movement	Positive by PRF

PRF, positive regulation factor.

membrane. In the cytosol, actin polymerization, induced by the actA gene product, occurs and the bacteria move to the surface into a protrusion which is then phagocytosed by the neighboring cell (MENGAUD et al. 1991; KOCKS et al. 1992). The metalloprotease and the lecithinase are responsible for destruction of the cellular double membrane, so that cell to cell spread occurs and the neighboring cell cytosol can be reached by *L. monocytogenes*, without contact to the extracellular space (DOMANN et al. 1991; VAZQUEZ-BOLAND et al. 1992). The relevant genes are all organized in a gene cluster and controlled by a positive regulation factor (PRF), which is encoded by the gene prfA upstream from this gene cluster (for review see PORTNOY et al. 1992).

Among all of the virulence factors of *L. monocytogenes*, listeriolysin is probably the most important. Genetically engineered mutant strains which carry a mutation either in the structural hly gene *(L. monocytogenes* strain M20) or in the promotor region (*L. monocytogenes* strain M3) and a spontaneously occurring ahemolytic mutant (*L. monocytogenes* strain SLCC53), with a deletion in the regulatory gene prfA, are avirulent in vivo and in vitro (LEIMEISTER-WÄCHTER et al. 1989; SZALAY et al. 1994) (Fig. 1). Thus, listeriolysin promotes evasion of *L. monocytogenes* organisms from the phagosome to the cytosol and is essential for virulence (GAILLARD et al. 1987; KATHARIOU et al. 1987).

The appearance of *L. monocytogenes* in the cytosol has consequences for both the bacteria and for the host immune response. The presence of *L. monocytogenes* organisms in the cytosol enables the host cell to process and present listerial antigens not only through the MHC class II pathway but also through that of MHC class I. The intracytosolic presence of antigens is generally thought to be essential for association with MHC class I molecules and recognition by CD8 T cells. Extracellular proteins, when introduced into the cytosol by osmotic shock, are processed through the MHC class I pathway and elicit a CD8 T cell response (MOORE et al. 1988; STEINHOFF et al. 1989). The classical presentation of bacterial peptides by MHC class II molecules allows the cellular

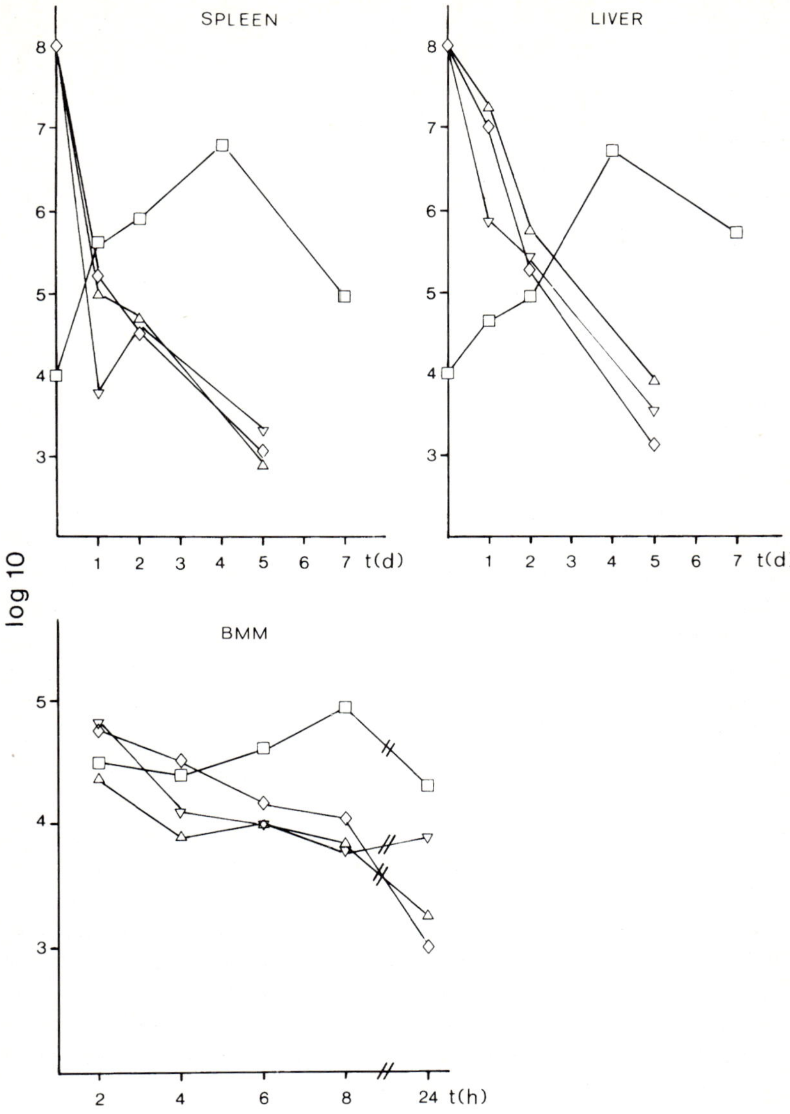

Fig. 1. In vivo and in vitro replication of hly-expressing and hly-deficient *Listeria monocytogenes* strains. In vivo replication of *L. monocytogenes* in spleen and liver:Mice were infected with 1×10^4hly–expressing *L. monocytogenes* strain EGD (□), or with 1×10^8 hly-deficient strain M3 (△), M2O (▽) or SLCC53 (◇). At the time points indicated, CFU were determined. In vitro replication in bone marrow-derived macrophages (BMM). Cell were infected with strain EGD (□), strain M3 (△), strain M2O (▽) or strain SLCC53 (◇) at a ratio of 20:1. At the time points indicated, cells were lysed, CFU were determined and calculated for 1×10^5 BMM. (From Szalay et al. 1994)

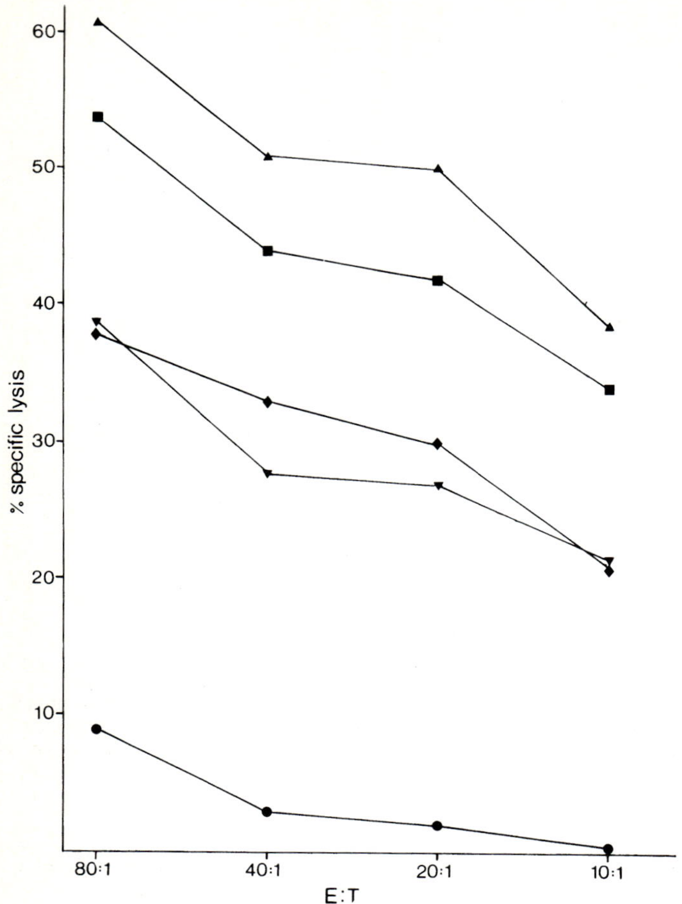

Fig. 2. Listeriolysin-indepepndent target cell recognition by *Listeria monocytogenes* reactive cytotoxic T lymphocytes (CTLS). Bone marrow-derived macrophasges (BMM) were infected overnight with listeriolysin positive *L. monocytogenes* EGD (■), or with listeriolysin negative M3 (▼), M20 (▲) or SLCC53 (◆). For control, BMM remained uninfected (●). Cells were used as targets for listeria-reactive CTL at indicated ratios. CTL were derived from bulk culture and restimulated with HKL as the specific antigen. (From SZALAY et al. 1994)

immune system to express CD4 T cell–mediated functions, including IL-2 and IFNγ production. Whilst IL-2 stimulates T cell proliferation, IFNγ, together with TNFα from MPS, activates antilisterial effector functions of MPS (KIDERLEN et al. 1984; GREEN et al. 1990; LANGERMANS et al. 1992; HUANG et al. 1993). Beside these typical helper functions, listeria-specific CD4 T cells also express cytolytic function in vitro and confer adoptive immunity (KAUFMANN et al. 1987). Moreover, in H2-IAß gene deficient mice infected with *L. monocytogenes,* the infection is exacerbated (LADEL et al. 1994). Nonetheless the major and more effective T cell subpopulation in listeriosis is the CD8 T cell, as demonstrated by adoptive

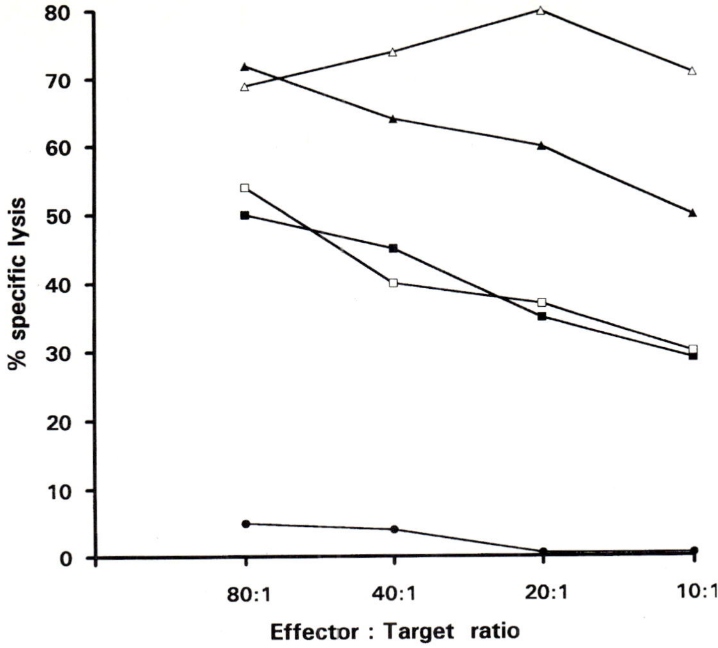

Fig. 3. P60-independent target cell recognition by *Listeria monocytogenes* reactive cytotoxic T lymphocytes (CTL). BMM were infected overnight with *L. monocytogenes* EGD (wild type) (■), RIII (p60-deficient) (▲), RIIIp60 (strain RIII reconstitued with p60 gene) (△), EGDp60 (p60 overexpressing wild type)(□), or remained uninfected (●). BMM were used as target cells for listeria-specific CTL at indicated ratios (Szalay et al., unpublished data)

transfer of CD8 T cells and infection of β_2m gene deficient mice (KAUFMANN et al. 1979, 1988; HARTY et al. 1992; ROBERTS et al. 1993; LADEL et al. 1994).

A link between listerial virulence and aquired cellular resistance has been revealed by showing that persistence of listeriae and strength of aquired resistance both depend on the degree of virulence of *L. monocytogenes* (KAUFMANN 1984). However, controversy exists as to whether: (1) listeriolysin alone mediates cytosolic localization of listeriae and (2) bacterial localization in the cytosol is necessary for association with MHC class I molecules. Moreover, the question whether listeriolysin serves as the major antigen recognized by CD8 T cells remains to be answered. Numerous studies have shown that listeriolysin is necessary for presentation of listerial antigens to CD8 T cells either as an antigen itself or as the mediator which promotes listerial escape from the phagosome to the cytosol (BRUNT et al. 1990; PAMER et al. 1991; BOUWER et al. 1992). Controverting these results are experiments showing that CD8 T cells also recognize and lyse bone marrow-derived macrophages (BMMS) infected with listeriolysin-negative *L. monocytogenes* mutants (Fig. 2) (BROWN et al. 1992; SZALAY et al. 1994). Blocking studies with monoclonal anti-CD8 antibodies and the fact that BMMS express only MHC class I molecules constitutively and not MHC class II molecules, together show that this cytolytic activity is due to CD8 T cells and not to CD4 T cells (KAUFMANN et al. 1987; SZALAY et al. 1994).

The invasion factor p60 is also recognized by CD8 T cells (PAMER 1994). Yet, target cells infected with the *L. monocytogenes* mutant RIII, lacking the p60 protein, and target cells infected with the p60 reconstituted RIII mutant (RIIIp60$^+$) or the p60 overexpressing mutant EGDp60$^+$ were lysed equally well by CD8 T cells (Fig. 3). These results question the essential participation of listeriolysin and p60 in antigen recognition by CD8 T cells.Furthermore, these data show that listerial antigens other than listeriolysin and p60 are recognized by *L. monocytogenes* reactive CD8 T cells. Finally, these experiments indicate that growth and persistence inside hostcells is not necessary for association of listerial peptides with MHC class I molecules.

The question how antigens from listeriolysin negative *L. monocytogenes* strains encounter the MHC class I processing pathway has not yet been resolved. Two alternatives are possible. First, listeriolysin-independent mechanisms which promote translocation of whole *L. monocytogenes* organisms from the endosomal to the cytosolic compartment need to be considered. Second, it is possible that *L. monocytogenes* remains in the endosome and that bacterial peptides gain access to the cytosol (PFEIFER et al. 1993).

5 MHC Class Ia and MHC Class Ib Restriction

The listeria-specific CD8 T cells so far described are all MHC class Ia-restricted (KAUFMANN et al. 1986; DELIBERO and KAUFMANN 1986; PAMER et al. 1991). In addition, MHC class Ib molecules are also involved in the presentation of listerial antigens (KAUFMANN et al. 1988; PAMER et al. 1992; KURLANDER et al. 1992). MHC class Ia or classical MHC class I molecules comprise the gene loci K, D, and L on chromosome 17 of mice. The nonclassical MHC class Ib molecules are encoded by the murine MHC genes Q, T, and M. MHC class Ib molecules differ from MHC class Ia molecules in their low degree of polymorphism and their partial tissue-specific expression (for overview see STROYNOWSKI and FISCHER LINDAHL 1994). Listerial peptides have been shown to be presented by H-2M3 molecules and this presentation can be blocked by artifical formylated peptides (PAMER et al. 1992; KURLANDER et al. 1992). Formylation at the initiation of protein synthesis is a property of prokaryotic and mitochondrial ribosomes. Recently a second nonclassical MHC class I molecule, H2-T, has been shown to present listerial peptides to CD8 T cells (BOUWER et al. 1994). Although in neither case has the sequence of the peptides yet been identified, it is assumed that MHC class Ib molecules focus on formylated peptides. Preferential association of MHC class Ib molecules with formylated peptides would favor presentation of bacterial peptides to CD8 T cells. The restricted polymorphism of MHC class Ib molecules could also facilitate rational design of peptide vaccines, because allelic variations of MHC molecules in the popu ation could be largely ignored and a few common peptides might be sufficient.

6 γ/δ T Cells

In addition to CD4 and CD8 T cells, $CD4^-CD8^-$ γ/δ T cells contribute to resistance against *L. monocytogenes* infection. In the listerial system direct evidence exists that γ/δ T cells confer early protection, though not as efficiently as α/β T cells (Hirumatsu et al. 1992; Mombaerts et al. 1993; Seen and Ziegler 1993; Ladel et al. 1996). Granuloma formaticn, characteristic for listerial and for mycobacterial infections, is markedly alterec in *L. monocytogenes*-infected, γ/δ T cell deficient mice. While in the presence of both T cell subpopulations distinct graulomas are formed in livers, the absence of γ/δ T cells leads to abscess-like lesion (Mombaerts et al. 1993). Such abscesses, however, do not develop in *M. bovis* BCG infected γ/δ deficient mice (Ladel et al. 1995b). Only indirect evidence has been obtained for participation of γ/δ T cells in protection against mycobacterial infections. In SCID mice, but not in nude mice, infection with *M. bovis* BCG leads to progressive infection (Izzo and North 1992). Since nude mice still habor relatively high levels of γ/δ T cells, this finding may be taken as evidence for γ/δ T cell contribution to protection against *M. bovis* BCG infection. The γ/δ T cells alone, however, are insufficient, because α/β T cell deficient mice succumb to *M. bovis* BCG infection within few months (Ladel et al. 1995b). The γ/δ T cell also accumulate at the side of *M. bovis* BCG infection (Augustin et al 1989; Janis et al. 1989; Inoue et al. 1991). In the human system γ/δ T cells are found in infected lesions, and in vitro γ/δ T cells respond to mycobacterial components (Modlin et al. 1989; Munk et al. 1990; Pfeffer et al. 1990; Schoel et al. 1994). However defined antigens, the mode of function and the MHC restriction of γ/δ T cells still remain largely unclear.

7 Immunization with Heat-killed Listeriae

Recent in vitro findings have shown that presentation of exogenous proteins by MHC class I molecules and recognition by CD8 T cells become possible after denaturation of the protein (Schirmbeck et al. 1994; Bachmann et al. 1994) with detergents or with heat (Schirmbeck et al. 1994; Martinez-Kinader et al. 1994). These results also indicate that recognition of exogenous proteins by CD8 T cells does not depend on viable organisms capable of egressing into the cytosol. Studies on specific recognition of listerial proteins by CD8 T cells have been extended to heat-killed bacteria. Earlier experiments with killed *L. mono-*

Table 2. Protection of mice from a lethal challenge with Listeria monocytogenes after HKL vaccination[a]

Mice	Proportion of protected mice
HKL vaccinated C57BL/6 mice	5/5
Naive C57BL/6 mice	0/5
HKL vaccinated $ß_2m^{-/-}$mice	5/5
Naive $ß_2m^{-/-}$mice	0/5
HKL vaccinated H-21a $ß^{-/-}$mice	5/5
Naive H-21a $ß^{-/-}$mice	1/5

[a]Mice were vaccinated with 1×10^9 HKL three times at 5 day intervals and after another 5 days challenged with 1.4×10^4 viable L.monocytogenes ($1.5 \times LD_{50}$). In the case of C57BL/6 mice, or 5×10^3 viable L.monocytogenes in the case of mutants. Survival was monitored over 10 days. Experimental groups were comprised of five animals and experiments were done twice(Szalay et al. 1995).

cytogenes as a vaccine led to contradictory results regarding the efficacy of vaccine-induced protection against listeriosis (VANDIJK et al. 1980; WIRSING VON KÖNIG et al. 1982). Based on the more recent observations mentioned above, we have reevaluated this issue. Mice were immunized with heat–killed. *L. monocytogenes* (HKL) by repeated intraveneous injections (SZALAY et al. 1995). Five days after the last injection, these mice were challenged with a high dose of viable *L. monocytogenes* organisms. The listerial burden in livers and spleens of vaccinated mice was significantly reduced compared to that in nonimmunized animals. Moreover, vaccinated mice survived an otherwise lethal challenge with *L. monocytogenes*. Vaccination was also effective in MHC class I (CD8 T cell) and MHC class II (CD4 T cell) deficient mice: $ß_2m$ and H-2I Aß gene deficient mice were protected from a lethal challenge with *L. monocytogenes* (Table 2). These data emphasize that specific CD8 T cells are activated by vaccination with dead bacteria. Moreover, these results show that *L. monocytogenes*-specific CD4 and CD8 T cells can develop independent from each other. In vitro characterization of these T cells revealed that listeria-specific CD8 T cells, activated by killed *L. monocytogenes*, recognize and lyse infected target cells. Finally, adoptive transfer of immune CD8 T cells to naive mice conferred protection against subsequent *L. monocytogenes* infection. These findings suggest that virulence and viability are not essential prerequistes for activation of CD8 T cell-mediated protective immunity. Hence, killed bacterial or even subunit vaccines against intracellular bacterial infections may reach the realm of feasibility.

8 Consequences for Antigen Presentation and Recognition by Cytolytic T Cells

Recognition of bacterial antigens by CD8 T cells has far-reaching consequences. First, immune surveillance is not restricted to the few target cells – B cells, dendritic cells, MPS – which express MHC class II molecules and are recognized by CD4 T cells. Rather, recognition of antigen/MHC class I complexes by CD8 T cells extents the scope of targets to virtually every cell type. Second, in contrast to CD4 T cells, which function indirectly via cytokine production, CD8 T cells directly lyse infected target cells. Thus, the way to destroy the niche of *L. monocytogenes* in order to promote bacterial uptake by highly activated professional phagocytes is an important function of CD8 T cells.

9 What Do We Learn from These Results for *Mycobacterium tuberculosis* Infection?

M. bovis BCG infection has been employed as the widely used model for studying mycobacterial infection. But experimental infection of mice with *M. bovis* BCG leads to an immune response dominated by CD4 T cells with only a small contribution by CD8 T cells. In contrast, recent analyses suggest that CD8 T cells are also central to immunity to tuberculosis (Flynn et al. 1992). In tuberculosis, participation of CD8 T cells seems to be primarily related to their cytolytic capacity rather than to their restriction by MHC class I molecules, since *M. tuberculosis* infects MPS only. Because *M. tuberculosis* deactivates MPS, destruction of infected MPS by CD8 cytotoxic T lympocytes (CTL) gains major relevance. A model system for infection with intracellular bacteria in which protective immunity largely depends on CD8 T cells is provided by *L. monocytogenes*. Moreover, this type of infection allows rapid analysis of the relevant immune mechanisms, due to the acute course of infection and the rapid development of specific T cells.

In listeriosis, the activation of specific CD8 T cells is possible because the pathogen egresses into the cytosol. This is in contrast to *M. bovis* BCG, which remains in the endosomal compartment. During infection with *M. tuberculosis* the exact localization of bacteria remains unclear. Two reports suggest that *M. tuberculosis*, but not *M. bovis* BCG, is translocated to the cytosol (Myrvik et al. 1984; McDounough et al. 1993). Expression by *M. tuberculosis* of cytolysins equivalent to listeriolysin might favor translocation. This notion is supported by the identification of a cytolysin which is only expressed by a virulent *M. tuberculosis* strain (King et al. 1993). Studies with hly$^-$ *L. monocytogenes* strains and with *E.coli* have raised the possibility that an alternate pathway exists which promotes translocation of mycobacterial proteins or peptides into the cytosol

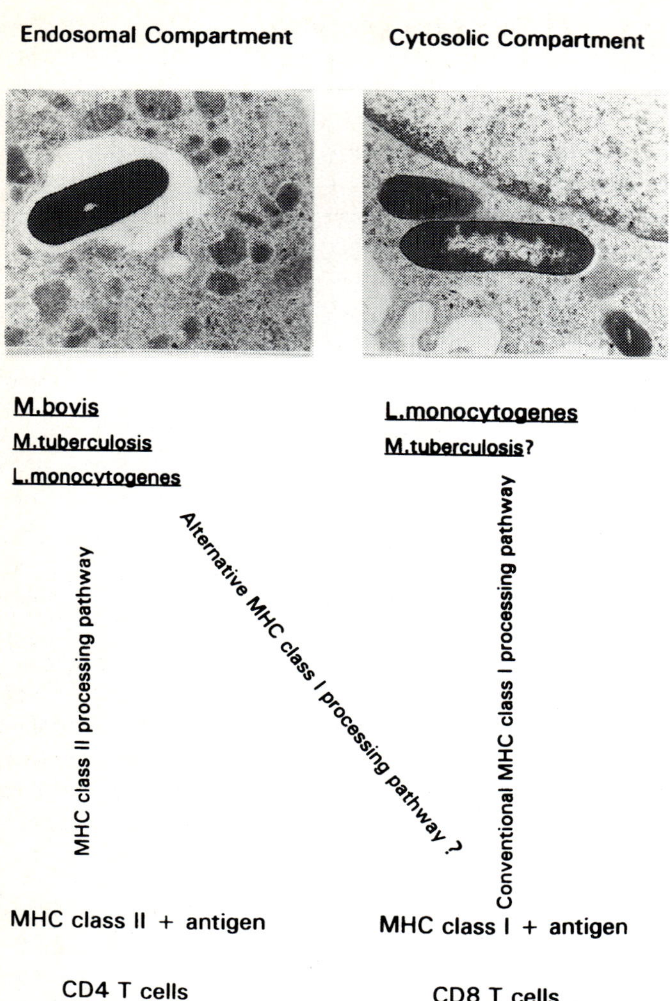

Fig. 4. Consequences of antigen compartmentalization for T cell recognition

with the pathogen remaining in the endosome (SZALAY et al. 1994; PFEIFER et al. 1993). Differential loading of MHC class I molecules and, as a consequence, differential action of CD8 T cells may help to explain the low efficacy of *M. bovis* BCG vaccination. An additional explanation for the low vaccination efficacy of *M. boris* BCG could be related to the differential metabolism of *M. bovis* BCG are available as *M.* bovis BCG and *M. tuberculosis* inside the phagosome. *M. tuberculosis* organisms largely circumvent macrophage killing and therefore secreted proteins may predominate over somatic proteins as the antigen source for CD8 T cells. In contrast, both secreted and somatic antigens of *M. bovis* BCG are available as *M. bovis* BCG cannot deactivate MPS; hence, a higher proportion of organisms is killed. This means that the equivalent proportion of the "protective" secreted antigens for CD8 T cell recognition may not be available in sufficient amounts during *M. bovis* BCG infection.

The question, however, remains, as to how mycobacterial antigens associate with MHC class I molecules so that they can be presented to CD8 T cells. We have learned from the listerial system that bacterial growth is not necessary for association with MHC class I molecules. Moreover, not only peptides derived from virulence factors function as listeria-specific antigens for CD8 T cells, and the cytosolic localization is not an absolute prerequisite for CD8 T cell recognition. We assume that the importance of CD8 T cells in infections with intracellular bacteria is primarily related to the cytolytic functions of CD8 T cells rather than to IFNγ secretion. Adoptive transfer of CD8 T cells protects mice in an apparently IFNγ-independent fashion, and mice with a gene defect in perforin suffer more severely from listerial infection (Harty et al. 1992; Kägi et al. 1994b). So, clearance of an infection with intracellular bacteria probably depends on CD8 CTL. While the cytoplasmic localization of virulent *L. monocytogenes* favors association with MHC class I molecules and recognition by CD8 T cells, *M. bovis* BCG organisms remain in the endosomal compartment and assoication with MHC class II molecules occurs. Recent findings, however, have revealed a contribution of CD8 T cells to the optimum defense against *M. bovis* BCG infection, as $ß_2$m gene deficient mice are more susceptible than their heterozygous littermates (Ladel et al. 1995). The definite localization of *M. tuberculosis*, *M. bovis* BCG and avirulent *L. monocytogenes* organisms needs to be clarified, but even if, by an alternative MHC class I processing pathway, the organisms remain in the endosome, translocation of antigenic peptides to the cytosol appears possible (Fig. 4).

10 Conclusions

As demonstrated in the listerial system, the advantage for the host of association of bacterial antigens with MHC class I molecules is undeniable, because: (1) MHC class I molecules are expressed by virtually all nucleated cells and therefore, even if *M. tuberculosis* has a second, yet unknown, habitat beside MPS, they can be recognized by specific T cells. (2) Presentation of bacterial antigens by MHC class Ib molecules is probably not restricted to *L. monocytogenes* infection, but can be generalized to other bacterial infections. Since at least some MHC class Ib molecules preferentially present formylated peptides and protein formylation is characteristic of prokaryotic cells, association of bacterial peptides and MHC class Ib molecules favors specific presentation. Moreover, MHC class Ib molecules seem to present bacterial antigens to γ/δ T cells, which contribute to antibacterial protection. Finally, CD8 T cells are typically cytolytic. The advantage of CD8 CTL is the destruction of infected, quiescent or deactivted host cells, thus promoting bacterial release and uptake by more efficiently activated macrophages.

Note Added in Proof. Since submission of this manuscript several findings relevant to the issues discussed in the chapter have been reported. These include: First, nonproteinacious ligands for human γ/δ T cells have been identified. They include nucleotide-, carbohydrate- and alkyl-derivatives. As a common feature all these ligands contain phosphate (Tanaka Y et al. [1995] Nature 375: 155–158; Schoel B et al. [1994] Eur J Immunol 24: 1886-1892; Constant P et al. [1994] Science 264: 267–270). Second, evidence has also been presented that listeria-specific MHC class 1 b-restricted CD8 T cells recognize nonproteinacious ligands (Nataraj et al. [1996] Int Immunol 8: 367–378). Third, the essential role of γ/δ T cells in resistance against murine tuberculosis has been formally proven (Ladel et al. [1995] Eur J Immunol 25: 2877–2881). Fourth, evidence is now overwhelming that microbial egression from the endosome into the cytosol is not essential for antigen introduction into the MHC class I pathway (e.g., Kovacsovics- Bankowski and Rock [1995] Science 267: 243–246). This pathway is independent of peptide regurgltation and exogenous MHC loading. Rather, it involves an alternative intracellular translocation pathway. Fifth, a superior role of secreted over somatic antigens has been demonstrated in a murine model using recombinant Salmonella vaccine carriers displaying listerial antigens in somatic or secreted form. Salmonella vaccine carriers secreting listerial antigens provided significantly higher protection against subsequent listerial infection as compared to carriers expressing the same antigens in somatic form (Hess et al. [1996] Proc Natl Acad Sci USA 93: 1458–1463).

Acknowledgements. This work received financial support from the Sonderforschungsbereich 322 "Lympho-Hamopoese", the German Science Foundation, project Ka 573/3-1/2 and the BMBF Verbundprojekt "Mykobakterielle Infektionen".

References

Andrew PW, Jackett PS, Lowrie DB (1985) Killing and degradation of microorganisms by macrophages. In: Dean RT, Jessop W (eds) Elsevier Biomedical, Amsterdam, pp 311–335

Arruda S, Bomfim G, Knights R, Huima-Byron T, Riley LW (1993) Cloning of a M. tuberculosis DNA fragment associated with entry and survival inside cells. Science 261: 1454–1457

Augustin A, Kubo RT, Sim GK (1989) Resident pulmonary lymphocytes expressing the γ/δ T cell receptor. Nature 340: 239–241

Babior, BM (1984) The respiratory burst of phagocytes. J. Clin Invest 73: 599–601

Bachmann MF, Küudig TM, Freer G, Li Y, Kang CY, Bishop DH, Hengartner H, Zinkernagel RM (1994) Induction of protective cytotoxi T cells with viral proteins. Eur J Immunol 24: 2228–2236

Boom Wh, Husson RN, Young RA, David JR, Piessens WF (1987) *In vivo* and In vitro characterization of murine T cell clones reactive to *Mycobacterium tuberculosis*. Infect Immun 55: 2223–2229

Bouwer HGA, Nelson CS, Gibbins BL, Portnoy DA, Hinrichs DJ (1992) Listeriolysin O is a target of the immune response to *Listeria monocytogenes*. J Exp Med 175: 1467–1471

Bouwer HGA, Fischer Lindahl K, Baldridge JR, Wagner CR, Barry RA, Hinrichs DJ (1994) An H2-T MHC class Ib molecule presents *Listeria monocytogenes*-derived antigen to immune $CD8^+$ cytotoxic T cells. J Immunol 152: 5352–5360

Britton Wj, Roche PW, Winter N (1994) Mechanisms of persistence of mycobacteria. Trends Microbiol 2: 284–288

Brown ML, Fields PE, Kurlander RJ (1992) Metabolic requirements for macrophage presentation of *Listeria monocytogenes* to immune CD8 Cells. J Immunol 148: 555–561

Brunt LM, Portnoy DA, Unanue ER (1990) Presentation of *Listeria monocytogenes* to $CD8^+$ T cells requires secretion of hemolysin and intracellular bacterial growth. J Immunol 145: 3540–3546

Chan J, Kaufmann SHE (1994) Immune mechanisms of protection tuberculosis. In: Bloom BR (ed) Pathogenesis, protection and control. American Society for Microbiologiy, Washington DC, pp 389–415

Colditz GA, Brewer TF, Berkey CS, Wilson ME, Burdick E, Fineberg HV, Mosteller F (1994) Efficacy of BCG vaccine in the prevention of tuberculosis. Meta-analysis of the published literature. JAMA 271: 698–702

DeLibero G, Kaufmann SHE (1986) Antigen-specific Lyt2+ cytolytic T lymphocytes from mice infected with the intracellular bacterium *Listeria monocytogenes*. J Immunol 137: 2688–2694

Delibero G, Flesch I, Kaufmann SHE (1988) Mycobacteria reactive Lyt2+ T cell lines. Eur J Immunol 18: 59–66

Domann E, Leimeister-Wachter M, Goebel W, Chakraborty T (1991) Molecular cloning, sequencing, and identification of a metalloprotease gene from *Listeria monocytogenes* that is species specific and physically linked to the listeriolysin gene. Infect Immun 59: 65–72

Emmerich F, Thole J, Van Embden JDA, Kaufmann SHE (1986) A recombinant 64 kiloDalton protein of *Mycobacterium bovis* BCG specifically stimulates human T4 clones reactive to mycobacterial antigens. J Exp Med 163: 1024–1029

Flynn JL, Goldstein MM, Triebold KJ, Koller B, Bloom BR (1992) Major histocompatibility complex class I-restricted T cells are required for resistance to *Mycobacterium tuberculosis* infection. Proc Natl Acad Sci USA 89: 12013–12017

Gaillard J-L, Berche P, Sansonetti P (1986) Transposon mutagenesis as a tool to study the role of hemolysin in the virulence of *Listeria monocytogenes*. Infect Immun 52: 50–55

Gaillard J-L, Berche P, Mounier J, Richard S, Sansonetti P (1987) *In vitro* model of penetration and intracellular growth of *Listeria monocytogenes* in the human enterocyte-like cell line Caco-2. Infect Immun 55: 2822–2829

Gaillard J-L, Berche P, Frehel C, Gouin E, Cossart P (1991) Entry of *L monocytogenes* into cells is mediated by internalin, a repeat protein reminiscent of surface antigens from gram-positive cocci. Cell 65: 1127–1141

Germain RN, Margulies DH (1993) The biochemistry and cell biology of antigen processing and presentation. Annu Rev Immunol 11: 403–450

Green SJ, Mellouk S, Hoffman SL, Meltzer MS, Nacy CA (1990) Cellular mechanisms of nonspecific immunity to intracellular infection: cytokine-induced synthesis of toxic nitrogen oxides from L-arginine by macrophages and hepatocytes. Immunol Lett 25: 15–20

Hahn H, Kaufmann SHE (1981) The role of cell mediated immunity in bacterial infections. Rev infect Dis 3: 1221–1250

Harty JT Bevan MJ (1992) CD8+ T cells specific for a single nonamer epitope of *Listeria monocytogenes* are protective *in vivo*. J Exp Med 175: 1531–1538

Harty JT, Schreiber RD, Bevan MJ (1992) CD8+ T cells can protect against an intracellular bacterium in an interferon γ-independent fashion. Proc Natl Acad Sci USA 89: 11612–11616

Hibbs JB Jr, Taintor RR, Vavrin Z, Rachlin EM (1988) Nitric oxide: a cytotoxic activated macrophage effector molecule. Biochem Biophys Res Commun 1987: 87–94

Hiromatsu K, Yoshikai Y, Matsuzaki G, Ohga S, Muramori K, Matsumoto K, Bluestone JA, Nomoto K (1992) A protective role of γ/δ T cells in primary infection with *Listeria monocytogenes* in mice. J Exp Med 175: 49

Horwitz MA (1988) Intracellular parasitism. Cur Opin Immunol 1: 41–36

Huang S, Hendriks W, Althage A, Hemmi S, Bluethmann H, Kamijo R, Vilcek J, Zinkernagel RM, Aguet M (1993) Immune response in mice that lack the interferon-γ receptor. Science 259: 1742–1745

Huygen K, Abramowicz D, Vandenbussche P, Jacobs F, DeBruyn J, Kentos A, Drowart A, Van Vooren J-P, Goldman M (1992) Spleen cell cytokine secretion in *Mycobacterium bovis* BCG-infected mice. Infect Immun 60: 2880–2886

Innoue T, Yoshikai Y, Matsuzaki G, Nomoto K (1991) Early appearing γ/δ-bearing T cells during infection with Calmette Guerin bacillus. J Immunol 146: 2754–2762

Izzo AA, North RJ (1992) Evidence for α/β T cell-independent mechanism of resistance to mycobacteria. Bacillus-Calmette-Guerin Causes progressive infection in severe combined immunodificient mice, but not in nude mice or in mice depleted of CD4+ and CD8+ T cells. J Exp Med 176: 581–586

Janis EM, Kaufmann SHE, Schwartz RH, Pardoll AM (1989) Activation of γ/δ T cells in the primary immune response to *Mycobacterium tuberculosis*. Science 244: 713–717

Kägi D, Ledermann B, Bürki K, Seiler P, Odermatt B, Olsen KJ, Podack ER, Zinkernagel RM, Hengartner H (1994a) Cytotoxicity mediated by T cells and natural killer cells is greatly impaired in perforin-deficient mice. Nature 369: 31–37

Kägi D, Ledermann B, Bürki K, Hengartner H, Zinkernagel RM (1994b) CD8+ T cell-mediated protection against an intracellular bacterium by perforin-dependent cytotoxicity. Eur J Immunol 24: 3068–3072

Kathariou S, Metz P, Hof H, Goebel W (1987) Tn916-induced mutations in the hemolysin determinant affecting virulence of *Listeria monocytogenes*. J Bacteriol 169: 1291–1297

Kaufmann SHE (1984) Acquired resistance to facultative intracellular bacteria: relationship between persistence, crossreactivity on the T cell level and capacity to stimulate cellular immunity of different *Listeria* strains Infect Imun 45: 234–241

Kaufmann SHE (1988) CD8+ T lymphocytes in intracellular microbial infections. Immunol Today 9: 168–174

Kaufmann SHE (1993) Immunity to intracellular bacteria Annu Rev Immunol 11: 129–163

Kaufmann SHE, Flesch IAE (1986) Function and antigen recognition pattern of L3T4+ T cell clones from *Mycobacterium tuberculosis* immune mice. Infect Immun 54: 291–296

Kaufmann SHE, Simon MM, Hahn H (1979) Specific Lyt 123 T cells are involved in protection against *Listeria monocytogenes* and in delayed-type hypersensitivity to listerial antigens. J Exp Med 150: 1033-1038

Kaufmann SHE, Hug E, DeLibero G (1986) *Listeria monocytogenes*-reactive T lymphocyte clones with cytolytic activity infected target cells. J Exp Med 164: 363–368

Kaufmann SHE, Hug E, Väth U, DeLibero G (1987) Specific lysis of *Listeria* monocytogenes-infected macroghages by class II-restricted L3T4+ T Cells. Eur J Immunol 17: 237–246

Kaufmann SHE, Rodewald H-R, Hug E, DeLibero G (1988) Cloned *Listeria monocytogenes* specific non-MHC-restricted Lyt-2+ T cells with cytolytic and protective activity. J Immunol 140: 3173–3179

Kawamura I, Tsukada H, Yoshikawa H, Fujita M, Nomoto K, Mitsuyama M (1992) IFN-γ producing ability as apossible marker for the protective T cells against *Mycobacterium bovis* BCG in mice. J Immunol 148: 2887–2893

Kiderlen AF, Kaufmann SHE, Lohmann-Matthes M-L (1984) Protection of mice against the intracellular bacterium *Listeria monocytogenes* by recombinant immune interferon. Eur J Immunol 14: 964–967

King C, Sathish M, Crawford JT, Shinnick TM (1993) Expression of contact-dependent cytolytic activity of *Mycobacterium tuberculosis* and isolation of the locus encoding the activity. Infect Immun 61: 2708–2712

Kocks C, Gouin E, Tabouret M, Berche P, Ohayon H, Cossart P (1992) *L. monocytogenes*-induced actin assembly requires the actA gene product, a surface protein. Cell 68: 521–531

Köhler S, Bubert A, Vogel M, Goebel W (1991) Expression of the iap gene coding for protein p60 of *Listeria monocytogenes* is controlled on the posttranscriptional level. J Bacteriol 173: 4668–4674

Kuhn M, Goebel W (1989) Identification of an extracellular protein of *Listeria monocytogenes* possibly involved in intracellular uptake by mammalian cells. Infect Immun 57: 55–61

Kurlander RJ, Shawar SM, Brown ML, Rich RR (1992) Specialized role for a murine class I-b MHC molecule in prokaryotic host defenses. Science 257: 678–679

Ladel CH, Flesch IAE, Arnoldi J, Kaufmann SHE (1994) Studies with MHC deficient knock-out mice reveal impact of both MHC I and MHC II dependent T cell responses on *Listeria monocytogenes* infection. J Immunol 153: 3116–3122

Ladel CH, Daugelat S, Kaufmann SHE (1995) Immune response to *Mycobacterium bovis* bacille Calmette Guerin infection in major histocompatibility complex I and II deficient knock-out mice: contribution of CD4 and CD8 T cells to aquired resistance. Eur J Immunol 25: 377–384

Ladel CH, Hess J, Daugelat S, Mombaerts P, Tonegawa S, Kaufmann SHE (1995b) Contribution of α/β and γ/δT lymphocytes to immunity against Mycobacterium bovis Bacillus Calmette Guerin: studies with T cell receptor-deficient mutant mice. Eur J Immun. 838-846

Ladel CH, Blum C, Kaufmann SHE (in press) Control of natural killer cell mediated innate resistance against the intracellular pathogen Listeria monocytogenes by γ/δ T lymphoctes. Infect Immun 64

Langermans JAM, Van der Hulst MEB, Nibbering PH, Van Furth R (1992) Endogenous tumor necrosis factor alpha is required for enhanced antimicrobial activity against *Toxoplasma gondii* and *Listeria monocytogenes* in recombinant gamma interferon-treated mice. Infect Immun 60: 5107–5112

Lehrer RI, Ganz T, Selsted ME (1991) Defensins: Endogenous antibiotic peptides of animal cells. Cell 64: 229–230

Leimeister-Wächter M, Goebel W, Chakraborty T (1989) Mutations affecting hemolysin production in *Listeria monocytogenes* located outside the listeriolysin gene. FEMS Microbiol Lett 65: 23–30

Leimeister-Wächter M, Domann E, Chakraborty T (1991) Detection of a gene encoding a phosphatidylinositol-specific phospholipase C that is co-ordinately expressed with listeriolysin in *Listeria monocytogenes*. Mol Mic·obiol 5: 361–366

Liew FY, Cox FEG (1990) Non-specific defence mechanism: the role of nitric oxide Immunol Today 12: A17–A21

Mackaness GB (1962) Cellular resistance to infection. J Exp Med 116: 381–406

Magee DM, Wing EJ (1988) Clcned L3T4+ T lymphocytes protect mice against *Listeria monocytogenes* by secreting IFN-γ. J Immunol 141: 3202–3207

Martinez-Kinader B, Wagner H, Heeg K (1994) Denaturated ovalbumin enters class I-restricted pathway of amtigen presentation and allows in vivo sensititization of cytolytic CD8+ T lymphocytes. Immunobiology 19·: 161

McDounough KA, Kress Y, Bloom BR (1993) Pathogenesis of tuberculosis: interaction of *Mycobacterium tuberculosis* witr macrophages Infect Immun 61: 2763–2773

Mengaud J, Chenevert J, Geoffroy C, Gaillard J-L, Cossart P (1987) Identification of the structural gene encoding the SH-activated hemolysin of *Listeria monocytogenes*: listeriolysin O is homologous to streptolysin O and pneumolysin. Infect Immun 55: 3225–3227

Mengaud J, Barun-Breton C, Cossart P (1991) Identification of a phosphatidylinositol-specific phospholipase C in *Listeria monocytogenes*: a novel type of virulence factor? Mol Microbiol 5: 367–372

Modlin RL, Pirmez C, Hofmann FM, Torigian V, Uyemura K, Rea TH, Bloom BR, Brenner MB (1989) Lymphocytes bearing antigen-specific γ/δ T-cell receptors accumulate in human infectious disease lesions. Nature 339: 544–548

Mombaerts P, Arnoldi J, Russ F, Tonegawa S, Kaufmann SHE (1993) Different roles of α/β and γ/δ T cells in immunity against an intracellular bacterial pathogen. Nature 365: 53–56

Moore MW, Carbone FR, Bevan MJ (1988) Introduction of soluble protein into the class I pathway of antigen processing and presentation. Cell 54: 777

Mosmann TR, Coffman RL (1989) TH1 and TH2 cells: different patterns of lymphokine secretion lead to different functional properties. Annu Rev Immunol 7: 145–173

Moulder JW (1985) Comparative biology of intracellular parasitism. Microbiol Rev 49: 298–337

Müller I, Cobbold SP, Waldmann H, Kaufmann SHE (1987) Impaired resistance against *Mycobacterium tuberculosis* infection after selective in-vivo depletion of $L3T4^+$ and $Lyt2^+$ T cells. Infect Immun 55: 2037–2041

Munk ME, Gatrill A, Kaufmann SHE (1990) Antigen-specific target cell lysis and interleukin-2 secretion by *Mycobacterium tuberculosis*-activated γ/δ T cells. J Immunol 145: 2434–2439

Myvrik QN, Leake ES, Wright MJ (1984) Disruption of phagosomal membranes of normal alveolar macrophages by the H37Rv strain of *Mycobacterium tuberculosis*. Am Rev Respir Dis 129: 322–328

Orme IM (1987) The kinetics of emergence and loss of mediator T lymphocytes acquired in response to infection with *Mycobacterium tuberculosis*. J Immunol 138: 293–298

Orme IM, Collins FM (1983) Protection against *Mycobacterium tuberculosis* infection by adoptive transfer. J Exp Med 158: 74–83

Orme IM, Miller ES, Roberts AD, Furney SK, Griffin JP, Dobos KM, Chi D, Rivoire B, Brennan PJ (1992) T lymphocytes mediating protection and cellular cytolysis during the course of *Mycobacterium tuberculosis* infection. J Immunol 148: 189–196

Ottenhoff THM, Kale Ab B, Van Embden JDA, Thole JER, Kiessling R (1988) The recombinant 65 kD heat shock protein of *Mycobacterium bovis* BCG/M tuberculosis. J Exp Med 168: 1947

Pamer EG (1994) direct sequence identification and kinetic analysis of an MHC class I-restricted *Listeria monocytogenes* CTL epitope. J Immunol 152: 686–694

Pamer EG, Harty JT, Bevan MJ (1991) Precise prediction of a dominant class I MHC-restricted epitope of *Listeria monocytogenes*. Nature 353: 852–855

Pamer EG, Wang C-R, Flaherty L, Fischer Lindahl K, Bevan MJ (1992) H-2M3 presents a *Listeria monocytogenes* peptide to cytotoxic T lymphocytes. Cell 70: 215–223

Peddrazini T, Louis JA (1986) Function analysis *in vitro* and *in vivo* of *Mycobacterium bovis* strain BCG-specific T cell clones. J Immunol 136: 1828–1834

Peddrazini T, Hug K, Louis JA (1987) Importance of L3T4+ and Lyt-2+ cells in the immunologic control of infection with *Mycobacterium bovis* strain bacillus Calmette-Guerin in mice. Assessment by elimination of T cell subsets *in vivo*. J Immunol 139: 2032–2037

Pfeffer K, Schoel B, Gulle H, Kaufmann SHE, Wagner H (1990) Primary responses of human T cells to mycobacteria: a frequent set of γ/δ T cells are stimulated by protease-resistant ligands. Eur J Immunol 20: 1175-1179

Pfeifer JD, Wick MJ, Roberts RL, Findlay K, Normark SJ, Harding CV (1993) Phagocytic processing of bacterial antigens for class I MHC presentation to T cells. Nature 361: 359–362

Portnoy DA, Schreiber RD, Connelly P, Tilney LG (1989) Gamma interferon limits access of *Listeria monocytogenes* to the macrophage cytoplasm. J Exp Med 170: 2141–2146

Portnoy DA, Chakraborty T, Goebel W, Cossart P (1992) Molecular determinants of *Listeria monocytogenes* pathogenesis. Infect Immun 60: 1263–1267

Roberts AD, Ordway DJ, Orme IM (1993) *Listeria monocytogenes* infection on ß2 microglobulin-deficient mice. Infect Immun 61: 1113–1116

Rosen H, Gordon S, North RJ (1989) Exacerbation of murine listeriosis by a monoclonal antibody specific for the type 3 complement receptor of myelomonocytic cells. J Exp Med 170: 27–37

Schirmbeck R, Böhm W, Reimann J (1994) Injection of detergent-denatured ovalbulmin primes murine class I-restricted cytotoxic T cells in vivo. Eur J Immunol 24: 2068–2072

Schoel B, Sprenger S, Kaufmann SHE (1994) Phosphate is essential for stimulation of Vγ9Vδ2 T lymphocytes by mycobacterial low molecular weight ligand. Eur J Immunol 24: 1886–1892

Sher A, Coffman RL (1992) Regulation of immunity to parasites by T cells and T cell-derived cytokines. Annu Rev Immunol 10: 385–410

Skeen MJ, Ziegler HK (1993) Induction of murine peritioneal γ/δ T cells and their role in resistance to bacterial infection. J Exp Med 178: 971–984

Steinhoff U, Schoel B, Kaufmann SHE (1989) Lysis of interferon-γ activated Schwann cell by cross-reactive CD8 $^{+}$ α/β T cells with specificity for the mycobacterial 65 kd heat shock protein. Int Immunol 2: 279–284

Stroynowski I, Fischer Lindahl K (1994) Antigen presentation by non-classical class I molecules. Curr Opin Immunol 6: 38–44

Szalay G, Hess J, Kaufmann SHE (1994) Presenation of *Listeria monocytogenes* antigens by major histocompatibility complex class I molecules to CD8 cytotoxic T lymphocytes independent of listeriolysin secretion and virulence. Eur J Immunol 24: 1471–1477

Szalay G, Ladel CH, Kaufmann SHE (1995) Stimulation of protective CD8 T lymphocytes by vaccination with nonliving bacteria. Proc Natl Acad Sci 92: 12389-12392

Van Dijk H, Hofhuls FMA Berna EMJJ, van der Mear C, Willers JM (1980) Killed *Listeria monocytogenes* vaccine is protective in C3H/HeJ mice without addition of adjuvants. Nature 286: 713–714

Vazquez-Boland J-A, Kocks C, Dramsi S, Ohayon H, Geoffroy C, Mengaud J, Cossart P (1992) Nucleotide sequence of the lecithinase operon of *Listeria monocytogenes* and possible role of lecithinase in cell-cell spread. Infect Immun 60: 219–230

Weinberg ED (1992) Iron depletion: a defense against intracellular infection and neoplasia. Life Science 50: 1289–1297

Wirsing von König CH, Finger H, Hof H (1982) Failure of killed *Listeria monocytogenes* vaccine to produce protective immunity. Nature 297: 233–234

Subject Index

Current Topics in Microbiology and Immunology

Volumes published since 1989 (and still available)

Vol. 175: **Aktories, Klaus (Ed.):** ADP-Ribosylating Toxins. 1992. 23 figs. IX, 148 pp. ISBN 3-540-54598-0

Vol. 176: **Holland, John J. (Ed.):** Genetic Diversity of RNA Viruses. 1992. 34 figs. IX, 226 pp. ISBN 3-540-54652-9

Vol. 177: **Müller-Sieburg, Christa; Torok-Storb, Beverly; Visser, Jan; Storb, Rainer (Eds.):** Hematopoietic Stem Cells. 1992. 18 figs. XIII, 143 pp. ISBN 3-540-54531-X

Vol. 178: **Parker, Charles J. (Ed.):** Membrane Defenses Against Attack by Complement and Perforins. 1992. 26 figs. VIII, 188 pp. ISBN 3-540-54653-7

Vol. 179: **Rouse, Barry T. (Ed.):** Herpes Simplex Virus. 1992. 9 figs. X, 180 pp. ISBN 3-540-55066-6

Vol. 180: **Sansonetti, P. J. (Ed.):** Pathogenesis of Shigellosis. 1992. 15 figs. X, 143 pp. ISBN 3-540-55058-5

Vol. 181: **Russell, Stephen W.; Gordon, Siamon (Eds.):** Macrophage Biology and Activation. 1992. 42 figs. IX, 299 pp. ISBN 3-540-55293-6

Vol. 182: **Potter, Michael; Melchers, Fritz (Eds.):** Mechanisms in B-Cell Neoplasia. 1992. 188 figs. XX, 499 pp. ISBN 3-540-55658-3

Vol. 183: **Dimmock, Nigel J.:** Neutralization of Animal Viruses. 1993. 10 figs. VII, 149 pp. ISBN 3-540-56030-0

Vol. 184: **Dunon, Dominique; Mackay, Charles R.; Imhof, Beat A. (Eds.):** Adhesion in Leukocyte Homing and Differentiation. 1993. 37 figs. IX, 260 pp. ISBN 3-540-56756-9

Vol. 185: **Ramig, Robert F. (Ed.):** Rotaviruses. 1994. 37 figs. X, 380 pp. ISBN 3-540-56761-5

Vol. 186: **zur Hausen, Harald (Ed.):** Human Pathogenic Papillomaviruses. 1994. 37 figs. XIII, 274 pp. ISBN 3-540-57193-0

Vol. 187: **Rupprecht, Charles E.; Dietzschold, Bernhard; Koprowski, Hilary (Eds.):** Lyssaviruses. 1994. 50 figs. IX, 352 pp. ISBN 3-540-57194-9

Vol. 188: **Letvin, Norman L.; Desrosiers, Ronald C. (Eds.):** Simian Immunodeficiency Virus. 1994. 37 figs. X, 240 pp. ISBN 3-540-57274-0

Vol. 189: **Oldstone, Michael B. A. (Ed.):** Cytotoxic T-Lymphocytes in Human Viral and Malaria Infections. 1994. 37 figs. IX, 210 pp. ISBN 3-540-57259-7

Vol. 190: **Koprowski, Hilary; Lipkin, W. Ian (Eds.):** Borna Disease. 1995. 33 figs. IX, 134 pp. ISBN 3-540-57388-7

Vol. 191: **ter Meulen, Volker; Billeter, Martin A. (Eds.):** Measles Virus. 1995. 23 figs. IX, 196 pp. ISBN 3-540-57389-5

Vol. 192: **Dangl, Jeffrey L. (Ed.):** Bacterial Pathogenesis of Plants and Animals. 1994. 41 figs. IX, 343 pp. ISBN 3-540-57391-7

Vol. 193: **Chen, Irvin S. Y.; Koprowski, Hilary; Srinivasan, Alagarsamy; Vogt, Peter K. (Eds.):** Transacting Functions of Human Retroviruses. 1995. 49 figs. IX, 240 pp. ISBN 3-540-57901-X

Vol. 194: **Potter, Michael; Melchers, Fritz (Eds.):** Mechanisms in B-cell Neoplasia. 1995. 152 figs. XXV, 458 pp. ISBN 3-540-58447-1

Vol. 195: **Montecucco, Cesare (Ed.):** Clostridial Neurotoxins. 1995. 28 figs. XI., 278 pp. ISBN 3-540-58452-8

Vol. 196: **Koprowski, Hilary; Maeda, Hiroshi (Eds.):** The Role of Nitric Oxide in Physiology and Pathophysiology. 1995. 21 figs. IX, 90 pp. ISBN 3-540-58214-2

Vol. 197: **Meyer, Peter (Ed.):** Gene Silencing in Higher Plants and Related Phenomena in Other Eukaryotes. 1995. 17 figs. IX, 232 pp. ISBN 3-540-58236-3

Vol. 198: **Griffiths, Gillian M.; Tschopp, Jürg (Eds.):** Pathways for Cytolysis. 1995. 45 figs. IX, 224 pp. ISBN 3-540-58725-X

Vol. 199/I: **Doerfler, Walter; Böhm, Petra (Eds.):** The Molecular Repertoire of Adenoviruses I. 1995. 51 figs. XIII, 280 pp. ISBN 3-540-58828-0

Vol. 199/II: **Doerfler, Walter; Böhm, Petra (Eds.):** The Molecular Repertoire of Adenoviruses II. 1995. 36 figs. XIII, 278 pp. ISBN 3-540-58829-9

Vol. 199/III: **Doerfler, Walter; Böhm, Petra (Eds.):** The Molecular Repertoire of Adenoviruses III. 1995. 51 figs. XIII, 310 pp. ISBN 3-540-58987-2

Vol. 200: **Kroemer, Guido; Martinez-A., Carlos (Eds.):** Apoptosis in Immunology. 1995. 14 figs. XI, 242 pp. ISBN 3-540-58756-X

Vol. 201: **Kosco-Vilbois, Marie H. (Ed.):** An Antigen Depository of the Immune System: Follicular Dendritic Cells. 1995. 39 figs. IX, 209 pp. ISBN 3-540-59013-7

Vol. 202: **Oldstone, Michael B. A.; Vitković, Ljubiša (Eds.):** HIV and Dementia. 1995. 40 figs. XIII, 279 pp. ISBN 3-540-59117-6

Vol. 203: **Sarnow, Peter (Ed.):** Cap-Independent Translation. 1995. 31 figs. XI, 183 pp. ISBN 3-540-59121-4

Vol. 204: **Saedler, Heinz; Gierl, Alfons (Eds.):** Transposable Elements. 1995. 42 figs. IX, 234 pp. ISBN 3-540-59342-X

Vol. 205: **Littman, Dan R. (Ed.):** The CD4 Molecule. 1995. 29 figs. XIII, 182 pp. ISBN 3-540-59344-6

Vol. 206: **Chisari, Francis V.; Oldstone, Michael B. A. (Eds.):** Transgenic Models of Human Viral and Immunological Disease. 1995. 53 figs. XI, 345 pp. ISBN 3-540-59341-1

Vol. 207: **Prusiner, Stanley B. (Ed.):** Prions Prions Prions. 1995. 42 figs. VII, 163 pp. ISBN 3-540-59343-8

Vol. 208: **Farnham, Peggy J. (Ed.):** Transcriptional Control of Cell Growth. 1995. 17 figs. IX, 141 pp. ISBN 3-540-60113-9

Vol. 209: **Miller, Virginia L. (Ed.):** Bacterial Invasiveness. 1996. 16 figs. IX, 115 pp. ISBN 3-540-60065-5

Vol. 210: **Potter, Michael; Rose, Noel R. (Eds.):** Immunology of Silicones. 1996. 136 figs. XX, 430 pp. ISBN 3-540-60272-0

Vol. 211: **Wolff, Linda; Perkins, Archibald S. (Eds.):** Molecular Aspects of Myeloid Stem Cell Development. 1996. 98 figs. XIV, 298 pp. ISBN 3-540-60414-6

Vol. 212: **Vainio, Olli; Imhof, Beat A. (Eds.):** Immunology and Developmental Biology of the Chicken. 1996. 43 figs. IX, 281 pp. ISBN 3-540-60585-1

Vol. 213/I: **Günthert, Ursula; Birchmeier, Walter (Eds.):** Attempts to Understand Metastasis Formation I. 1996. 35 figs. XV, 293 pp. ISBN 3-540-60680-7

Vol. 213/II: **Günthert, Ursula; Birchmeier, Walter (Eds.):** Attempts to Understand Metastasis Formation II. 1996. 33 figs. XV, 288 pp. ISBN 3-540-60681-5

Vol. 213/III: **Günthert, Ursula; Schlag, Peter M.; Birchmeier, Walter (Eds.):** Attempts to Understand Metastasis Formation III. 1996. 14 figs. XV, 262 pp. ISBN 3-540-60682-3

Vol. 214: **Kräusslich, Hans-Georg (Ed.):** Morphogenesis and Maturation of Retroviruses. 1996. 34 figs. XI, 344 pp. ISBN 3-540-60928-8

Springer Verlag and the environment

We at the Springer-Verlag firmly believe that an international science publisher has a special obligation to the environment, and our corporate policies consistently reflect this conviction. We also expect our business partners – paper mills, printers, packaging manufacturers, etc. – to commit themselves to using environmentally friendly materials and production processes. The paper in this book is made from low- or no-chlorine pulp and is acid free, in conformance with international standards for paper permanency.

Printing: Saladruck, Berlin
Binding: Buchbinderei Lüderitz & Bauer, Berlin